T0335934

Fundamentals of Nanotransistors

Lessons from Nanoscience: A Lecture Note Series

ISSN: 2301-3354

"Lessons from Nanoscience" aims to present new viewpoints that help understand, integrate, and apply recent developments in nanoscience while also using them to re-think old and familiar subjects. Some of these viewpoints may not yet be in final form, but we hope this series will provide a forum for them to evolve and develop into the textbooks of tomorrow that train and guide our students and young researchers as they turn nanoscience into nanotechnology. To help communicate across disciplines, the series aims to be accessible to anyone with a bachelor's degree in science or engineering.

More information on the series as well as additional resources for each volume can be found at: http://nanohub.org/topics/LessonsfromNanoscience

---

Published:

Vol. 1 Lessons from Nanoelectronics: A New Perspective on Transport
by Supriyo Datta

Vol. 2 Near-Equilibrium Transport: Fundamentals and Applications
by Mark Lundstrom, Changwook Jeong and Raseong Kim

Vol. 3 Thermal Energy at the Nanoscale
by Timothy S Fisher

Vol. 4 Fundamentals of Atomic Force Microscopy, Part I: Foundations
by Ronald Reifenberger

Vol. 5 Lessons from Nanoelectronics: A New Perspective on Transport (Second Edition)
Part A: Basic Concepts
by Supriyo Datta

Vol. 6 Fundamentals of Nanotransistors
by Mark Lundstrom

Lessons from Nanoscience:
A Lecture Note Series

Vol. 6

Fundamentals of Nanotransistors

Mark Lundstrom

Purdue University, USA

NEW JERSEY • LONDON • SINGAPORE • BEIJING • SHANGHAI • HONG KONG • TAIPEI • CHENNAI • TOKYO

Published by

World Scientific Publishing Co. Pte. Ltd.
5 Toh Tuck Link, Singapore 596224
USA office: 27 Warren Street, Suite 401-402, Hackensack, NJ 07601
UK office: 57 Shelton Street, Covent Garden, London WC2H 9HE

Library of Congress Cataloging-in-Publication Data
Names: Lundstrom, Mark, author.
Title: Fundamentals of nanotransistors / Mark Lundstrom, Purdue University, USA.
Other titles: Lessons from nanoscience ; v. 6.
Description: Singapore ; Hackensack, NJ : World Scientific Publishing Co. Pte. Ltd., [2017] |
Series: Lessons from nanoscience: a lecture notes series ; vol. 6 |
Includes bibliographical references and index.
Identifiers: LCCN 2017010454| ISBN 9789814571722 (hardcover ; alk. paper) |
ISBN 9814571725 (hardcover ; alk. paper) | ISBN 9789814571739 (pbk ; alk. paper) |
ISBN 9814571733 (pbk ; alk. paper)
Subjects: LCSH: Metal oxide semiconductor field-effect transistors. | Nanostructured materials.
Classification: LCC TK7871.99.M44 L86 2017 | DDC 621.3815/284--dc23
LC record available at https://lccn.loc.gov/2017010454

British Library Cataloguing-in-Publication Data
A catalogue record for this book is available from the British Library.

Printed in Singapore

To

Will and Nick,

who make their father proud

Preface

The transistor is the basic component from which electronic systems are built. The discovery of the transistor in 1947 set the stage for a revolution in electronics. The invention of the integrated circuit in 1958–59 launched the revolution by providing a way to mass produce monolithic circuits of interconnected transistors. As semiconductor technology developed, the number of transistors on an integrated circuit chip doubled each year. This doubling of the number of transistors per chip, driven by continuously downscaling the size of transistors, has continued at about the same pace for more than 50 years. The continuous increase in the capabilities of electronic systems and the continuous decrease in the cost per function that resulted from downscaling have shaped the world we live in today.

The theory of the MOSFET (the most common type of transistor) was formulated in the 1960's when transistor channels were about 10 micrometers (10,000 nanometers) long. As semiconductor technology matured, transistor dimensions shrank, new physics became important, and the models evolved. By the end of the 20th century, transistor dimensions had reached the nanoscale, and the transistor became the first active, nanoscale device in high-volume manufacturing. Today, transistor channel lengths are approaching 10 nm – 1000 times shorter than the first MOSFETs. My goal in these lectures is to discuss the physical operating principles of modern, nanoscale MOSFETs.

MOSFETs operate by controlling energy barriers for charge carriers with voltages that are applied to the transistors' terminals. This basic principle applies to all commonly-used transistors. Only MOSFETs (Metal-Oxide-Semiconductor Field-Effect Transistors) will be discussed in these lectures, but the same principles apply to Junction Field-Effect Transistors

(JFETs), Metal-Semiconductor Field-Effect Transistors (MESFETs), High Electron Mobility FETs (HEMTs, which are also called MOdulation-Doped FETs, MODFETs), and even to bipolar junction transistors (BJTs). The manipulation of energy barriers is described by electrostatics. The electrostatics of modern MOSFETs is much like that for earlier MOSFETs.

When the energy barrier is pushed down, charge carriers flow across the device. Electrons and holes in modern, nanoscale transistors flow much differently than they did 50 years ago when transistor models were first developed for 10,000 nm channel length MOSFETs. Understanding how electrons and holes flow at the nanoscale is essential to understanding the physics of modern transistors. Carrier transport at the nanoscale is quite different from transport at the microscale, but the principles are remarkably simple and easy to understand. The approach is based on a new understanding of electron transport that has emerged from research on molecular and nanoscale electronics [1]. Although carrier transport at the nanoscale is much different, the theory of the modern MOSFET retains much of the original theory of the MOSFET because of the central role of MOS electrostatics. Finally, in addition to describing a specific device, these notes should serve as an example of how other nanodevices might be understood and modeled.

The MOSFET models used by designers of today's billion transistor integrated circuits are very sophisticated and describe many features of transistor operation that are not discussed in these brief notes. My goal is to present a sound, but very basic treatment of the physics of modern nanoscale MOSFETs and, by extension, of most other nanoscale transistors. For most readers, this treatment may be sufficient. For those pushing the frontiers of transistor science and technology, these notes are a starting point that aims to convey some important fundamentals that can be built upon.

To follow these lecture notes, a reader should understand basic semiconductor physics, but no familiarity with transistors is assumed. Part 1 introduces the transistor and discusses some fundamental concepts. Part 2 discusses MOS electrostatics, a critically important component of MOSFET theory. Readers with a strong background in traditional MOSFET theory may wish to skim Parts 1 and 2 and go directly to Parts 3 and 4 where the new approach is presented. Online versions of these lectures are also available, along with an extensive set of additional resources for self-learners at nanoHUB-U [2]. These lecture notes are an elaboration of the ideas presented succinctly in [3].

These notes are part of the *Lessons from Nanoscience* Lecture Note Series, which aims to bring new approaches and new ways of thinking to the field of electronic materials and devices. They draw from research in nanoscience, but they are not research monographs that discuss the latest "hot topics". Rather, the goal is to produce lectures notes that have lasting value by re-thinking the way we teach electronic materials and devices so that working from the nanoscale to the macroscale is seamless and intuitive. I hope that readers find these notes a useful introduction to a topic that is both scientifically interesting and technologically important.

Mark Lundstrom
Purdue University
May 2017

[1] Supriyo Datta, *Lessons from Nanoelectronics*, 2nd Ed., PART A: Basic Concepts, World Scientific Publishing Company, Singapore, 2017.

[2] "nanoHUB-U: Online courses broadly accessible to students in any branch of science or engineering," `http://nanohub.org/u`, 2016.

[3] Mark S. Lundstrom and Dimitri A. Antoniadis, "Compact Models and the Physics of Nanoscale FETs", *IEEE Trans. Electron Devices*, **61**, pp. 225-233, 2014.

Acknowledgments

Thanks to World Scientific Publishing Corporation and our series editor, Zvi Ruder, for their support with this lecture notes series. Special thanks to the U.S. National Science Foundation, the Intel Foundation, and Purdue University for their support of the Network for Computational Nanotechnology's "Electronics from the Bottom Up" initiative, which laid the foundation for this series.

My understanding of the physics of nanoscale transistors has evolved over many years during which I have had numerous opportunities to work with and learn from a remarkable group of students and colleagues. Former students who contributed specifically to this understanding are Drs. Farzin Assad, Zhibin Ren, Ramesh Venugopal, Jung-Hoon Rhew, Jing Guo, Jing Wang, Anisur Rahman, Sayed Hasan, Himadri Pal, Yang Liu, Raseong Kim, Changwook Jeong, Xingshu Sun, Piyush Dak, and Evan Witkoske. Professor Supriyo Datta of Purdue University and Professor Dimitri Antoniadis of the Massachusetts Institute of Technology are two of the many colleagues I've been fortunate to work with. Datta's approach to carrier transport at the nanoscale provides a clear, simple, and sound way to understand transport in nanoscale transistors. The "virtual source" model of Antoniadis captures the essential ideas discussed in these lectures and embodies them in a useful compact model. His careful analysis of nanoscale transistor measurements has done much to clarify my own understanding of these remarkable devices.

Acknowledgments

Constants Used in These Lectures

Magnitude of electronic charge	$q = 1.6 \times 10^{-19}$ C (Coulomb)
Unit of energy	$1 \text{ eV} = 1.6 \times 10^{-19}$ J (Joule)
Boltzmann's constant	$k_B = 1.38 \times 10^{-23}$ J/K
Planck's constant	$h = 6.626 \times 10^{-34}$ J-s
Reduced Planck's constant	$\hbar = h/2\pi = 1.055 \times 10^{-34}$ J-s
Free electron mass	$m_0 = 9.109 \times 10^{-31}$ kg
Permittivity of free space	$\varepsilon_0 = 8.854 \times 10^{-12}$ F/m

Some Symbols Used

β	Velocity saturation parameter	Dimensionless
δ	DIBL parameter	V/V
ε_{ins}	Permittivity of gate insulator	F/m
η_F	Dimensionless Fermi energy	Dimensionless
κ	Device scaling parameter	Dimensionless
κ_{ox}	Relative dielectric constant of gate oxide	Dimensionless
κ_S	Relative dielectric constant of semiconductor	Dimensionless
λ	Mean-free-path (MFP) for backscattering	m
λ_0	Near-equilibrium MFP for backscattering	m
Λ	MFP or geometric screening length	m
μ	Mobility	m^2/V-s
μ_{app}	Apparent mobility	m^2/V-s
μ_{ball}	Ballistic mobility	m^2/V-s
μ_n	Electron mobility	m^2/V-s
ρ	Space charge density	C/m^3
τ	Scattering time or device delay metric	s
τ_E	Energy relaxation time	s
τ_m	Momentum relaxation time	s
Φ_M	Metal (gate) work function	J
Φ_S	Semiconductor work function	J
Φ_{MS}	Metal-semiconductor work function difference	J
$\psi(y)$	Electrostatic potential vs. position	V
ψ_B	Fermi potential	V
ψ_S	Surface potential	V
Ω	Normalization volume	m^3

A_v	Self-gain	Dimensionless
CET	Capacitance equivalent thickness	m
C_D	Depletion layer capacitance per unit area	$\mathrm{F/m^2}$
C_{ox}	Oxide capacitance per unit area	$\mathrm{F/m^2}$
C_Q	Quantum capacitance per unit area	$\mathrm{F/m^2}$
C_S	Semiconductor capacitance per unit area	$\mathrm{F/m^2}$
$\vec{D}$	Electric displacement vector	$\mathrm{C/m^2}$
$D(E)$	Density of states	$\mathrm{J^{-1}m^{-3}}$
$D_{2D}(E)$	2D Density of states	$\mathrm{J^{-1}m^{-2}}$
D_n	Electron diffusion coefficient	$\mathrm{m^2}/s$
$DIBL$	Drain-induced barrier lowering	mV/V
E_C	Conduction band minimum	J
E_F	Fermi level	J
$E_{FS,D}$	Fermi level in source (drain)	J
$\mathcal{E}$	Electric field	V/m
$\mathcal{E}_{cr}$	Critical electric field for velocity saturation	V/m
$\mathcal{E}_S$	Surface electric field	V/m
$f(E)$	Fermi function	Dimensionless
$f_{S,D}(E)$	Fermi function in source (drain)	Dimensionless
G_{CH}	Channel conductance	S
G_B	Ballistic conductance	S
g_m	Transconductance	S or S/m
g_v	Valley degeneracy	Dimensionless
k	Electron wavenumber	$\mathrm{m^{-1}}$
I_{ON}	On-current	A or A/μm
I_{OFF}	Off-current	A or A/μm
I_{DSAT}	Drain current in saturation region	A or A/μm
I_{DLIN}	Drain current in linear region	A or A/μm
L	Channel length	m
L_{eff}	Effective channel length	m
L_D	Debye length	m
ℓ	Critical length for backscattering	m
m	Body effect parameter	Dimensionless
$M(E)$	Number of channels at energy, E	Dimensionless
M_{2D}	Number of channels per unit width	$\mathrm{m^{-1}}$
n_i	Intrinsic carrier density	$\mathrm{m^{-3}}$
n_S	Sheet electron density	$\mathrm{m^{-2}}$
n_0	Equilibrium electron density	$\mathrm{m^{-3}}$

N_A	Acceptor doping density	m^{-3}
N_A^-	Ionized acceptor doping density	m^{-3}
N_D	Donor doping density	m^{-3}
N_D^+	Ionized donor doping density	m^{-3}
N_C	Conduction band effective density of states	m^{-3}
N_V	Valence band effective density of states	m^{-3}
N_{2D}	2D effective density of states	m^{-2}
P_D	Power dissipation	W
p_0	Equilibrium hole density	m^{-3}
Q_D	Depletion charge	$\mathrm{C/m}^2$
Q_F	Fixed charge at insulator/semiconductor interface	$\mathrm{C/m}^2$
Q_n	Mobile (electron) sheet charge	$\mathrm{C/m}^2$
Q_S	Sheet charge in the semiconductor	$\mathrm{C/m}^2$
R_{ch}	Channel resistance	Ω or $\Omega-\mu\mathrm{m}$
$R_{S,D}$	Series resistance at source (drain)	Ω or $\Omega-\mu\mathrm{m}$
r_o	Output resistance	Ω or $\Omega-\mu\mathrm{m}$
SS	Subthreshold swing	mV/decade
$S(\vec{p}\rightarrow\vec{p}')$	Scattering transition rate from $\vec{p}$ to $\vec{p}'$	s^{-1}
t_{inv}	Inversion layer thickness	m
t_{ins}	Gate insulator thickness	m
t_{ox}	Gate oxide thickness	m
T	Temperature	K
$\mathcal{T}$	Transmission	Dimensionless
υ_{inj}	Injection velocity	m/s
υ_{sat}	High-field saturation velocity	m/s
υ_T	Uni-directional thermal velocity	m/s
$\langle v_x^+(E)\rangle$	Angle-averaged velocity in +x direction	m/s
$\langle\langle v_x^+\rangle\rangle$	Energy and angle-averaged velocity in +x direction	m/s
$\vdots$		
V_{FB}	Flat band voltage	V
V_{bi}	Built-in potential	V
V_G	Applied gate voltage	V
V_{DSAT}	Drain saturation voltage	V
V_G'	Applied gate voltage assuming $\Phi_{MS}=0$	V
V_R	Reverse bias between source and body	V

V_T	Threshold voltage	V
W	Channel width	m
W_D	Depletion layer thickness	m
W_T	Maximum depletion layer thickness	m
$W(E)$	Fermi window function	J^{-1}

Contents

List of Figures

PART 1
MOSFET Fundamentals

Lecture 1

Overview

1.1 Introduction

The transistor has been called "the most important invention of the 20th century" [1]. Transistors are everywhere; they are the basic building blocks of electronic systems. As transistor technology advanced, their dimensions were reduced from the micrometer (μm) to the nanometer (nm) scale, so that more and more of them could be included in electronic systems. Today, billions of transistors are in our smartphones, tablet and personal computers, supercomputers, and the other electronic systems that have shaped the world we live in. In addition to their economic importance, transistors are scientifically interesting nano-devices. These lectures aim to present a clear treatment of the essential physics of the nanotransistor. This first lecture introduces the topics we'll discuss and gives a roadmap for the remaining lectures.

Figure 1.1 shows the most common transistor in use today, the Metal-Oxide-Semiconductor Field-Effect Transistor (MOSFET). On the left is the schematic symbol we use when drawing transistors in a circuit, and on the right is a scanning electron micrograph (SEM) of a silicon MOSFET circa 2000. The transistor consists of a *source* by which electrons enter the device, a *gate*, which controls the flow of electrons from the source and across the *channel*, and a *drain* through which electrons leave the device. The gate insulator, which separates the gate electrode from the channel, is less than 2 nm thick (less than the diameter of a DNA double helix). The length of the channel was about 100 nm at the turn of the century, and is about 20 nm today. The operation of a nanoscale transistor is interesting scientifically, and the technological importance of transistors is almost impossible to overstate.

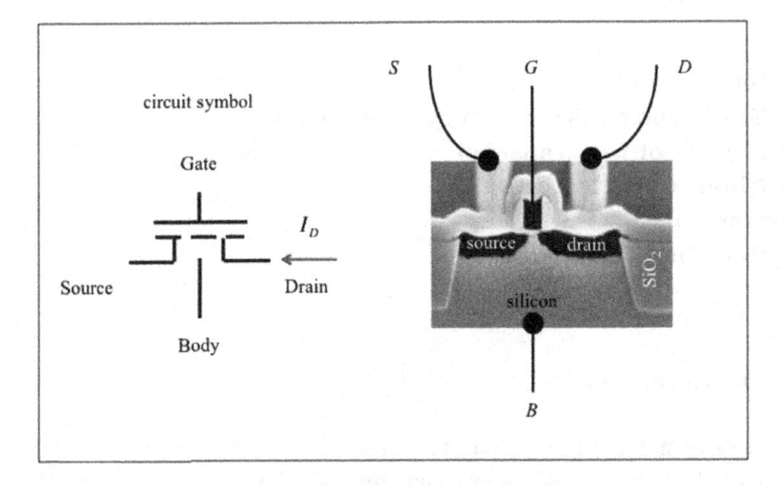

Fig. 1.1 The silicon MOSFET. Left: The circuit schematic of an enhancement mode MOSFET showing the source, drain, gate, and body contacts. The dashed line represents the conductive channel, which is present when a large enough gate voltage is applied. Right: An SEM cross-section of a silicon MOSFET circa 2000. The source, drain, gate, silicon body, and gate insulator are all visible. The channel is the gap between the source and the drain. (Source: Texas Instruments, circa 2000.)

Figure 1.2 shows the current-voltage (IV) characteristics of a MOSFET. Electrons flow from the source to the drain when the gate voltage is large enough. Devices with IV characteristics like this are useful in electronic circuits. They can operate as digital switches, either on or off, or as analog amplifiers of input signals. The shape of the IV characteristic and the magnitude of the current are controlled by the physics of the device. My goal in these lectures is to relate the IV characteristic of a nanotransistor to its internal physics.

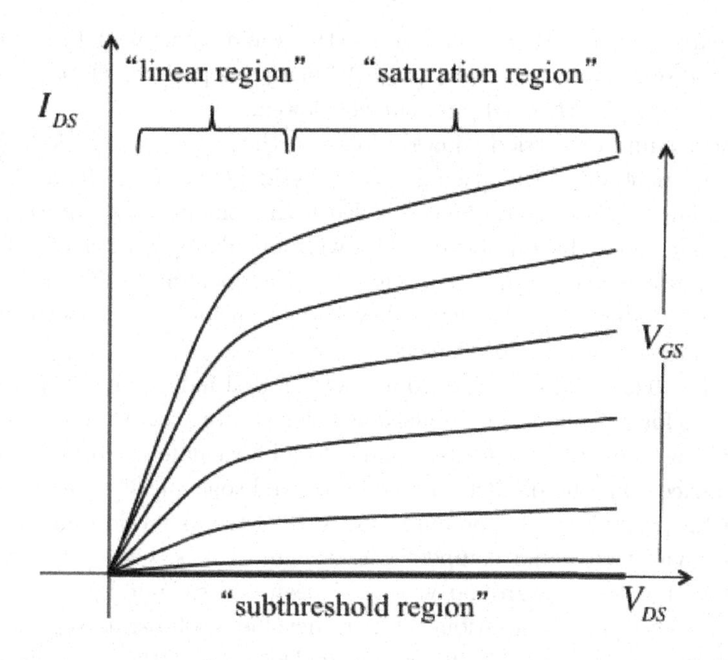

Fig. 1.2 The common source output *IV* characteristics of an N-channel MOSFET. The vertical axis is the current that flows between the drain and source terminals, and the horizontal axis is the voltage between the drain and source. Each line corresponds to a different gate voltage. The two regions of operation, to be discussed later, are also labeled. The maximum voltage applied to the gate and drain terminals is the power supply voltage, V_{DD}. (The small leakage current in the subthreshold region is not visible on a linear scale for I_{DS}.)

1.2 Electronic devices: A very brief history

Electronic systems are circuits of interconnected electronic devices. Resistors, capacitors and inductors are very simple devices, but most electronic systems rely on non-linear devices, the simplest being the diode, which allows conduction for one polarity of applied voltage but not for the other. The first use of diodes was for detecting radio signals. In the early 1900's, semiconductor diodes were demonstrated as were vacuum tube diodes. Semiconductor diodes were metal-semiconductor junctions consisting of a metal wire (the "cat's whisker") placed in a location on the crystal that gave the best performance. Because they were finicky, these crystal detectors were soon replaced with vacuum tube detectors, which consisted of a heated filament that boiled off electrons and a metal plate inside an evacuated bulb. When the voltage on the plate was positive, electrons from the filament were attracted, and current flowed.

The vacuum tube triode quickly followed the vacuum tube diode (and, later, the pentode). By placing a metal grid between the filament and plate, a large current could be controlled with a small voltage on the grid, and signals could be amplified. The widespread application of vacuum tubes transformed communications and entertainment and enabled the first digital computers, but vacuum tubes had problems — they were large, fragile, and consumed a lot of power.

In the 1920's, Julius Lilienfeld and Oskar Heil independently patented a concept for a "solid-state" replacement for the vacuum tube triode. By eliminating the need for a heated filament and a vacuum enclosure, a solid-state device would be smaller, more reliable, and consume less power. Semiconductor technology was too immature at the time to develop this concept into a device that could compete with vacuum tubes, but by the end of World War II, enough ground work had been laid to spur Bell Telephone Laboratories to mount a serious effort to develop a solid-state replacement for the vacuum tube [2]. The result, in December 1947, was the transistor – not the field-effect transistor (FET) of Lillenfeld and Heil but a point contact bipolar transistor (something like the original cat's whisker crystal rectifier). Over the years, however, the technological problems associated with FETs were solved, and today, the Metal-Oxide-Semiconductor Field-Effect Transistor (MOSFET) is the mainstay of electronic systems [3]. These lectures are about the MOSFET, but the basic principles apply to several different types of transistors.

By 1960, technologists learned how to manufacture several transistors in one, monolithic piece of semiconductor and to wire them up in circuits as part of the manufacturing process, instead of first making transistors and then wiring them up individually by hand. Gordon Moore noticed in 1965 that the number of transistors on these integrated circuit "chips" was doubling every technology generation (about one year then, about 1.5 years now) [4]. He predicted that this doubling of the number of transistors per chip would continue for some time, but even he must have been surprised to see it continue for more than 50 years [5].

The doubling of the number of transistors per chip each technology generation (now known as *Moore's Law*) was accomplished by down-scaling the size of transistors. Because transistor dimensions were first measured in micrometers, electronics technology became known as "microelectronics."

Device physicists developed simple, mathematical models for transistors [6-9], which succinctly described the operation of the device in a way that designers could use for circuit and system design. Over the years, these models were refined and extended to describe evolving transistor technology [10, 11]. Each technology generation, the lateral dimensions of transistors shrunk by a factor of $\sqrt{2}$, which reduced the area by a factor of two and doubled the number of transistors on a chip. About the year 2000, the length of the transistor channel reached 100 nanometers, microelectronics became nanoelectronics, and the nanotransistor became a high-impact success of the nanotechnology revolution. It now seems clear that transistor channel lengths will shrink to the 10 nm scale, and the question today is: "how far below 10 nanometers can transistor technology be pushed?"

As the size of transistors crossed the nanometer threshold, the characteristics of the device as measured at its terminals did not change dramatically (indeed, if they had, we would no longer have what we call a "transistor"). But something did change; the internal physics that controls the transport of charge carriers from the source to the drain in a transistor changed in a very significant way. Understanding electronic transport in nanoscale transistors in a simple, but physically sound way is the goal of these lectures.

1.3 Physics of the transistor

The vast majority of transistors operate by controlling the height of an energy barrier with an applied voltage. An energy barrier in the channel prevents electrons from flowing form the source to the drain. As voltages are applied to the gate and drain electrodes, the height of this energy barrier can be manipulated, and the flow of electrons from the source to the drain can be controlled. In Lecture 3, I will discuss this energy band view of the MOSFETs in more detail; it contains most of the physics that we will later use to develop mathematical models to describe transistors.

The mathematical analysis of a MOSFET often begins with the equation,

$$I_{DS} = W |Q_n (V_{GS}, V_{DS})| \langle \upsilon \rangle \ , \tag{1.1}$$

where W is the width of the transistor in the direction normal to the current flow, Q_n is the mobile sheet charge in the device (C/m^2), and $\langle \upsilon \rangle$ is the average velocity at which it flows. When "doing the math" it is important to keep the physical picture in mind. The charge comes from electrons surmounting the energy barrier, and the velocity represents the velocity at which they then move. Understanding MOSFETs boils down to understanding electrostatics (Q_n) and transport ($\langle\upsilon\rangle$). While the electrostatic design principles of MOSFETs have not changed much for the past few decades, the nature of electron transport in transistors has changed considerably as transistors have been made smaller and smaller. A proper treatment of transport in nanoscale transistors is essential to understanding and designing these devices.

The *drift-diffusion equation* is the cornerstone of traditional semiconductor device physics. It states that the current in a uniform semiconductor is proportional to the electric field, $\mathcal{E}$, and that in the absence of an electric field, the current is carried by electrons diffusing down a concentration gradient. In general, both processes occur at the same time, and we add the two to find the current carried by electrons in the conduction band as

$$J_n = nq\mu_n q\mathcal{E} + qD_n dn/dx \,, \tag{1.2}$$

where n is the density of electrons in the conduction band, q is the magnitude of the charge on an electron, μ_n is the electron *mobility*, and D_n is the *diffusion coefficient*. Although most semiconductor textbooks still begin with eqn. (1.2), it is not at all clear that the approximations necessary to derive eqn. (1.2) are valid for the small devices that are now being manufactured. Indeed, sophisticated computer simulations show that electron transport in nanoscale transistors is quite complex [12, 13]. For our purposes in these lectures, we need a simple description of transport designed to work at the nanoscale.

The Landauer approach describes carrier transport at the nanoscale. Instead of eqn. (1.2), we compute the current from [14, 15]:

$$I = \frac{2q}{h}\int M(E)\mathcal{T}(E)\left[f_1(E) - f_2(E)\right]dE \,, \tag{1.3}$$

where q is the magnitude of the charge on an electron, h is Planck's constant, $M(E)$ is the number of channels at energy, E, that are available for conduction, $\mathcal{T}(E)$ is the transmission, $f_1(E)$ the equilibrium Fermi function of contact one and $f_2(E)$, the Fermi function for contact two. The number of channels is analogous to the number of lanes on a highway, and the transmission is a number between zero and one; it is the probability that an electron injected from contact one exits from contact two. For large devices, eqn. (1.3) reduces to eqn. (1.2), but eqn. (1.3) can also be applied to nanodevices for which it is not so clear how to make use of eqn. (1.2).

The transport effects discussed so far are semiclassical — they consider electrons to be particles with the quantum mechanics being embedded in the band structure or effective mass, but as devices continue to shrink, it is becoming important to consider explicitly the quantum mechanical nature of electrons. We should expect that quantum mechanical effects will become important when the potential energy changes rapidly on the scale of the electron's de Broglie wavelength. A simple estimate of the de Broglie wavelength of thermal equilibrium electrons in Si gives about 10 nm, which is not much less than present day channel lengths. During the past decade or two, powerful techniques to treat the quantum mechanical transport of electrons in transistors have been developed [13]. As channel lengths shrink below 10 nm, it is becoming increasingly necessary to describe electron transport quantum mechanically, but for channel lengths above about 10 nm, the semiclassical picture works well.

A significant research effort over the past few decades has been devoted to understanding transport at the nanoscale and at developing techniques to simulate it on computers. The essential physics of transport at the nanoscale is readily understood, and this simple understanding is useful for interpreting experiments and detailed simulations as well as for for designing and optimizing transistors. This simple, intuitive, "essential only" approach to transport in nanotransistors is the subject of these lecture notes.

1.4 About these lectures

The lectures presented in this volume are divided into four parts.

Part 1: MOSFET Fundamentals
Part 2: MOS Electrostatics
Part 3: The Ballistic Nanotransistor
Part 4: Transmission Theory of the Nanotransistor

Part 1: MOSFET Fundamentals

Part 1 introduces the transistor. The lecture that follows this overview treats transistors as "black boxes" and describes their electrical characteristics and key performance metrics. A lecture on the MOSFET as a barrier controlled device shows how simple it is to understand the MOSFET in terms of energy band diagrams. One lecture then presents the traditional derivation of the MOSFET *IV* characteristics. The final lecture in Part 1 introduces the "Virtual Source" (VS) model, a semi-empirical model for MOSFETs [16] that will serve as an overall framework for the subsequent lectures.

Part 2: MOS Electrostatics

Part 2 discusses the most important physics of a MOSFET — MOS electrostatics — how the potential barrier between the source and drain is controlled by the gate and drain voltages. Five lectures discuss one-dimensional MOS electrostatics (the dimension normal to the channel) much as it is presented in traditional textbooks. The effects of two-dimensional electrostatics (i.e. the role of the drain voltage) are then described. In the final lecture of Part 2, we return to the VS model and show how to improve it with a better treatment of MOS electrostatics.

Part 3: The Ballistic MOSFET

Part 3 is about the ballistic MOSFET, a device for which electrons in the channel do not scatter. The section begins with an introduction to the Landauer approach to transport and then continues by applying this approach in the ballistic limit to MOSFETs. Modern MOSFETs operate quite close to the ballistic limit. The ballistic MOSFET model looks much different than the traditional MOSFET model, but when we relate it to the VS model, we'll find that it can be expressed in the traditional language of MOSFET analysis.

Part 4: Transmission Theory of the MOSFET

Part 4 adds carrier scattering to the model. A transmisson theory of MOSFETs that includes electron transport from the no-scattering (ballistic) to strong-scattering (diffusive) regimes is developed. Part 4 begins with a lecture on the fundamentals of carrier scattering and the relation of transmission to the mean-free-path. The transmission theory of the nano-MOSFET is then presented and related to traditional MOSFET theory via the VS model. The use of the Transmission/VS model to experimentally characterize nanotransistors is discussed, and Part 4 concludes with a lecture that examines the limits of transistors and some of the limitations of the transmission approach to nano-MOSFETs.

1.5 Summary

My objectives in writing these lectures are to present a simple, physical way to understand the operation of nanoscale MOSFETs and to relate this new understanding to the traditional theory of the MOSFET. Transistor science and technology is a complex, but readily understood subject. I am only able in these lectures to touch upon a few, important concepts. The goal is to develop a firm understanding of a few key principles will provide a starting point that can be filled in and extended as needed. Since the nano-MOSFET is the first high-volume, high-impact, active nano-device, understanding the MOSFET as a nano-device also provides a case study in developing models for other nano devices.

To follow these lectures, only a basic understanding of semiconductor physics is necessary – e.g. concepts like bandstructure, effective mass, mobility, doping, etc. The first two parts of the lectures are for those with little or no background in transistors and MOSFETs, and the last two parts present a novel approach to understanding nanotransistors. Those with a good background in transistors and MOSFETs may want to skip (or just skim) Parts 1 and 2. For those with little or no background in transistors and MOSFETs, Parts 1 and 2 will provide the necessary background for understanding Parts 3 and 4. The reader will notice that some words are *italicized.* This is done to alert the reader when important terms that should be remembered are first encountered. Finally, an extensive set of online materials that supplement and extend these lectures can be found on the author's home page: https://nanohub.org/groups/mark_lundstrom/.

1.6 References

To learn about the interesting history of the transistor, see:

[1] Ira Flatow, *Transistorized!*, `http://www.pbs.org/transistor/`, 1999.

[2] Michael Riordan and Lillian Hoddeson, *Crystal Fire: The Birth of the Information Age*, W.W. Norton & Company, Inc., New York, 1997.

[3] Bo Lojek, *History of Semiconductor Engineering*, Springer, New York, 2007.

The famous paper that predicted what is now known as "Moore's Law," the doubling of the number of transistors on an integrated circuit chip each technology generation is:

[4] G.E. Moore, "Cramming more components onto integrated circuits," *Electronics Magazine*, pp. 4-7, 1965.

For a 2003 perspective on the future of Moore's Law, see:

[5] M. Lundstrom, "Moore's Law Forever?" an Applied Physics Perspective, *Science*, **299**, pp. 210-211, January 10, 2003.

The mathematical modeling of transistors began in the 1960's. Some of the first papers on the type of transistor that we'll focus on are listed below.

[6] S.R. Hofstein and F.P. Heiman, "The Silicon Insulated-Gate Field-Effect Transistor, *Proc. IEEE*, pp. 1190-1202, 1963.

[7] C.T. Sah, "Characteristics of the Metal-Oxide-Semiconductor Transistors, *IEEE Trans. Electron Devices*, **11**, pp. 324-345, 1964.

[8] H. Shichman and D. A. Hodges, "Modeling and simulation of insulated-gate field-effect transistor switching circuits," *IEEE J. Solid State Circuits*, **SC-3**, 1968.

[9] B.J. Sheu, D.L. Scharfetter, P.-K. Ko, and M.-C. Jeng, "BSIM: Berkeley Short-Channel IGFET Model for MOS Transistors," *IEEE J. Solid-State Circuits*, **SC-22**, pp. 558-566, 1987.

For comprehensive, authoritative treatments of the state-of-the-art in MOSFET device physics and modeling, see:

[10] Y. Tsividis and C. McAndrew, *Operation and Modeling of the MOS Transistor*, 3rd Ed., Oxford Univ. Press, New York, 2011.

[11] Y. Taur and T. Ning, *Fundamentals of Modern VLSI Devices*, 2nd Ed., Oxford Univ. Press, New York, 2013.

The following references are examples of physically detailed MOSFET device simulation — the first semiclassical and the second quantum mechanical.

[12] D. Frank, S. Laux, and M. Fischetti, "Monte Carlo simulation of a 30 nm dual-gate MOSFET: How short can Si go?," Intern. Electron Dev. Mtg., pp. 553-556, Dec., 1992.

[13] Z. Ren, R. Venugopal, S. Goasguen, S. Datta, and M.S. Lundstrom "nanoMOS 2.5: A Two-Dimensional Simulator for Quantum Transport in Double-Gate MOSFETs," *IEEE Trans. Electron. Dev.,* **50**, pp. 1914-1925, 2003.

The Landauer approach to carrier transport at the nanoscale is discussed in Vols. 5 and 2 of this series.

[14] Supriyo Datta, *Lessons from Nanoelectronics*, 2nd Ed., PART A: Basic Concepts, World Scientific Publishing Company, Singapore, 2017.

[15] Mark Lundstrom, *Near-Equilibrium Transport: Fundamentals and Applications*, World Scientific Publishing Company, Singapore, 2012.

The MIT Virtual Source Model, which provides a framework for these lectures, is described in:

[16] A. Khakifirooz, O.M. Nayfeh, and D.A. Antoniadis, "A Simple Semiempirical Short-Channel MOSFET Current Voltage Model Continuous Across All Regions of Operation and Employing Only Physical Parameters," *IEEE Trans. Electron. Dev.,* **56**, pp. 1674-1680, 2009.

Lecture 2

The Transistor as a Black Box

2.1 Introduction

The goal for these lectures is to relate the internal physics of a transistor to its terminal characteristics; *i.e.* to the currents that flow through the external leads in response to the voltages applied to those leads. This lecture will define the external characteristics that subsequent lectures will explain in terms of the underlying physics. We'll treat a transistor as an engineer's "black box," as shown in Fig. 2.1. A large current flows through terminals 1 and 2, and this current is controlled by the voltage on (or, for some transistors the current injected into) terminal 3. Often there is a fourth terminal too. There are many kinds of transistors [1], but all transistors have three or four external leads like the generic one sketched in Fig. 2.1. The names given to the various terminals depends on the type of transistor. The IV characteristics describe the current into each lead in terms of the voltages applied to all of the leads.

Before we describe the IV characteristics, we'll begin with a quick look at the most common transistor — the field-effect transistor (FET). In these lectures, our focus is on a specific type of FET, the silicon Metal-Oxide-Semiconductor Field-Effect Transistor (MOSFET). A different type of FET, the High Electron Mobility Transistor (HEMT), finds use in radio frequency (RF) applications. Bipolar junction transistors (BJTs) and heterojunction bipolar transistor (HBTs) are also used for RF applications. Most of the transistors manufactured and used today are one of these four types of transistors. Although our focus is on the Si MOSFET, the basic principles apply to these other transistors as well.

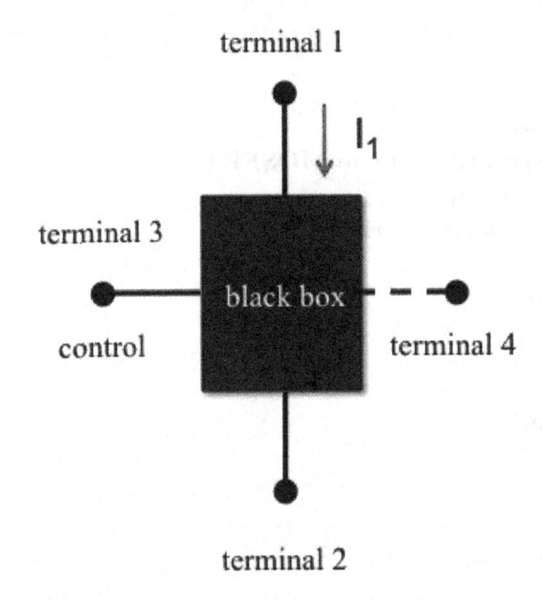

Fig. 2.1 Illustration of a transistor as a black box. The currents that flow in the four leads of the device are controlled by the voltages applied to the four terminals. The relation of the currents to the voltages is determined by the internal device physics of the transistor. These lectures will develop simple, analytical expressions for the current vs. voltage characteristics and relate them to the underlying device physics.

2.2 Physical structure of the MOSFET

Figure 2.2 (same as Fig. 1.1) shows a scanning electron micrograph (SEM) cross section of a Si MOSFET circa 2000. The drain and source terminals (terminals 1 and 2 in Fig. 2.1 are clearly visible, as are the gate electrode (terminal 3 in Fig. 2.1) and the Si body contact (terminal 4 in Fig. 2.1).

Note that the gate electrode is separated from the Si substrate by a thin, insulating layer that is less than 2 nm thick. In present-day MOSFETs, the gap between the source and drain (the channel) is only about 20 nm long.

Also shown in Fig. 2.2 is the schematic symbol used to represent MOSFETs in circuit diagrams. The dashed line represents the channel between the source and drain. It is dashed to indicate that this is an *enhancement mode* MOSFET, one that is only "on" with a channel present when the magnitude of the gate voltage exceeds a critical value known as the *threshold voltage.*

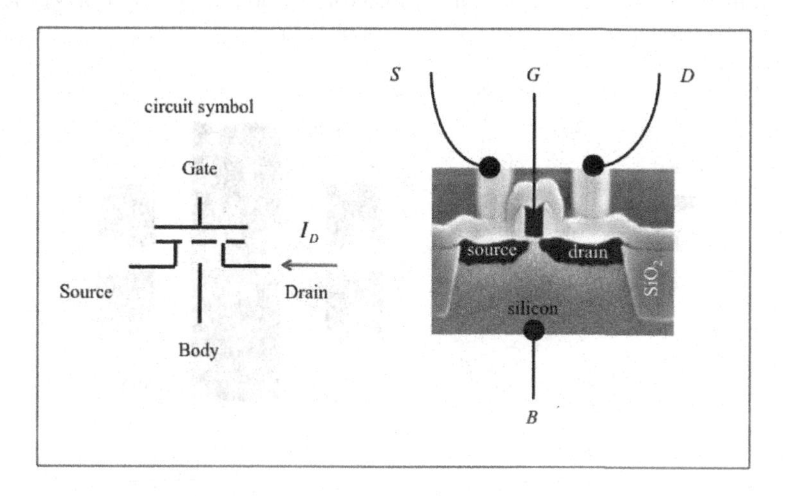

Fig. 2.2 The n-channel silicon MOSFET. Left: The circuit schematic of an enhancement mode MOSFET showing the source, drain, gate, and body contacts. The dashed line represents the channel, which is present when a large enough gate voltage is applied. Right: An SEM cross-section of a silicon MOSFET circa 2000. The source, drain, gate, silicon body, and gate insulator are all visible. (This figure is the same as Fig. 1.1.)

Figure 2.3 compares the cross-sectional and top-views of an n-channel, silicon MOSFET. On the left is a "cartoon" illustration of the cross-section, similar to the SEM in Fig. 2.2. An n-channel MOSFET is built on a p-type Si substrate. The source and drain regions are heavily doped n-type regions; the transistor operates by controlling conduction across the channel that separates the source and drain. On the right side of Fig. 2.3 is a top view of the same transistor. The large rectangle is the transistor itself. The black squares on the two ends of this rectangle are contacts to the source and drain regions, and the black rectangle in the middle is the gate electrode. Below the gate is the gate oxide, and under it, the p-type silicon channel.

The channel length, L, is a critical parameter; it sets the overall "footprint" (size) of the transistor, and determines the ultimate speed of the transistor (the shorter L is, the faster the ultimate speed of the transistor). The width, W, determines the magnitude of the current that flows. For a given technology, transistors are designed to be well-behaved for channel lengths greater than or equal to some minimum channel length. Circuit designers specify the lengths and widths of transistors to achieve the desired circuit performance. For the past several decades, the minimum channel length (and, therefore, the minimum size of a transistor) has steadily shrunk, which has allowed more and more transistors to be placed on an integrated circuit "chip" [2, 3].

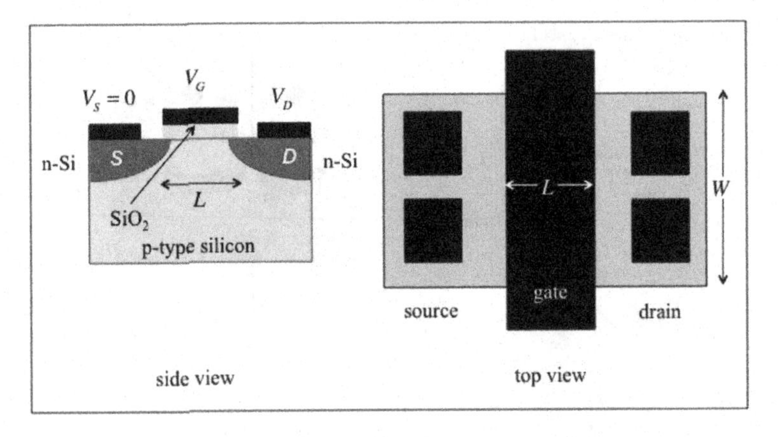

Fig. 2.3 Comparison of the cross-sectional, side view (left) and top view (right) of an n-channel, silicon MOSFET.

In the n-channel MOSFET shown in Fig. 2.3, conduction is by electrons in the conduction band. As shown in Fig. 2.4, it is also possible to make the complementary device in which conduction is by electrons in the valence band (which can be visualized in terms of "holes" in the valence band). A p-channel MOSFET is built on an n-type substrate. The source and drain regions are heavily doped p-type; the transistor operates by controlling conduction across the n-type channel that separates the source and drain. Note that $V_{DS} < 0$ for the p-channel device and that $V_{GS} < 0$ to turn the device on. Also note that the drain current flows out of the drain, rather than into the drain as for the n-channel device. Modern electronics is largely built with CMOS (or complementary MOS) technology for which every n-channel device is paired with a p-channel device.

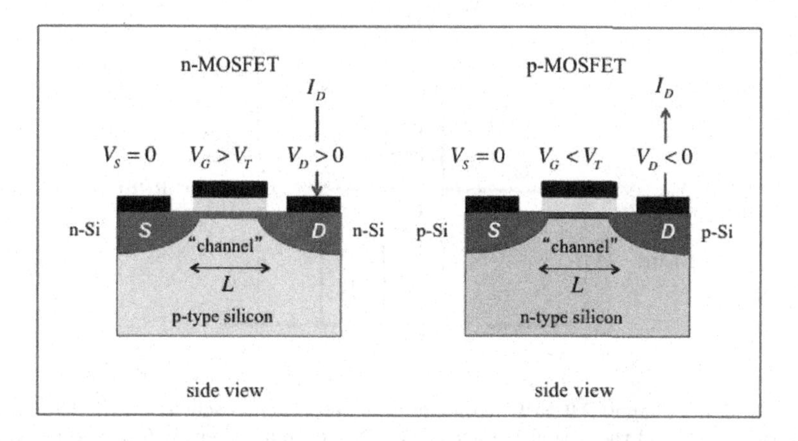

Fig. 2.4 Comparison of an n-channel MOSFET (left) and a p-channel MOSFET (right). Note that $V_{DS}, V_{GS} > 0$ for the n-channel device and $V_{DS}, V_{GS} < 0$ for the p-channel device. The drain current flows in the drain of an n-channel MOSFET and out the drain of a p-channel MOSFET.

For circuit applications, transistors are usually configured to accept an input voltage and to operate at a certain output voltage. The input voltage is measured across the two input terminals and the output voltage across the two output terminals. The input current is the current that flows into one of the two input terminals and out of the other, and the output current is the current that flows into one of the two output terminals and out of the other. (By convention, the "circuit convention," the current is considered to be positive if it flows into a terminal, so the drain current of an n-channel MOSFET is positive, and the drain current of a p-channel MOSFET is negative.) Since we only have three terminals (the body contact is special — it tunes the operating characteristics of the MOSFET), one of terminals must be connected in common to both the input and the output. Possibilities are *common source*, *common drain*, and *common gate* configurations.

Figure 2.5 shows an n-channel MOSFET connected in the common source configuration. In this case, the DC output current is the drain to source current, I_{DS}, and the DC output voltage is the drain to source voltage, V_{DS}. The DC input voltage is the gate to source voltage, V_{GS}. For MOSFETs, the DC gate current is typically very small and can usually be neglected.

Our goal in this lecture is to understand the general features of transistor IV characteristics and to introduce some of the terminology used. Two

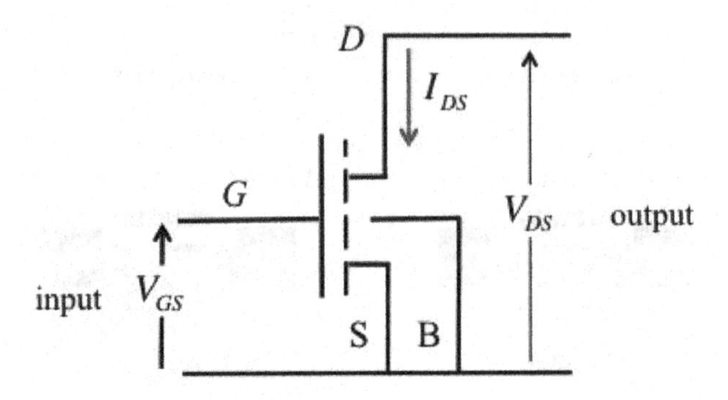

Fig. 2.5 An n-channel MOSFET configured in the common source mode. The input voltage is V_{GS}, and the output voltage, V_{DS} . The output current is I_{DS}, and the gate current is typically negligibly small, so the DC input current is assumed to be zero.

types of *IV* characteristics are of interest; the first are the *output characteristics*, a plot of the output current, I_{DS}, vs. the output voltage, V_{DS}, for a constant input voltage, V_{GS}. The second *IV* characteristic of interest is the *transfer characteristic*, a plot of the output current, I_{DS}, as a function of the input voltage, V_{GS} for a fixed output voltage, V_{DS}. In the remainder of this lecture, we treat the transistor as a black box, as in Fig. 2.1, and simply describe the *IV* characteristics and define some terminology. Subsequent lectures will relate these *IV* characteristics to the underlying physics of the device.

2.3 IV characteristics

Figure 2.6 shows the *IV* characteristics of a simple device, a resistor. For an ideal resistor, the current is proportional to the voltage according to $I = V/R$, where R is the resistance in Ohms. Figure 2.7 shows the *IV* characteristics of a current source. For an ideal current source, the current is independent of voltage, but real current sources show some dependence of the current on the voltage across the terminals. Accordingly, a real current source can be represented as an ideal current source in parallel with an ideal resistor, as shown in Fig. 2.7. The output characteristics of a MOSFET look like a resistor for small V_{DS} and like a current source for large V_{DS}.

The output characteristics of an n-channel MOSFET are shown in Fig. 2.8 (same as Fig. 1.2). Each line in the family of characteristics corre-

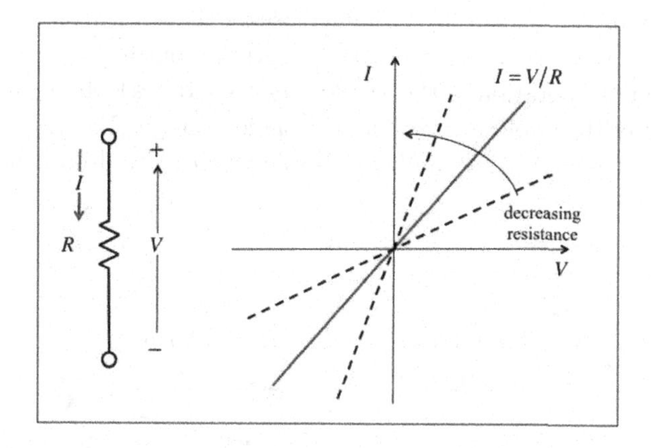

Fig. 2.6 The IV characteristics of an ideal resistor.

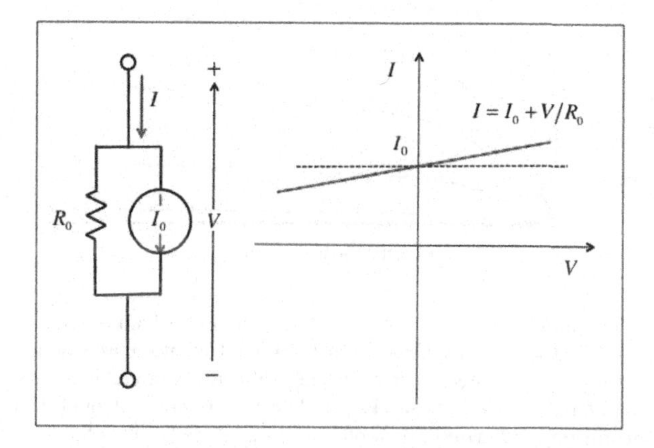

Fig. 2.7 The IV characteristics of a current source. The dashed line is an ideal current source, for which the current is independent of the voltage across the terminals. Real current sources show some dependence of the current on the voltage, which can be represented by a ideal current source in parallel with a resistor, R_O, as shown on the left.

sponds to a different input voltage, V_{GS}. For V_{DS} less than some critical value (called V_{DSAT}), the current is proportional to the voltage. In this small V_{DS} (*linear* or *ohmic*) region, a MOSFET operates like a resistor with the resistance being determined by the input voltage, V_{GS}.

For $V_{DS} > V_{DSAT}$, (the *saturation* or *beyond pinch-off* region), the MOSFET operates as a current source with the value of the current being determined by V_{GS}. The current increases a little with increasing V_{DS}, which shows that the current source has a finite output resistance, r_d. A third region of operation is the *subthreshold* region, which occurs for V_{GS} less than a critical voltage, V_T, the threshold voltage. For $V_{GS} < V_T$, the drain current is very small and not visible when plotted on a linear scale as in Fig. 2.8.

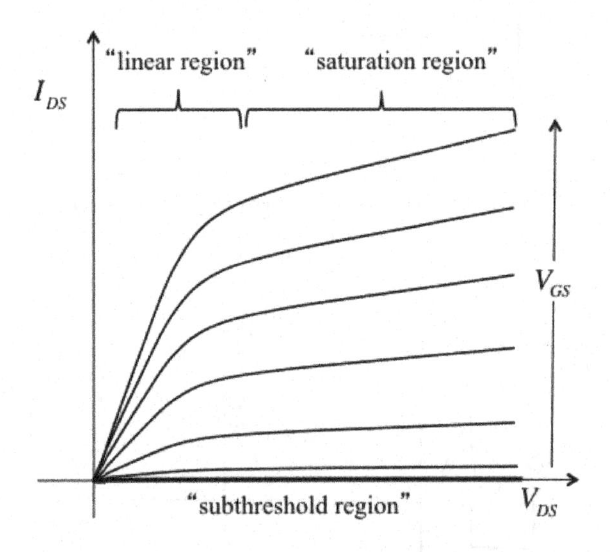

Fig. 2.8 The common source output IV characteristics of an n-channel MOSFET. The vertical axis is the current that flows between the drain and source, I_{DS}, and the horizontal axis is the voltage between the drain and source, V_{DS}. Each line corresponds to a different gate voltage, V_{GS}. The two regions of operation, linear (or ohmic) and saturation (or beyond pinch-off) are also labeled. (This figure is the same as Fig. 1.2.)

Figure 2.9 compares the output and transfer characteristics for an n-channel MOSFET. The output characteristics are shown on the left. Consider fixing V_{DS} to a small value and sweeping V_{GS}. This gives the line labeled V_{DS1} in the transfer characteristics on the right. If we fix V_{DS} to a large value and sweep V_{GS}, then we get the line labeled V_{DS2} in the transfer characteristic. The transfer characteristics also show that for $V_{GS} < V_T$, the current is very small. A plot of $\log_{10}(I_{DS})$ vs. V_{GS} is used to resolve the current in this subthreshold region (see Fig. 2.12).

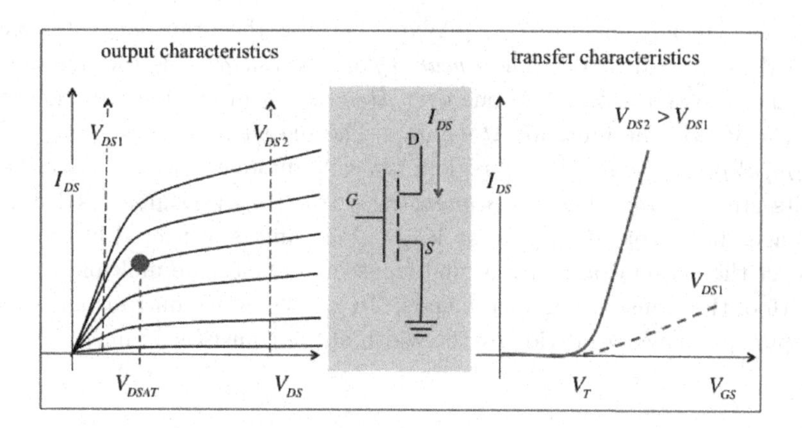

Fig. 2.9 A comparison of the common source output characteristics of an n-channel MOSFET (left) with the common source transfer characteristics of the same device (right). The line labeled V_{DS1} in the transfer characteristics is the low V_{DS} line indicated on the output characteristic on the left, and the line labeled V_{DS2} in the transfer characteristic corresponds to the high V_{DS} line indicated on the output characteristic.

2.4 MOSFET device metrics

The performance of a MOSFET can be summarized by a few device metrics as listed below.

on-current, I_{ON}, in μA/μm
linear region on resistance R_{ON}, in $\Omega - \mu$m
output resistance, r_d, in $\Omega - \mu$m
transconductance, g_m, in μS/μm.
off-current, I_{OFF}, in μA/μm
subthreshold swing, SS, in mV/decade
drain-induced barrier lowering, $DIBL$, in mV/V
threshold voltage, V_T(lin) and V_T(sat) in V
drain saturation voltage, V_{DSAT}, in V

The units listed above are those that are commonly used, which are not necessarily MKS units. For example, the transconductance is not typically quoted in Siemens per meter (S/m), but in micro-Siemens per micrometer, μS/μm or milli-Siemens per millimeter, mS/mm.

As shown in Fig. 2.10, several of the device metrics can be determined from the common source output characteristics. The *on-current* is the maximum drain current, which occurs at $I_{DS}(V_{GS} = V_{DS} = V_{DD})$, where V_{DD} is the power supply voltage. Note that the drain to source current,

I_{DS}, is typically measured in μA/μm, because the drain current scales linearly with width, W. The *linear region on-resistance* is the minimum channel resistance, which is one over dI_{DS}/dV_{DS} in the linear region for $V_{GS} = V_{DD}$. The units are $\Omega - \mu$m. The *output resistance* is one over dI_{DS}/dV_{DS} in the saturation region; typically quoted at $V_{GS} = V_{DD}$. The units are $\Omega - \mu$m. The *transconductance* is dI_{DS}/dV_{GS} at a fixed drain voltage. It is typically quoted at $V_{DS} = V_{DD}$ and is measured in μS/μm. To get the actual drain current and transconductance, we multiply by the width of the transistor in micrometers. To get the actual on-resistance and output resistance, we divide by the width of the transistor in micrometers.

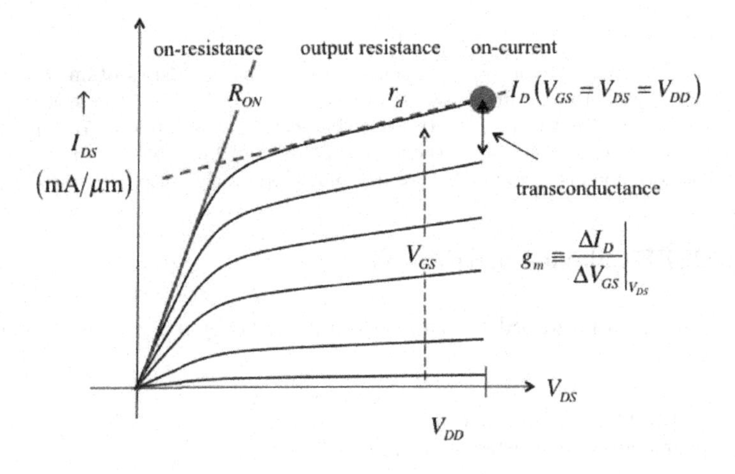

Fig. 2.10 The common source output characteristics of an n-channel MOSFET with four device metrics indicated.

As shown in Fig. 2.11, additional device metrics can be determined from the common source transfer characteristics with the current plotted on a linear scale. The two different IV characteristics are for low V_{DS} (linear region of operation) and for high V_{DS} (saturation region). The on-current noted in Fig. 2.10 is also shown in Fig. 2.11. If we find the point of maximum slope on the IV characteristics, plot a line tangent to the curve at that point, and read off the x-axis intercept, we find the threshold voltage. Note that there are two threshold voltages, one obtained from the linear region plot, V_T(lin) and another from the saturation region plot, V_T(sat) and that $V_T(\text{sat}) < V_T(\text{lin})$. Note that the off to on transition is gradual and V_T is approximately the point at which this transition is complete. Finally, the *off-current*, $I_{DS}(V_{GS} = 0, V_{DS} = V_{DD}$, is also indicated in Fig. 2.11, but it is too small to read from this plot.

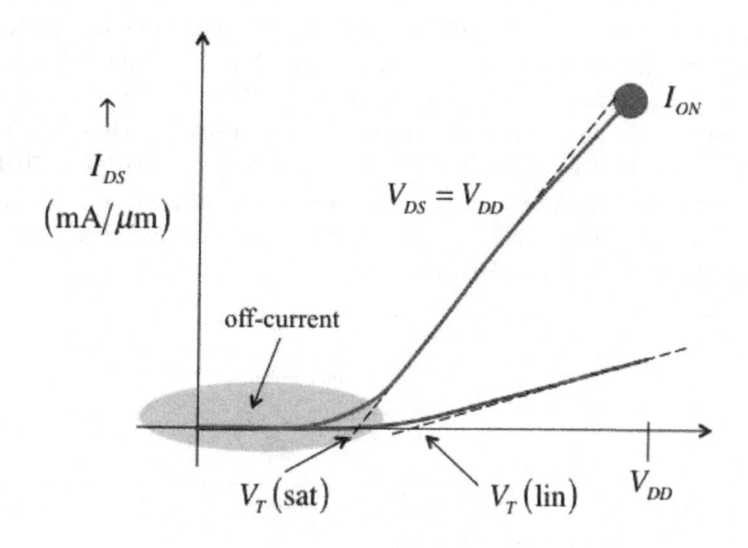

Fig. 2.11 The common source transfer characteristics of an n-channel MOSFET with three device metrics indicated, V_T(lin) and V_T(sat), and the on-current. The drain current, I_{DS}, is plotted on a linear scale in this plot.

To resolve the subthreshold characteristics, we should plot the drain current on a logarithmic scale, as shown in Fig. 2.12. Both the off-current, $I_{OFF} = I_{DS}(V_{GS} = 0, V_{DS} = V_{DD})$, and the on-current, $I_{ON} = I_{DS}(V_{GS} = V_{DS} = V_{DD})$, are identified in this figure. The subthreshold current in a well-behaved MOSFET increases exponentially with V_{GS}. The *subthreshold swing*, SS, is given by

$$SS = \left[\frac{d(\log_{10} I_{DS})}{dV_{GS}}\right]^{-1} \tag{2.1}$$

and is typically quoted in millivolts per decade. In words, the subthreshold swing is the change in gate voltage (typically quoted in millivolts) needed to change the drain current by a factor of 10. The smaller the SS, the lower is the gate voltage needed to switch the transistor from off to on. As we'll discuss in Sec. 2, the physics of subthreshold conduction dictate that $SS \geq 60$ mV/decade. In a well-behaved MOSFET, the subthreshold swing is the same for the low and high V_{DS} transfer characteristics. An increase of SS with increasing V_{DS} is often observed and is attributed to the influence of two-dimensional electrostatics (which will also be discussed in Part 2: MOS Electrostatics).

Finally, we note that the subthreshold IV characteristics are shifted to the left for increasing drain voltages. This shift is attributed to an effect known as *drain-induced barrier lowering* ($DIBL$) and is defined as the horizontal shift in the low and high V_{DS} subthreshold characteristics divided by the difference in drain voltages (typically $V_{DD} - 0.05$ V). DIBL is closely related to the two threshold voltages shown in Fig. 2.11. An ideal MOSFET has zero DIBL and a threshold voltage that does not change with drain voltage, *i.e.*, $V_T(\text{lin}) = V_T(\text{sat})$.

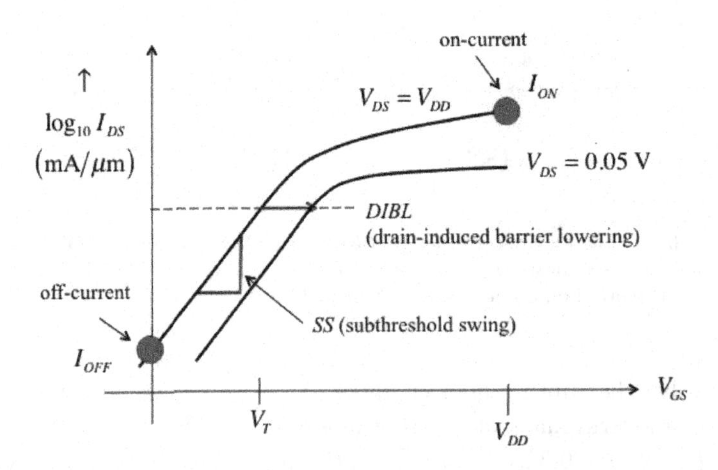

Fig. 2.12 The common source transfer characteristics of an n-channel MOSFET with two additional device metrics indicated, SS and $DIBL$. The drain current, I_{DS}, is plotted on a logarithmic scale in this plot.

As mentioned earlier, it is important to note that threshold voltage is not a precisely defined quantity. It is approximately the gate voltage at which significant drain current begins to flow, and there are different ways to specify this voltage. For example, it may be determined from the x-intercept of a plot of I_{DS} vs. V_{GS} as indicated in Fig. 2.11. Alternatively, one could specify a small drain current (e.g. perhaps 10^{-7}A/μm as in the horizontal dashed line in Fig. 2.12) and simply define V_T as the gate voltage needed to achieve this current. When a threshold voltage is quoted, one should, therefore, be sure to understand exactly how V_T was defined.

Finally, a word about notation. In Figs. 2.2 and 2.4, we define the current flowing into the drain of an n-MOSFET as I_D. Ideally, the same current flows out of the channel and $I_D = -I_S = I_{DS}$. In practice, there may be some leakage currents (i.e. some of the drain current may flow to the substrate), so that $I_D > I_S$. We'll not be concerned with these leakage currents in these notes and will assume that $I_D = I_S = I_{DS}$, where I_{DS} is the current that flows from the drain to the source.

2.5 Summary

In this lecture we described the shape of the IV characteristics of a MOSFET and defined several metrics that are commonly used to characterize the performance of MOSFETs. We briefly discussed the physical structure of a MOSFET, but did not discuss what goes on inside the black box to produce the IV characteristics we described. Subsequent lectures will focus on the physics that leads to these IV characteristics and on developing simple expressions for several of the key device metrics.

2.6 References

There are many types of transistors, for an incomplete list, see:

[1] Kwok K. Ng "A survey of semiconductor devices," *IEEE Trans, Electron Devices*, **43**, pp. 1760-1766, 1996.

For an introduction to Moore's Law and its implications for electronics, see:

[2] "Moore's law," `http://en.wikipedia.org/wiki/Moore's_law`, July 19, 2013.

[3] M. Lundstrom, "Moore's Law Forever?" an Applied Physics Perspective, *Science*, **299**, pp. 210-211, January 10, 2003.

Lecture 3

The MOSFET: A Barrier-Controlled Device

3.1 Introduction

Most transistors operate by controlling the height of an energy barrier with an applied voltage. This includes so-called field-effect transistors (FET's), such as MOSFETs, JFETs (junction FET's), HEMTs (high electron mobility transistors, which are also FET's) as well as BJT's (bipolar junction transistors) and HBT's (heterojunction bipolar transistors) [1, 2]. The operating principles of these transistors are most easily understood in terms of energy band diagrams, which provide a qualitative way to understand MOS electrostatics. The energy band view of a MOSFET is the subject of this lecture.

3.2 Equilibrium energy band diagram

As sketched in Fig. 3.1, the MOSFET is inherently a two-dimensional (or even three-dimensional) device. For a complete understanding of the device, we must understand multi-dimensional energy band diagrams, but most of the essential principles can be conveyed with 1D energy band diagrams. Accordingly, we aim to understand the energy vs. position plot along the surface of the MOSFET as indicated in Fig. 3.1.

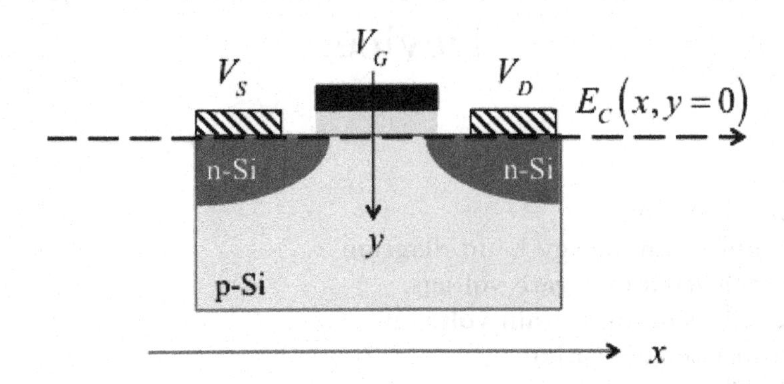

Fig. 3.1 Sketch of a MOSFET cross-section showing the line along the Si surface for which we will sketch the energy vs. position, $E_C(x, y = 0)$, from the source, across the channel, to the drain. The z-axis is into the page, in the direction of the width of the transistor, W.

The source and drain regions of the n-channel MOSFET are heavily doped n-type, and the channel is p-type. In a uniformly doped bulk semiconductor, the bands are independent of position with the Fermi level near E_c for n-type semiconductors and near E_v for p-type. The upper part of Fig. 3.2 shows separate n-type, p-type, and n-type regions. We conceptually put these three regions together to draw the energy band diagram. In equilibrium, we begin with the principle that the Fermi level (electrochemical potential) is constant. Far to the left, deep in the source, E_c must be near E_F, and far to the right, deep in the drain, the same thing must occur. In the channel, E_v must be near E_F. In order to line up the Fermi levels in the three regions, the source and drain energy bands must drop in energy until E_F is constant (or, equivalently, the channel must rise in energy). The alignment of the Fermi levels occurs because electrons flow from higher Fermi level to lower Fermi level (from the source and drain regions to the channel), which sets up a charge imbalance and produces an electrostatic potential difference between the two regions. The source and drain regions acquire a positive potential (the so-called *built-in potential*), which lowers the bands according to

$$E_c(x) = E_{co} - q\psi(x)$$
$$E_v(x) = E_{vo} - q\psi(x), \quad (3.1)$$

where the subscript "o" indicates the value in the absence of an electrostatic potential, ψ.

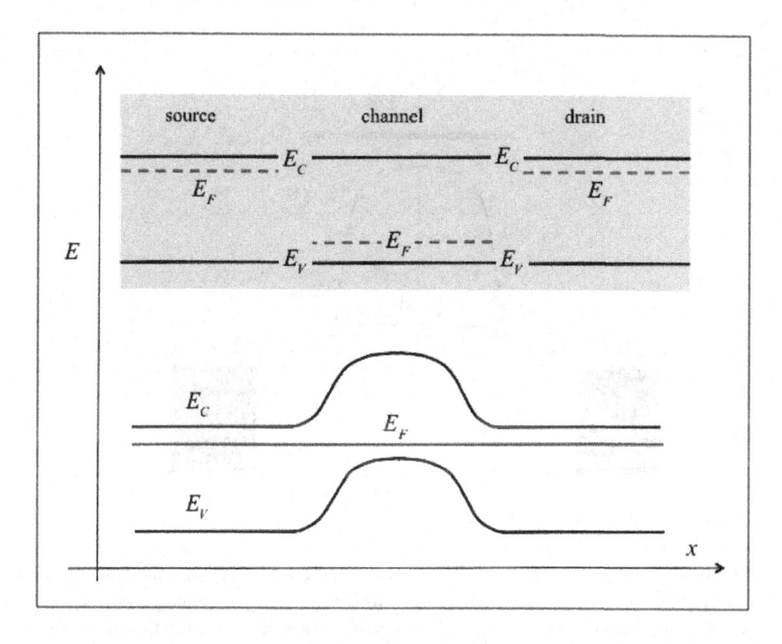

Fig. 3.2 Sketch of the equilibrium energy band diagram along the top surface of a MOSFET. Top: separate n-type, p-type, and n-type regions representing the source, channel, and drain regions. Bottom: The resulting equilibrium energy band diagram when all three regions are connected and $V_S = V_G = V_D = 0$.

Because the device of Fig. 3.2 is in equilibrium, no current flows. Note that there is a potential energy barrier that separates electrons in the source from electrons in the drain. This barrier will play an important role in our understanding of how transistors work. The next step is to understand how the energy bands change when voltages are applied to the gate and drain terminals.

3.3 Application of a gate voltage

Figure 3.3 shows what happens when a positive voltage is applied to the gate. In this figure, we show only the conduction band, because we are discussing an n-channel MOSFET for which the current is carried by electrons in the conduction band. Also shown in Fig. 3.3 are the metal source and drain contacts. (We assume ideal contacts, for which the Fermi levels in the metal align with Fermi level in the semiconductor in equilibrium.) Since $V_S = V_D = 0$, the Fermi levels in the source, device, and drain all align; the device is in equilibrium, and no current flows.

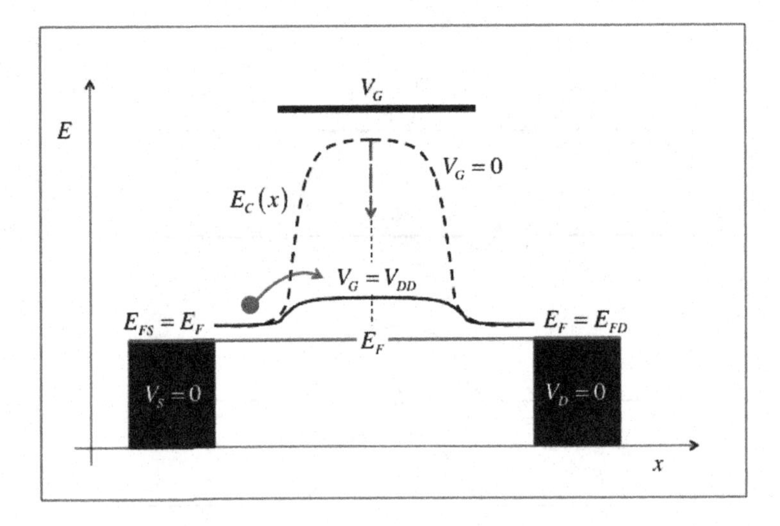

Fig. 3.3 Sketch of the equilibrium electron potential energy vs. position for an n-channel MOSFET for low gate voltage (dashed line) and for high gate voltage (solid line). The voltages on the source, drain, and gate electrodes are zero. The Fermi levels in the source and source contact, in the channel, and in the drain and drain contact are all equal, $E_{FS} = E_F = E_{FD}$ because the device is in equilibrium with no voltage applied to the source and drain contacts. The application of a gate voltage does not disturb equilibrium because the gate electrode is insulated from the silicon by the gate oxide insulator.

Also shown in Fig. 3.3 is what happens when a positive gate voltage is applied. The gate electrode is separated from the silicon channel by an insulating layer of SiO_2, but the positive potential applied to the gate influences the potential in the semiconductor. A positive gate voltage increases the electrostatic potential in the channel, which lowers the conduction band according to eqn. (3.1).

It is important to note that the application of a gate voltage does not affect the Fermi level in the underlying silicon. A positive gate voltage lowers the Fermi level in the gate electrode, but the gate electrode is isolated from the underlying silicon by the gate oxide. The Fermi level in the device can only change if the source or drain voltages change, because the source and drain Fermi levels are connected to the Fermi level in the device.

We conclude that the application of a gate voltage simply raises or lowers the potential energy barrier between the source and drain. The device remains in equilibrium, and no current flows. The fact that the device is in equilibrium even with a gate voltage applied simplifies the analysis of MOS electrostatics, which we will discuss in the next few lectures.

3.4 Application of a drain voltage

Figure 3.4 shows what happens when a large drain voltage is applied. The source is grounded, so the quasi-Fermi level (electrochemical potential) in the source does not change from equilibrium, but the positive drain voltage lowers the quasi-Fermi level in the drain. Lowering the Fermi level lowers E_c too, because $E_F - E_c$ determines the electron density. Electrostatics will attempt to keep the drain neutral, so $n \approx n_0 \approx N_D$, where N_D is the doping density in the drain. The resulting energy band diagrams under low and high gate voltages are shown in Fig. 3.4. Note that we have only shown the quasi-Fermi levels in the source and drain, but $F_n(x)$ varies smoothly across the device. In general, numerical simulations are needed to resolve $F_n(x)$, but it is clear that there will be a slope to $F_n(x)$, so current will flow. The device is no longer in equilibrium when the electrochemical potential varies with position.

Consider first the case of a large drain voltage and small gate voltage as shown in the dashed line of Fig. 3.4. In a well-designed transistor, the height of energy barrier between the source and the channel is controlled only (or mostly) by the voltage applied to the gate. If the gate voltage is low, the energy barrier is high, and very few electrons have enough energy to surmount the barrier and flow to the drain. The transistor is in the *off-state* corresponding to the $I_{DS} \approx 0$ part of the IV characteristic of Fig. 2.10. Current flows, but only the small leakage off-current, I_{OFF} (Fig. 2.12).

When a large gate voltage is applied in addition to the large drain voltage (shown as the solid line in Fig. 3.4), the gate voltage increases the electrostatic potential in the channel and lowers the height of the barrier.

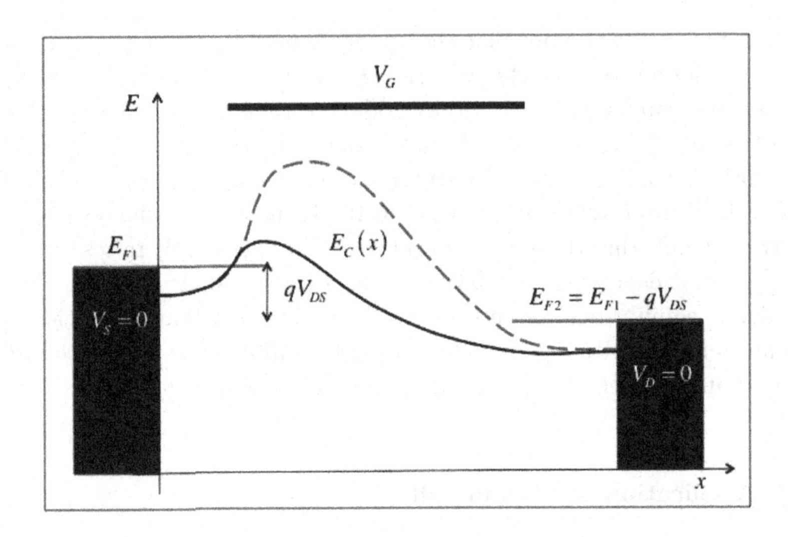

Fig. 3.4 Sketch of $E_c(x)$ vs. x along the channel of an n-channel MOSFET. Dashed line: Large drain voltage and small gate voltage. Solid line: Large drain voltage and large gate voltage.

If the barrier is low enough, a significant fraction of the electrons in the source can hop over the energy barrier and flow to the drain. The transistor is in the *on-state* with the maximum current being the on-current, I_{ON}, at $V_{GS} = V_{DS} = V_{DD}$ of Fig. 2.10.

3.5 Transistor operation

Figure 3.4 illustrates the basic operating principle of most transistors – controlling current by modulating the height of an energy barrier with an applied voltage. We have described the physics of the off-state and on-state of the *IV* characteristic of Fig. 2.10, but the entire characteristic can be understood with energy band diagrams. Figure 3.5 shows numerical simulations of the conduction band vs. gate voltage in the linear region of operation. Note that under high gate voltage, $E_c(x)$ varies linearly with position in the channel, which corresponds to a constant electric field, as expected in the linear region of operation where the device acts as a gate voltage controlled resistor.

Figure 3.6 shows simulations of the conduction band vs. gate voltage in the saturated region of operation. As the gate voltage pushes the potential

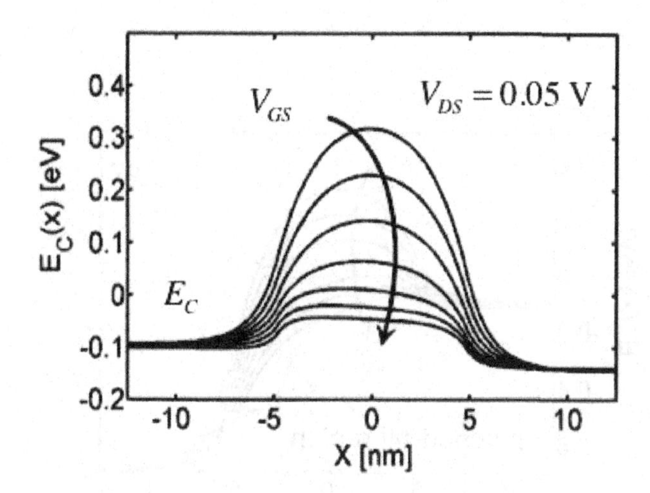

Fig. 3.5 Simulations of $E_C(x)$ vs. x for a short channel transistor. A small drain voltage is applied, so the device operates in the linear region. Each line corresponds to a different gate voltage, with the gate voltage increasing from the top down. (©IEEE 2002. Reprinted, with permission, from: Mark Lundstrom and Zhibin Ren, "Essential Physics of Carrier Transport in Nanoscale MOSFETs," *IEEE Trans. Electron Dev.*, **49**, pp. 133-141, 2002.)

energy barrier down, electrons in the source hop over the barrier and then flow down hill to the drain. This figure also illustrates why the drain current saturates with increasing drain voltage. It is the barrier between the source and channel that limits the current. Electrons that make it over the barrier flow down hill and out the drain. Increasing the drain voltage (assuming that it does not lower the source to channel barrier) should not increase the current. Note also that even under very high gate voltage, a small barrier remains. Without this barrier and its modulation by the gate voltage, we would not have a transistor.

3.6 IV characteristic

The mathematical form of the *IV* characteristic of a transistor can also be understood with the help of energy bandindexMOSFET!thermionic emission model diagrams and a simple, *thermionic emission* model. Consider first the common source characteristic of Fig. 2.10. The net drain current is the current from the left to right (from the source, over the barrier, and out the drain) minus the current form the right to left (from the drain, over

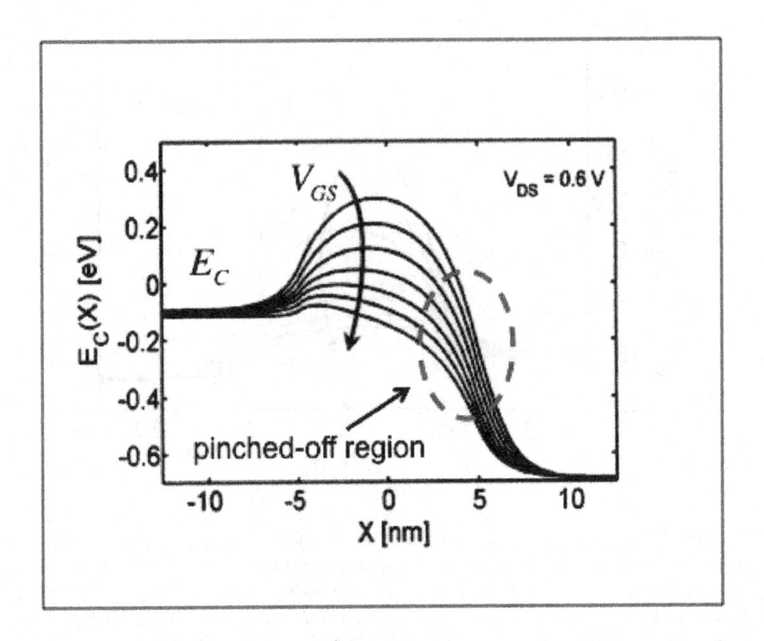

Fig. 3.6 Simulations of $E_c(x)$ vs. x for a short channel transistor. A large drain voltage is applied, so the device operates in the saturation region. Each line corresponds to a different gate voltage, with the gate voltage increasing from the top down. (The pinch-off region will be discussed in Sec. 4.6.) (©IEEE 2002. Reprinted, with permission, from: Mark Lundstrom and Zhibin Ren, "Essential Physics of Carrier Transport in Nanoscale MOSFETs," *IEEE Trans. Electron Dev.*, **49**, pp. 133-141, 2002.)

the barrier, and out the source):

$$I_{DS} = I_{LR} - I_{RL} \,. \tag{3.2}$$

The probability that an electron can surmount the energy barrier and flow from the source to the drain is $\exp(-E_{SB}/k_BT)$, where E_{SB} is the barrier height from the source to the top of the barrier, so the current from the left to the right is

$$I_{LR} \propto e^{-E_{SB}/k_BT} \,. \tag{3.3}$$

The probability that an electron can surmount the barrier and flow from the drain to the source is $\exp(-E_{DB}/k_BT)$, where E_{DB} is the barrier height from the drain to the top of the barrier. The current from the right to left is, therefore,

$$I_{RL} \propto e^{-E_{DB}/k_BT} \,. \tag{3.4}$$

Because the drain voltage pulls the conduction band in the drain down, $E_{DB} > E_{SB}$. When there is no DIBL, $E_{DB} = E_{SB} + qV_{DS}$, so $I_{RL}/I_{LR} = \exp(-qV_{DS}/k_BT)$, and we can write the net drain current as

$$I_{DS} = I_{LR} - I_{RL} = I_{LR}\left(1 - e^{-qV_{DS}/k_BT}\right) . \tag{3.5}$$

At the top of the barrier, there are two streams of electrons, one moving to the right and one to the left. They have the same kinetic energy, so their velocities, υ_T, are the same. Current is charge times velocity. For a MOSFET, the charge flows in a two-dimensional channel, so it is the charge per area in C/cm^2 that is important. The left to right current is $I_{LR} = WQ_n^+(x = 0)\upsilon_T$, where $Q_n^+(x = 0)$ is the charge in C/cm^2 at the top of the barrier due to electrons with positive velocities, and W is the width of the MOSFET. Similarly, $I_{RL} = WQ_n^-(x = 0)\upsilon_T$. We find the total charge by adding the charge in the two streams,

$$\begin{aligned} Q_n(x=0) &= \frac{I_{LR} + I_{RL}}{W\upsilon_T} \\ &= \frac{I_{LR}}{W\upsilon_T}\left(1 + I_{RL}/I_{LR}\right) \\ &= \frac{I_{LR}}{W\upsilon_T}\left(1 + e^{-qV_{DS}/k_BT}\right) . \end{aligned} \tag{3.6}$$

Finally, if we solve eqn. (3.6) for I_{LR} and insert the result in eqn. (3.5), we find the IV characteristic of a ballistic MOSFET as

$$I_{DS} = W|Q_n(x=0)|\upsilon_T \frac{\left(1 - e^{-qV_{DS}/k_BT}\right)}{\left(1 + e^{-qV_{DS}/k_BT}\right)} . \tag{3.7}$$

In Lecture 13, we will derive eqn. (3.7) more formally, learn some of its limitations, and define the velocity, υ_T. The general form of the ballistic IV characteristic is, however, easy to understand in terms of thermionic emission in a barrier controlled device.

Now let's examine the general result, eqn. (3.7) under low and high drain bias. For small drain bias, a Taylor series expansion of the exponentials gives

$$I_{DS} = W|Q_n(x=0)|\frac{\upsilon_T}{2k_BT/q}V_{DS} = G_{CH}V_{DS} = V_{DS}/R_{CH} , \tag{3.8}$$

where G_{CH} (R_{CH}) is the channel conductance (resistance). Equation (3.8) is a ballistic treatment of the the linear region of the IV characteristic in Fig. 2.10.

Consider next the high V_{DS}, saturated region of the common source characteristic of Fig. 2.10. In this case, $I_{RL} \ll I_{LR}$ and the drain current saturates at $I_{DS} = I_{LR}$. In the limit, $V_{DS} \gg k_BT/q$, eqn. (3.7) becomes

$$I_{DS} = W|Q_n(x=0)|\upsilon_T \,. \tag{3.9}$$

The high V_{DS} current is seen to be independent of V_{DS}, but we will see later that DIBL causes $Q_n(x=0)$ to increase with drain voltage, so I_{DS} does not completely saturate.

Having explained the common source IV characteristic, we now turn to the transfer characteristic of Fig. 2.12. The transfer characteristic is a plot of I_{DS} vs. V_{GS} for a fixed V_{DS}. Let's assume that we fix the drain voltage at a high value, so the current is given by eqn. (3.9) and the question is: "How does $Q_n(x=0)$ vary with gate voltage?"

For high drain voltage, $I_{RL} = 0$, so eqn. (3.6) gives

$$|Q_n(x=0)| = \frac{I_{LR}}{W\upsilon_T} \,. \tag{3.10}$$

The current, I_{LR} is due to thermionic emission over the source to channel barrier. Application of a gate voltage lowers this barrier, so we can write:

$$I_{LR} \propto e^{-E_{SB}/k_BT} = e^{-(E_{SB}^0 - qV_{GS}/m)/k_BT} \,, \tag{3.11}$$

where E_{SB}^0 is the barrier height from the source to the top of the barrier at $V_{GS} = 0$, and $1/m$ is the fraction of the gate voltage that gets to the semiconductor surface (some of the gate voltage is dropped across the gate oxide). From eqns. (3.11) and (3.10), we find

$$Q_n(V_{GS}) = Q_n(V_{GS}=0)\; e^{qV_{GS}/mk_BT} \,. \tag{3.12}$$

From eqns. (3.12) and (3.9), we see that the current increases exponentially with gate voltage,

$$I_{DS} = W|Q_n(V_{GS}=0)|\upsilon_T\; e^{qV_{GS}/mk_BT} \,. \tag{3.13}$$

In fact, it is easy to show that to increase the current by a factor of ten (a decade), the gate voltage must increase by $2.3mk_BT/q \geq 0.060$ V at room temperature. This 60 mV per decade is characteristic of thermionic emission over a barrier.

According to eqn. (3.13), the drain current is independent of drain voltage; in practice, there is a small increase in drain current with increasing drain voltage because the drain voltage "helps" the gate pull down the source to channel barrier. This is the physical explanation for DIBL — it is due to the two-dimensional electrostatics that we will discuss in Lecture 10.

Equation (3.13) describes the exponential increase of I_{DS} with V_G observed in Fig. 2.12 below threshold, but above threshold, the drain current of a MOSFET does not increase exponentially with gate voltage; it increases approximately linearly with gate voltage. Above threshold, eqn. (3.3) still applies, it is just that the decrease in the height of the potential barrier is no longer proportional to V_{GS} above threshold; there is a lot of charge in the semiconductor, which screens the charge on the gate, and makes it difficult for the gate voltage to push the barrier down. The parameter, m, becomes very large. The same considerations apply to the charge as well. Below threshold, eqn. (3.12) shows that the charge varies exponentially with gate voltage, but above threshold, we will find that it varies linearly with gate voltage.

When we discuss MOS electrostatics in Lectures 8 and 9, we will show that above threshold, the charge increases linearly with gate voltage as in eqn. (3.14) below.

$$\begin{aligned} Q_n(V_{GS}, V_{DS}) &= -C_{ox}\left(V_{GS} - V_T\right) \\ V_T &= V_{T0} - \delta V_{DS} \end{aligned} , \qquad (3.14)$$

where Q_n is the mobile electron charge, $C_{ox} = \kappa_{ox}\epsilon_0/t_{ox}$, where t_{ox} is the oxide thickness, is the gate capacitance per unit area. Also in eqn. (3.14), V_T is the threshold voltage, and δ is the drain-induced barrier lowering (DIBL) parameter. (We'll see later that the appropriate capacitance to use is a little less than C_{ox}.)

This discussion shows that the IV characteristics of a ballistic MOSFET can be easily understood in terms of thermionic emission over a gate controlled barrier. When we return to this problem in Lecture 13, we will learn a more formal and more comprehensive way to treat ballistic MOSFETs, but the underlying physical principles will be the same.

3.7 Discussion

Transistor physics boils down to electrostatics and transport. The energy band diagram is a qualitative illustration of transistor electrostatics. In practice, most of transistor design is about engineering the device so that the energy barrier is appropriately manipulated by the applied voltages. The design challenges have increased as transistors have gotten smaller and smaller, and we understand transistor electrostatics better now, but the basic principles are the same as they were in the 1960's.

Figure 3.7 illustrates the key principles of a well-designed short channel MOSFET. The *top of the barrier* is a critical point; it marks the beginning of the channel and is also called the *virtual source.* In a well-designed MOSFET, the height of the source to channel energy barrier is strongly controlled by the gate voltage and only weakly dependent on the drain voltage.

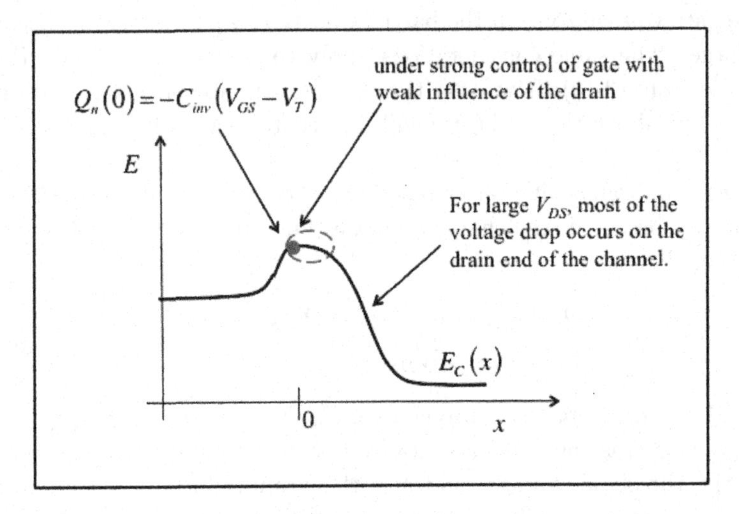

Fig. 3.7 Sketch of a well-designed short channel MOSFET under high gate and drain bias. In a well-designed short channel MOSFET, the charge at the top of the barrier is very close to the value it would have in a long channel device, for which the lateral electric field could be neglected. In a well-designed MOSFET, there is a low lateral electric field near the beginning of the channel and under high V_{DS}, the drain voltage has only a small influence on the region near the top of the barrier.

Under low V_{DS} and high V_{GS}, the potential drops approximately linearly in the channel, so the electric field is approximately constant. Under high drain and gate bias, the electric field is high and varies non-linearly with position. Near the beginning of the channel (near the top of the barrier) the electric field is low, but near the drain, the electric field is very large. In the saturation region, increases in drain voltage increase the potential drop in the high field part of the channel but leave the region near the top of the barrier relatively unaffected (if DIBL is small). Since the region near the top of the barrier controls the current, the drain current is relatively insensitive to the drain voltage in the saturation region.

Note from Fig. 3.7 that electrons that surmount the barrier and flow to the drain gain lot of kinetic energy. Some energy will be lost by electron-phonon scattering, but in a nanoscale transistor, there is not enough time for electrons to shed their kinetic energy as they flow to the drain. Accordingly, the velocity is very high in the part of the channel where the lateral potential drop (electric field) is high. Because current is the product of charge times velocity, the electron charge density will be very low in the region where the velocity is high. The part of the channel where the lateral potential drop is large and the electron density low is known in classical MOS theory as the *pinch-off* region. In a short channel device, the pinch-off region can be a substantial part of the channel, but for an electrostatically well-designed MOSFET, there must always be a small region near the source where the potential is largely under the control of the gate, and the lateral potential drop is small.

Figure 3.8 is a sketch of a long channel transistor under high gate and drain bias. Compared to the short channel transistor sketched in Fig. 3.7, we see that the low-field region under the control of the gate is a very large part of the channel, but there is still a short, pinch-off region near the drain. The occurrence of the pinch-off region under high drain bias is what causes the current to saturate. In the saturation or *beyond pinch-off* region, the current is mostly determined by transport across the low-field part of the channel, which is near the source, but most of the potential drop across the channel occurs in the high-field portion of the channel, which is near the drain. Once electrons enter the pinch-off region, they are quickly swept out to the drain.

In a well-designed MOSFET, the region near the top of the barrier is under the strong control of the gate voltage and only weakly affected by the drain voltage. The goal in transistor design is to achieve this performance as channel length scaling brings the drain closer and closer to the source. Once electrons hop over the source to channel barrier, they can flow to the drain. The electrostatic design of MOSFETs has gotten more challenging as device dimensions have been scaled down over the past five decades, but the principles have not changed. The nature of electron transport in transistors has, however, changed considerably as transistors have become smaller and smaller. A proper treatment of transport in nanoscale transistors is essential to understanding and designing these devices and will be our focus beginning in Lecture 14.

We have discussed 1D energy bands for a MOSFET by sketching $E_c(x)$ for $z = 0$, the surface of the silicon. Figure 3.9 shows these energy band

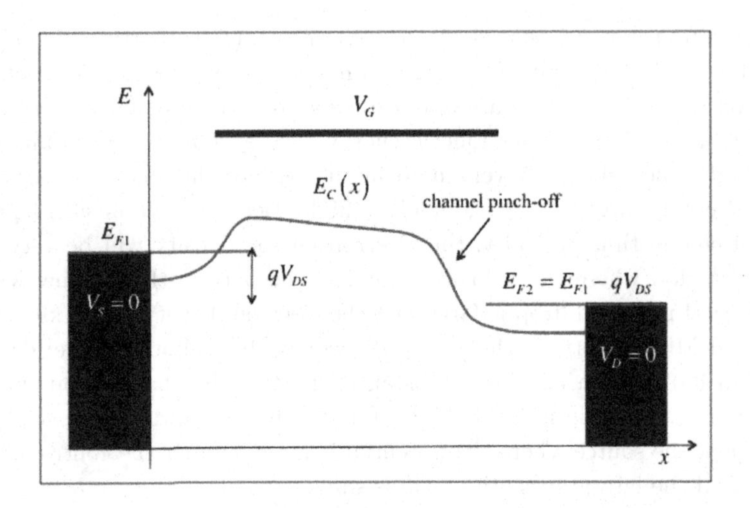

Fig. 3.8 Sketch of a long channel MOSFET under high gate and drain bias. In this case, the low lateral electric occupies a substantial part of the channel and the pinch-off region is short. Additional increases in V_{DS}, lengthen the pinch-off region a bit, but in a long channel transistor, it occupies a small portion of the channel.

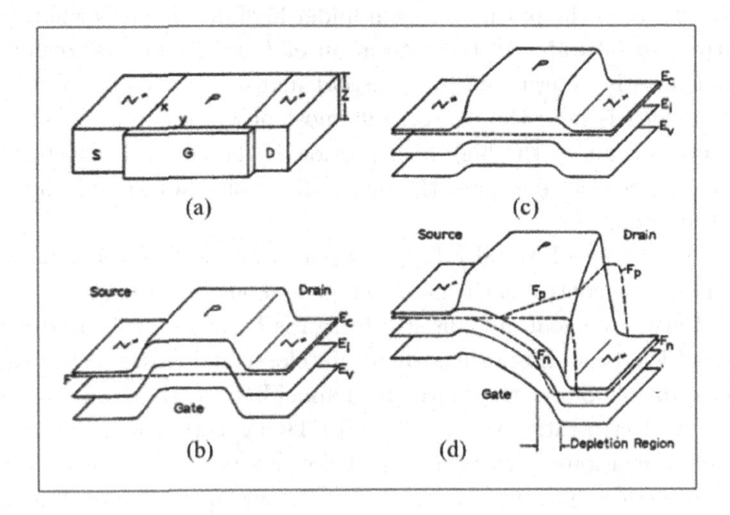

Fig. 3.9 Two dimensional energy band diagram for an n-channel MOSFET. (a) the device structure, (b) the equilibrium energy band diagram, (c) an equilibrium energy band diagram with a large gate voltage applied, and (d) the energy band diagram with large gate and drain voltages applied. (Reprinted from *Solid-State Electron.*, **9**, pp. 927-937, H.C Pao and C.T. Sah, "Effects of Diffusion Current on the Characteristics of Metal-Oxide (Insulator)-Semiconductor Transistors," ©1966, with permission from Elsevier.)

diagrams in two dimensions. Figure 3.9a is a sketch of the device. Figure 3.9b shows a device in equilibrium with $V_S = V_D = 0$ and the gate voltage adjusted so that the bands are flat in the direction normal to the channel. Figure 3.9c shows the device with a large gate voltage applied, but with V_S and V_D still at zero volts. Note that E_c along the surface of the device is just like the solid line in Fig. 3.3. Figure 3.9d shows the energy band diagram with large gate and drain voltages applied. In this case, E_c along the surface is just like the solid line in Fig. 3.4.

Finally, we note that the energy band diagrams that we have sketched are similar to the energy band diagrams for a bipolar transistor [1, 2]. In fact, the two devices both operate by controlling current by manipulating the height of an energy barrier [3]. The source of the MOSFET is analogous to the emitter of the BJT, the channel to the base of the BJT, and the drain to the collector of a BJT. This close similarity will prove useful in understanding the operation of short channel MOSFETs.

3.8 Summary

The MOSFET operates by controlling current through the manipulation of an energy barrier with a gate voltage. Understanding this gives a clear, physical understanding of how long and short channel MOSFETs operate. The control of current by a energy barrier is what gives a transistor its characteristic shape.

We can write the drain current as

$$I_{DS} = W|Q_n(V_{GS}, V_{DS})| \langle v \rangle \, . \tag{3.15}$$

This equation simply says that the drain current is proportional to the amount of charge in the channel and how fast that charge is moving. (The sign of Q_n is negative and because the current is defined to be positive when it flows into the drain, the absolute value is taken.) The charge, Q_n, flows into the channel to balance the charge on the gate electrode. While the shape of the IV characteristic is determined by MOS electrostatics, the magnitude of the current depends on how fast that charge flows.

3.9 References

Most of the important kinds of transistors are discussed in these texts:

[1] Robert F. Pierret *Semiconductor Device Fundamentals, 2nd Ed.*, Addison-Wesley Publishing Co, 1996.

[2] Y. Taur and T. Ning, *Fundamentals of Modern VLSI Devices*, 2nd Ed., Oxford Univ. Press, New York, 2013.

Johnson describes the close relation of bipolar and field-effect transistors.

[3] E.O. Johnson, "The IGFET: A Bipolar Transistor in Disguise," *RCA Review*, **34**, pp. 80-94, 1973.

Lecture 4

MOSFET IV: Traditional Approach

4.1 Introduction

The traditional approach to MOSFET theory was developed in the 1960's [1-4] and although they have evolved considerably, the basic features of the models used today are very similar to those first developed more than 50 years ago. My goal in this lecture is to briefly review the traditional theory of the MOSFET as it is presented in most textbooks (e.g. [5, 6]). Only the essential ideas of the traditional approach will be discussed. For example, we shall be content to compute the linear region current, and the saturated region current and not the entire IV characteristic. Only the above threshold IV characteristics will be discussed, not the subthreshold characteristics. Those interested in a full exposition of traditional MOSFET theory should consult standard texts such as [7, 8]. Later in these lectures, we will develop a much different approach to MOSFET theory — one better suited to the physics of nanoscale transistors, but we will also show, that it can be directly related to the traditional approach reviewed in this lecture.

4.2 Current, charge, and velocity

Figure 4.1 is a "cartoon" sketch of a MOSFET for which the drain to source current can be written as in eqn. (1.1),

$$I_{DS} = W|Q_n(x)|\langle\upsilon(x)\rangle , \tag{4.1}$$

where W is the width of the transistor in the y-direction, Q_n is the mobile sheet charge in the $x-y$ plane (C/m^2), and $\langle\upsilon\rangle$ is the average velocity at which the charge flows. We assume that the device is uniform in the z-direction (out of the page) and that current flows in the x-direction from the source to the drain. The quantity, Q_n, is called the *inversion layer charge* because it is an electron charge in a p-type material. The electron charge and velocity vary with position along the channel, but the current is constant if there is no electron recombination or generation. Accordingly, we can evaluate the current at the location along the channel where it is the most convenient to do so.

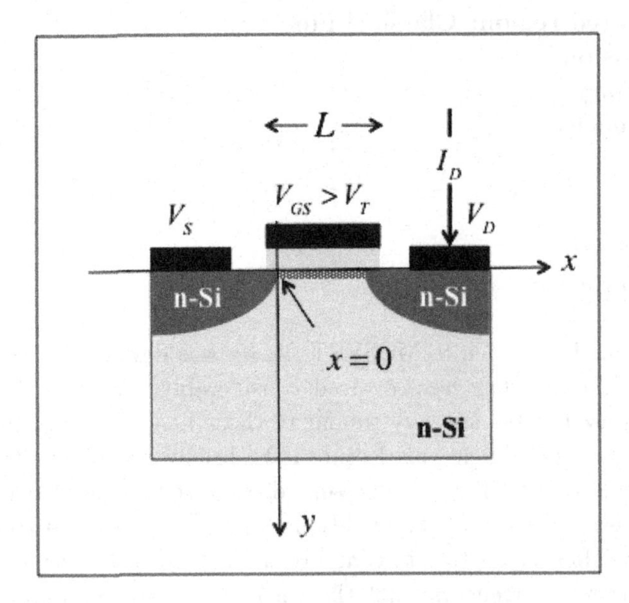

Fig. 4.1 Sketch of a simple, n-channel, enhancement mode MOSFET. The z-direction is normal to the channel, and the y-axis is out of the page. The beginning of the channel is located at $x = 0$. An inversion charge is present in the channel because $V_{GS} > V_T$; it is uniform between $x = 0$ and $x = L$ as shown here, if $V_S = V_D = 0$.

Consider the MOSFET of Fig. 4.1 with $V_S = V_D = 0$, but with $V_G > 0$. The MOSFET is in equilibrium and no current flows. In this case, the inversion layer charge is independent of x. As we will discuss in Part 2: MOS Electrostatics, there is very little charge when the gate voltage is less than a critical value, the threshold voltage, V_T. For $V_{GS} > V_T$, the charge is negative and proportional to $V_{GS} - V_T$,

$$Q_n(V_{GS}) = -C_{ox}\left(V_{GS} - V_T\right), \tag{4.2}$$

where C_{ox} is the gate oxide capacitance per unit area,

$$C_{ox} = \frac{\kappa_{ox}\epsilon_0}{t_{ox}} \quad \text{F/m}^2, \tag{4.3}$$

with the numerator being the dielectric constant of the oxide and the denominator the thickness of the oxide. (As we'll discuss in Lecture 8, the gate capacitance is actually somewhat less than C_{ox} when the oxide is thin.) For $V_{GS} \leq V_T$, the charge is assumed to be negligibly small.

When $V_D > V_S$, the inversion layer charge density varies with position along the channel, and so does the average velocity of electrons. As we shall see when we discuss MOS electrostatics, in a well-designed transistor, Q_n at the beginning of the channel is given by eqn. (4.2). Accordingly, we will evaluate I_{DS} at $x = 0$, where we know the charge, and we only need to deduce the average velocity, $\langle \upsilon\left(x = 0\right)\rangle$.

4.3 Linear region

In the small V_{DS}, or linear region of the output characteristics (Fig. 2.8), a MOSFET acts as a voltage controlled resistor. Above threshold, the electric field in the channel is constant, and we can write the average velocity as

$$\langle \upsilon \rangle = -\mu_n \mathcal{E} = -\mu_n V_{DS}/L. \tag{4.4}$$

Using eqns. (4.2) and (4.4) in (4.1), we find

$$\boxed{I_{DS} = \frac{W}{L}\mu_n C_{ox}\left(V_{GS} - V_T\right) V_{DS},} \tag{4.5}$$

which is the classic expression for the small V_{DS} drain current of a MOSFET. Note that we have labeled the mobility as μ_n, but in traditional MOS theory, this mobility is called the *effective mobility*, μ_{eff}. The effective mobility is the depth-averaged mobility in the inversion layer. It is smaller than the electron mobility in the bulk, because *surface roughness scattering* at the oxide-silicon interface lowers the mobility.

Equation (4.4) provides a straightforward way to determine the threshold voltage, V_T. For gate voltages well above V_T, I_{DS} increases linearly with V_{GS}. Extrapolating this straight line to $I_{DS} = 0$ gives the *extrapolated threshold voltage*, which is one of a few ways to define V_T (see [7], p. 232 for a discussion of various definitions of threshold voltage).

4.4 Saturated region: Velocity saturation

In the large V_{DS}, or saturated region of the output characteristics (Fig. 2.8), a MOSFET acts as a voltage controlled current source. For a relatively small drain to source voltage of about 1 V, the electric field in the channel of a modern short channel ($\approx$ 20 nm) MOSFET is very high — well above the $\approx$ 10 kV/cm needed to saturate the velocity in bulk Si (recall Fig. 4.5). If the electric field is large across the entire channel for $V_{DS} > V_{DSAT}$, then the velocity is constant across the channel with a value of υ_{sat}, and we can write the average velocity as

$$\langle \upsilon(x) \rangle = \upsilon_{sat} \approx 10^7 \text{ cm/s}. \tag{4.6}$$

Using eqns. (4.2) and (4.6) in (4.1), we find

$$\boxed{I_{DS} = W C_{ox} \upsilon_{sat} (V_{GS} - V_T)}, \tag{4.7}$$

which is the classic expression for the *velocity saturated* drain current of a MOSFET. Note that in practice, the current does not completely saturate, but increases slowly with drain voltage. In a well-designed Si MOSFET, the output conductance is primarily due to DIBL as described by eqn. (3.11).

Finally, we should note that it is now understood that in a short channel MOSFET, the maximum velocity in the channel does not saturate — even when the electric field is very high. Nevertheless, the traditional approach to MOSFET theory, still presented in most textbooks, assumes that the electron velocity saturates when the electric field in the channel is large.

4.5 Saturated region: Classical pinch-off

Consider next a long channel MOSFET under high drain bias. In this case, the electric field is moderate, and the velocity is not expected to saturate. Nevertheless, we still find that the drain current saturates, so it must be for a different reason. This was the situation in early MOSFET's for which the channel length was about 10 micrometers (10,000 nanometers), and the explanation for drain current saturation was *pinch-off* near the drain.

Under high drain bias, the potential in the channel varies significantly from V_S at the source to V_D at the drain end (See Ex. 4.2). Since it is the difference between the gate voltage and the Si channel that matters, eqn. (4.2) must be extended as

$$Q_n(V_{GS}, x) = -C_{ox}\left(V_{GS} - V_T - V(x)\right), \tag{4.8}$$

where $V(x)$ is the potential along the channel. According to eqn. (4.8), when $V_D = V_{GS} - V_T$, at the drain end, we find $Q_n(V_{GS}, L) = 0$. We say that the channel is pinched off at the drain. Of course, if $Q_n = 0$, then eqn. (4.1) states that $I_{DS} = 0$, but a large drain current is observed to flow. This occurs because in the pinched off region, carriers move very fast in the high electric field, so Q_n is finite, although very small. The current saturates for drain voltages above $V_{GS} - V_T$ because the additional voltage is dropped across the small, pinched off part of the channel. The voltage drop across the conductive part of the channel remains at about $V_{GS} - V_T$. We are now ready to compute the saturated drain current due to pinch-off.

Figure 4.2 is an illustration of a long channel MOSFET under high gate bias and for a drain bias greater than $V_{GS} - V_T$. Over most of the channel, there is a strong inversion layer, and $\upsilon(x) = -\mu_n \mathcal{E}(x)$. When carriers enter the pinched-off region, the large electric field quickly sweeps the carriers across and to the drain. (The energy band view of pinch-off was presented in Fig. 3.8.)

In the part of the channel where the inversion charge density is large, we can write the average velocity as

$$\langle \upsilon(x) \rangle = -\mu_n \mathcal{E}(x). \tag{4.9}$$

The voltage at the beginning of the channel is $V(0) = V_S = 0$, and the voltage at the end of the channel where it is pinched off off is $V_{GS} - V_T$. The electric field at the beginning of the channel is (see Ex. 4.2)

$$\mathcal{E}(0) = \frac{V_{GS} - V_T}{2L'}, \tag{4.10}$$

where the factor of two comes from a proper treatment of the nonlinear electric field in the channel and L' is the length of the part of the channel that is not pinched off. Using eqn. (4.10) in (4.9), we find

$$\langle \upsilon(0) \rangle = -\mu_n \mathcal{E}(0) = -\mu_n \frac{V_{GS} - V_T}{2L'}. \tag{4.11}$$

Finally, using eqns. (4.2) and (4.11) in (4.1), we find

$$\boxed{I_{DS} = \frac{W}{2L'} \mu_n C_{ox} (V_{GS} - V_T)^2 \,,} \tag{4.12}$$

the so-called *square law IV characteristic* of a long channel MOSFET. In practice, the current does not completely saturate, but increases slowly with drain voltage as the pinched-off region slowly moves towards the source, which effectively decreases the length of the conductive part of the channel, L'.

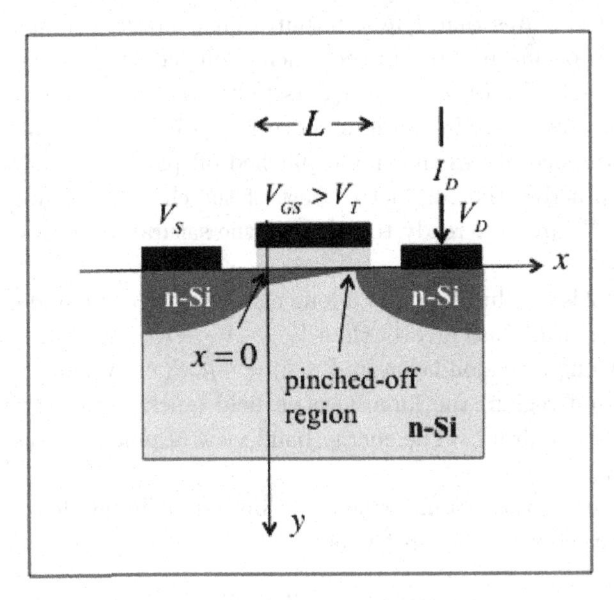

Fig. 4.2 Sketch of a long channel MOSFET showing the pinched-off region. Note that the thickness of the channel in this figure is used to illustrate the magnitude of the charge density (more charge near the source end of the channel than near the drain end). The channel is physically thin in the y-direction near the source end, where the gate to channel potential is large and physically thicker near the drain end, where the gate to channel potential is smaller. The length of the part of the channel where Q_n is substantial is $L' < L$.

Exercise 4.1: Linear to saturation square law IV characteristic.

Equations (4.5) and (4.12) describe the linear and saturation region currents as given by the traditional square law theory of the MOSFET. In this exercise, we'll compute the complete IV characteristic from the linear region to the saturation region. We begin with eqn. (4.1) for the drain current and use eqn. (4.4) for the velocity to write

$$I_{DS} = W|Q_n(x)|\langle v(x)\rangle = W|Q_n(x)|\mu_n \frac{dV}{dx}. \tag{4.13}$$

Next, we use eqn. (4.8) for the charge to write,

$$I_{DS} = W\mu_n C_{ox}(V_{GS} - V_T - V(x))\frac{dV}{dx}, \tag{4.14}$$

then separate variables and integrate across the channel to find,

$$I_{DS}\int_0^{L'} dx = W\mu_n C_{ox}\int_{V_S}^{V_D}(V_{GS} - V_T - V)\,dV, \tag{4.15}$$

where we have assumed that I_{DS} is constant (no recombination-generation in the channel) and that μ_n is constant as well. Integration gives us the IV characteristic of the MOSFET,

$$I_{DS} = \frac{W}{L'}\mu_n C_{ox}\left[(V_{GS} - V_T)V_{DS} - V_{DS}^2/2\right]. \tag{4.16}$$

Equation (4.16) gives the drain current for $V_{GS} > V_T$ and for $V_{DS} \le (V_{GS} - V_T)$. The charge in eqn. (4.8) goes to zero at $V_{DS} = V_{GS} - V_T$, which defines the beginning of the pinch-off region. The current beyond pinch-off is found by evaluating eqn. (4.16) for $V_{DS} = V_{GS} - V_T$ and is

$$I_{DS} = \frac{W}{2L'}\mu_n C_{ox}(V_{GS} - V_T)^2 \tag{4.17}$$

and only changes for increasing V_{DS} because of channel length shortening due to pinch-off (i.e. $L' < L$).

Equations (4.16) and (4.17) give the square law IV characteristics of the MOSFET — not just the linear and saturated regions, but the entire IV characteristics.

Exercise 4.2: Electric field vs. position in the channel.

In the development of the traditional model, we asserted that the electric field in the channel was V_{DS}/L under low drain bias and $(V_{GS} - V_T)/2L'$ under high drain bias in a long channel MOSFET. In this exercise, we will compute the electric field in the channel and show that these assumptions are correct.

Beginning with eqn. (4.14), we can use (4.16) for I_{DS} to find

$$\frac{1}{L'}\left[(V_{GS} - V_T)V_{DS} - V_{DS}^2/2\right] = (V_G - V_T - V(x))\frac{dV}{dx}, \tag{4.18}$$

then we separate variables and integrate from the source at $x = 0, V_S = 0$ to an arbitrary location, x, in the channel where $V = V(x)$. The result is

$$\left[(V_{GS} - V_T)\,V_{DS} - V_{DS}^2/2\right]\frac{x}{L'} = (V_{GS} - V_T)\,V(x) - V^2(x)/2\,, \quad (4.19)$$

which is a quadratic equation for $V(x)$ that can be solved to find

$$V(x) = (V_{GS} - V_T)\left[1 - \sqrt{1 - \frac{2(V_{GS} - V_T)V_{DS} - V_{DS}^2/2}{(V_{GS} - V_T)^2}\left(\frac{x}{L'}\right)}\,\right]. \quad (4.20)$$

Equation (4.20) can be differentiated to find the electric field. Let's examine the electric field for two cases. First, assume small V_{DS}, the linear region of operation, where eqn. (4.20) becomes

$$V(x) = (V_{GS} - V_T)\left[1 - \sqrt{1 - \frac{2V_{DS}}{(V_{GS} - V_T)}\left(\frac{x}{L'}\right)}\,\right], \quad (4.21)$$

and the square root can be expanded for small argument ($\sqrt{(1-x)} \approx 1 - x/2$) to find

$$V(x) = V_{DS}\frac{x}{L} \quad (4.22)$$

(Note that $L' = L$ for small V_{DS}.) Finally, differentiating eqn. (4.22), we find that the electric field for small V_{DS} is

$$-\frac{dV(x)}{dx} = \mathcal{E} = -\frac{V_{DS}}{L}\,, \quad (4.23)$$

which is the expected result.

Next, let's evaluate the electric field under pinched-off conditions, $V_{DS} = V_{GS} - V_T$. Equation (4.20) becomes

$$V(x) = (V_{GS} - V_T)\left[1 - \sqrt{1 - x/L'}\,\right], \quad (4.24)$$

and the electric field is

$$\mathcal{E}(x) = -\frac{dV}{dx} = -\frac{(V_{GS} - V_T)}{2L'}\left[\frac{1}{\sqrt{1 - x/L'}}\right]. \quad (4.25)$$

At $x = 0$, eqn. (4.25) gives the result, eqn. (4.10), which we simply asserted earlier. At $x = L'$, where the channel is pinched-off, we find $\mathcal{E}(L') \to \infty$. This result should be expected because in our model, $Q_n = 0$ at the pinch-off point, so it takes an infinite electric field to carry a finite current.

4.6 Discussion

i) velocity saturation and drain current saturation

Equations (4.5), (4.7), and (4.12) describe the linear and saturation region IV characteristics of MOSFETs according to traditional MOS theory. We have presented two different treatments of the saturated region current; in the first, drain current saturation was due to velocity saturation in a high channel field, and in the second, it was due to the development of a pinched-off region near the drain end of the channel. When the average electric field in the channel is much larger than the critical field for velocity saturation ($\approx$ 10 kV/cm) then we expect to use the velocity saturation model. We should use the velocity saturation model when

$$\frac{V_{GS} - V_T}{L} \gg \mathcal{E}_{cr} \approx 10 \text{ kV/cm}. \tag{4.26}$$

Putting in typical numbers of $V_{GS} = V_{DD} = 1$ V and $V_T = 0.2$ V, we find that the velocity saturation model should be used when $L \lesssim 1\,\mu$m. Indeed, velocity saturation models first began to be widely-used in the 1980's when channel lengths reached one micrometer [9].

Figure 4.3 shows the common source output characteristics of an n-channel Si MOSFET with a channel length of about 60 nm. It is clear from the results that $I_{DS} \propto (V_{GS} - V_T)$ under high drain bias, so that the velocity saturation model of eqn. (4.7) seems to describe this device. Indeed, the observation of a saturation current that varies linearly with gate voltage is taken as the "signature" of velocity saturation.

For the MOSFET of Fig. 4.3, $V_T \approx 0.4$ V. For the maximum gate voltage, the pinch-off model would give a drain saturation voltage of $V_{DSAT} = V_{GS} - V_T \approx 0.8$ V, which is clearly too high and tells us that the drain current is not saturating due to classical pinch-off. References [7] and [8] discuss V_{DSAT} in the presence of velocity saturation.

Although velocity saturation models seem to accurately describe short channel MOSFETs, there is a mystery. Detailed computer simulations of carrier transport in nanoscale MOSFETs show that the velocity **does not saturate** in the high electric field portion of a short channel MOSFET. There is simply not enough time for carriers to scatter enough to saturate the velocity; they traverse the channel and exit the drain too quickly. Nevertheless, the IV characteristic of Fig. 4.3 tell us that the velocity in the channel saturates. Understanding this is a mystery that we will unravel as we explore the nanoscale MOSFET.

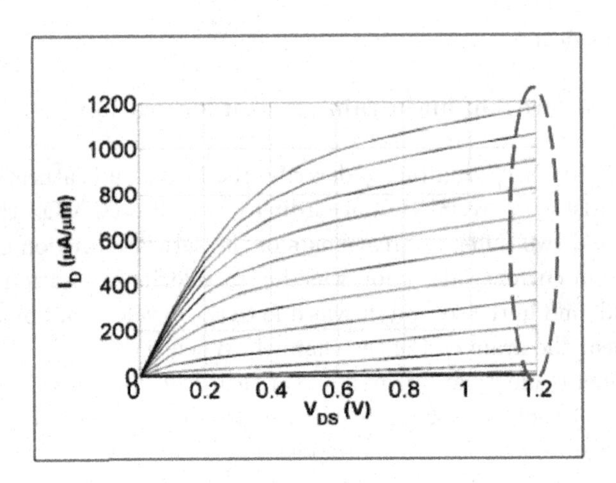

Fig. 4.3 Common source output characteristics of an n-channel. Si MOSFET with a gate length of $L \approx 60$ nm. The top curve is for $V_{GS} = 1.2$ V and the step is 0.1 V. Note that for large V_{DS}, the drain current increases linearly with gate voltage. This behavior is considered to be the signature of velocity saturation in the channel. The device is described in C. Jeong, D. A. Antoniadis and M.S. Lundstrom, "On Backscattering and Mobility in Nanoscale Silicon MOSFETs, *IEEE Trans. Electron Dev.*, **56**, pp. 2762-2769, 2009.

ii) device metrics

Equations (4.5) and (4.7) describe the IV characteristic of modern short channel MOSFETs and can be used to relate some of the device metrics listed in Sec. 2.4 to the underlying physics. Using these equations, we find:

$$\boxed{\begin{aligned}
&I_{ON} = WC_{ox}\upsilon_{sat}\left(V_{DD} - V_T\right) \qquad\qquad V_T = V_{T0} - \delta V_{DS} \\
&R_{ON} = \left(\left.\frac{\partial I_{DS}}{\partial V_{DS}}\right|_{V_{GS}=V_{DD},V_{DS}\approx 0}\right)^{-1} = \left(\frac{W}{L}\mu_n C_{ox}\left(V_{GS} - V_T\right)\right)^{-1} \\
&g_m^{sat} = \left.\frac{\partial I_{DS}}{\partial V_{GS}}\right|_{V_{GS}=V_{DS}=V_{DD}} = WC_{ox}\upsilon_{sat} \\
&r_d = \left(\left.\frac{\partial I_{DS}}{\partial V_{DS}}\right|_{V_{GS}=V_{DD},V_{DS}>V_{DSAT}}\right)^{-1} = \frac{1}{g_m^{sat}\delta} \\
&|A_v| = g_m^{sat} r_d = \frac{1}{\delta}
\end{aligned}} \tag{4.27}$$

The parameter, $|A_v|$ is the *self-gain*, an important figure of merit for analog applications.

Finally, we should discuss energy band diagrams. While energy bands did not appear explicitly in our discussion, they are present implicitly. The beginning of the channel, $x = 0$, is the top of the energy barrier in Figs. 3.5 and 3.6 (or close to the top of the barrier [10]). As we'll discuss later, in a well-designed MOSFET, the charge at the top of the barrier is given by eqn. (4.2). This charge comes from electrons in the source that surmount the energy barrier. The location at the beginning of the channel where eqn. (4.2) applies is also known as the *virtual source.*

The energy band view is especially helpful in understanding pinch-off. From Fig. 4.2, it can be confusing as to how carriers can leave the end of the channel and flow across the pinched-off region. Energy bands make it clear. As was shown in Fig. 3.6, the pinched-off region is the high electric field region near the drain, where the slope of $E_c(x)$ is the steepest. Electrons that enter this region from the channel simply flow downhill and out the drain. There is nothing to stop them when they enter the pinched-off region.

4.7 Summary

In this lecture, we reviewed traditional MOSFET *IV* theory. In practice, there are several complications to consider, such as the role of the depleted charge in eqn. (4.8), current for an arbitrary drain voltage, etc. [5-8], but the essential features of the traditional approach are easy to grasp, and will give us a point of comparison for the much different picture of the nanoscale MOSFET that will be developed in subsequent lectures.

According to eqn. (4.1), the drain current is proportional to the product of charge and velocity. The charge is controlled by MOS electrostatics (*i.e.* by manipulating the energy barrier between the source and the channel). The traditional approach to MOS electrostatics is still largely applicable, with some modifications due to quantum confinement. The lectures in Part 2 will review the critically important electrostatics of the MOSFET.

4.8 References

The mathematical modeling of transistors began in the 1960's. Some of the early papers MOSFET IV characteristics are listed below.

[1] S.R. Hofstein and F.P. Heiman, "The Silicon Insulated-Gate Field-Effect Transistor," *Proc. IEEE*, **51**, pp. 1190-1202, 1963.

[2] C.T. Sah, "Characteristics of the Metal-Oxide-Semiconductor Transistors," *IEEE Trans. Electron Devices*, **11**, pp. 324-345, 1964.

[3] H. Shichman and D. A. Hodges, "Modeling and simulation of insulated-gate field-effect transistor switching circuits," *IEEE J. Solid State Circuits*, **SC-3**, 1968.

[4] B.J. Sheu, D.L. Scharfetter, P.-K. Ko, and M.-C. Jeng, "BSIM: Berkeley Short-Channel IGFET Model for MOS Transistors," *IEEE J. Solid-State Circuits*, **SC-22**, pp. 558-566, 1987.

The traditional theory of the MOSFET reviewed in this chapter is the approach used in textbooks such as the two listed below.

[5] Robert F. Pierret, *Semiconductor Device Fundamentals, 2nd Ed.*, Addison-Wesley Publishing Co, 1996.

[6] Ben Streetman and Sanjay Banerjee, *Solid State Electronic Devices, 6th Ed.*, Prentice Hall, 2005.

For authoritative treatments of classical MOSFET theory, see:

[7] Y. Tsividis and C. McAndrew, *Operation and Modeling of the MOS Transistor*, 3rd Ed., Oxford Univ. Press, New York, 2011.

[8] Y. Taur and T. Ning, *Fundamentals of Modern VLSI Devices*, 2nd Ed., Oxford Univ. Press, New York, 2013.

As channel lengths shrunk to the micrometer scale, velocity saturation became important. The following paper from that era discusses the impact on MOSFETs and MOSFET circuits.

[9] C.G. Sodini, P.-K. Ko, and J.L. Moll, "The effect of high fields on MOS device and circuit performance," *IEEE Trans. Electron Dev.*, **31**, pp. 1386-1393, 1984.

The virtual source or beginning of the channel is not always exactly at the top of the energy barrier, as discussed by Liu.

[10] Y. Liu, M. Luisier, A. Majumdar, D. Antoniadis, and M.S. Lundstrom, "On the Interpretation of Ballistic Injection Velocity in Deeply Scaled MOSFETs," *IEEE Trans. Electron Dev.*, **59**, pp. 994-1001, 2012.

Lecture 5

MOSFET IV: The Virtual Source Model

5.1 Introduction

In Lecture 4, we developed expressions for the linear and saturation region drain currents as:

$$\begin{aligned} I_{DLIN} &= \frac{W}{L}\mu_n C_{ox}\left(V_{GS} - V_T\right) V_{DS} \\ I_{DSAT} &= W C_{ox} \upsilon_{sat}\left(V_{GS} - V_T\right) \end{aligned} . \tag{5.1}$$

These equations assume $V_{GS} > V_T$, so they cannot describe the subthreshold characteristics. As shown in Fig. 5.1, these equations provide a rough description of I_{DS} vs. V_{DS}, especially if we include DIBL as in eqn. (3.14), so that the finite output conductance is included. If we define the drain saturation voltage as the voltage where $I_{DLIN} = I_{DSAT}$, we find

$$V_{DSAT} = \frac{\upsilon_{sat} L}{\mu_n} . \tag{5.2}$$

For $V_{DS} \ll V_{DSAT}$, $I_{DS} = I_{DLIN}$, and for $V_{DS} \gg V_{DSAT}$, $I_{DS} = I_{DSAT}$.

Traditional MOSFET theory develops expressions for I_{DS} vs. V_{DS} that smoothly transition from the linear to saturation regions as V_{DS} increases from zero to V_{DD} [1, 2]. The goal in this lecture is to develop a simple, semi-empirical expression that describes the complete $I_{DS}(V_{DS})$ characteristic from the linear to saturated region. The approach is similar to the so-called *virtual source MOSFET model* that has been developed and successfully used to describe a wide variety of nanoscale MOSFETs [3]. We'll take a different approach to developing a virtual source model and begin with the traditional approach, and then use the VS model as a framework for subsequent discussions. As we extend and interpret the VS model in subsequent lectures, we'll develop a simple, physical model that provides an accurate quantitative descriptions of modern transistors.

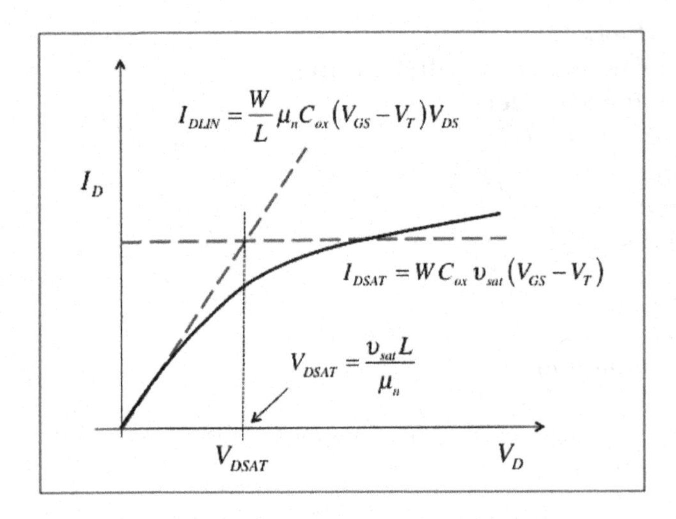

Fig. 5.1 Sketch of a common source output characteristic of an n-channel MOSFET at a fixed gate voltage (solid line). The dashed lines are the linear and saturation region currents as given by eqns. (5.1).

5.2 Channel velocity vs. drain voltage

The drain current is proportional to the product of charge at the beginning of the channel times the average carrier velocity at the beginning of the channel. From eqn. (4.1) at the beginning of the channel, we have

$$I_{DS}/W = |Q_n(x=0)|\upsilon(x=0) \; . \tag{5.3}$$

Equation (5.1) for the linear current can be re-written in this form as

$$
\begin{aligned}
I_{DLIN}/W &= |Q_n(V_{GS})|\, \upsilon(V_{DS}) \\
Q_n(V_{GS}) &= -C_{ox}\,(V_{GS} - V_T) \\
\upsilon(V_{DS}) &= \left(\mu_n \frac{V_{DS}}{L}\right) .
\end{aligned}
\tag{5.4}
$$

Similarly, eqn. (5.1) for the saturation current can be re-written as

$$
\begin{aligned}
I_{DSAT}/W &= |Q_n(V_{GS})|\, \upsilon(V_{DS}) \\
Q_n(V_{GS}) &= -C_{ox}\,(V_{GS} - V_T) \\
\upsilon(V_{DS}) &= \upsilon_{sat} .
\end{aligned}
\tag{5.5}
$$

If we can find a way for the average velocity to go smoothly from its value at low V_{DS} to υ_{sat} at high V_{DS}, then we will have a model that covers the complete range of drain voltages.

The VS model takes an empirical approach and writes the average velocity at the beginning of the channel as [3]

$$
\begin{aligned}
\upsilon(V_{DS}) &= F_{SAT}(V_{DS})\upsilon_{sat} \\
F_{SAT}(V_{DS}) &= \frac{V_{DS}/V_{DSAT}}{\left[1 + (V_{DS}/V_{DSAT})^{\beta}\right]^{1/\beta}} ,
\end{aligned}
\tag{5.6}
$$

where V_{DSAT} is given by eqn. (5.2) and β is an empirical parameter chosen to fit the measured IV characteristic.

The form of the drain current saturation function, F_{SAT}, is motivated by the observation that the lower of the two velocities in eqns. (5.4) and (5.5) should be the one that limits the current. We might, therefore, expect

$$
\frac{1}{\upsilon(V_{DS})} = \frac{1}{(\mu_n V_{DS}/L)} + \frac{1}{\upsilon_{sat}} ,
\tag{5.7}
$$

which can be re-written as

$$
\upsilon(V_{DS}) = \frac{V_{DS}/V_{DSAT}}{[1 + (V_{DS}/V_{DSAT})]}\upsilon_{sat} .
\tag{5.8}
$$

Equation (5.8) is similar to eqn. (5.6), except that (5.6) introduces the empirical parameter, β, which is adjusted to better fit data. Typical values of β for n- and p-channel Si MOSFETs are between 1.4 and 1.8 [3].

Equations (5.3), (4.2), and (5.6) give us a description of the above-threshold MOSFET for any drain voltage from the linear to the saturated regions.

5.3 Level 0 VS model

Our simple model for the above threshold MOSFET is summarized as follows:

$$\boxed{\begin{array}{l} I_{DS}/W = |Q_n(0)|\, \upsilon(0) \\ \\ Q_n(V_{GS}) = 0 \quad V_{GS} \leq V_T \\ Q_n(V_{GS}) = -C_{ox}\left(V_{GS} - V_T\right) \quad V_{GS} > V_T \\ V_T = V_{T0} - \delta V_{DS} \\ \\ \langle \upsilon(V_{DS}) \rangle = F_{SAT}(V_{DS})\upsilon_{sat} \\ F_{SAT}(V_{DS}) = \dfrac{V_{DS}/V_{DSAT}}{\left[1 + \left(V_{DS}/V_{DSAT}\right)^{\beta}\right]^{1/\beta}} \\ \\ V_{DSAT} = \dfrac{\upsilon_{sat} L}{\mu_n} \end{array}}, \tag{5.9}$$

With this simple model, we can compute reasonable MOSFET *IV* characteristics, and the model can be extended step by step to make it more and more realistic. There are only six device-specific input parameters to this model: C_{ox}, V_T, μ_n, υ_{sat}, L, and β. The level 0 model does not describe the subthreshold characteristics, but after discussing MOS electrostatics in the next few lectures, we will be able to include the subthreshold region. Series resistance is important in any real device, and can be readily included as discussed next.

5.4 Series resistance

As illustrated on the left of Fig. 5.2, we have developed expressions for the *IV* characteristic of an intrinsic MOSFET — one with no series resistance between the intrinsic source and drain and the metal contacts to which the voltages are applied. In practice, these series resistors are always there and must be accounted for.

The figure on the right in Fig. 5.2 shows how the voltages applied to the terminals of the device are related to the voltages on the intrinsic contacts.

Here, V'_D, V'_S, and V'_G refer to the voltages on the terminals and V_D, V_S, and V_G refer to the voltages on the intrinsic terminals. (No resistance is shown in the gate lead, because we are considering D.C. operation now.) Since the D.C. gate current is zero, a resistance in the gate has no effect. Gate resistance is, however, an important factor in the R.F. operation of transistors.)

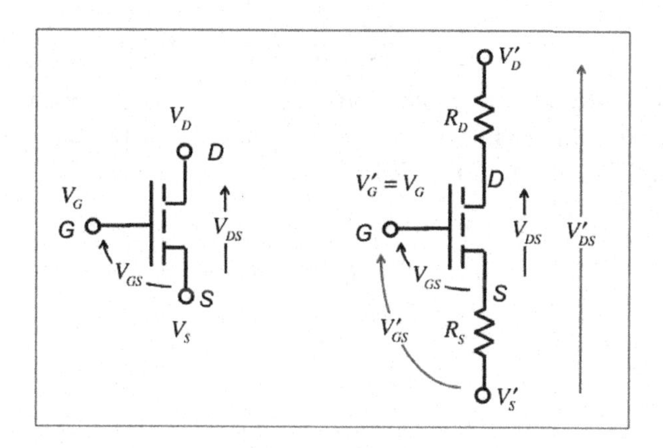

Fig. 5.2 Series resistance in a MOSFET. Left: the intrinsic device. Right: The actual, extrinsic device showing how the voltages applied to the external contacts are related to the voltages on the internal contacts.

From Fig. 5.2, we relate the internal (unprimed) voltages to the external (primed) voltages by

$$\begin{aligned} V_G &= V'_G \\ V_D &= V'_D - I_{DS}(V_G, V_S, V_D) R_D \, , \\ V_S &= V'_S + I_{DS}(V_G, V_S, V_D) R_S \end{aligned} \tag{5.10}$$

Since we know the *IV* characteristic of the intrinsic device, $I_{DS}(V_G, V_S, V_D)$, Equations (5.10) are two equations in two unknowns — the internal voltages, V_D and V_S. Given applied voltages on the gate, source, and drain, V'_G, V'_S, V'_D, we can solve these equations for the internal voltages, V_S and V_D, and then determine the current, $I_{DS}(V'_G, V'_S, V'_D)$.

Figure 5.3 illustrates the effect of series resistance on the *IV* characteristic. In the linear region, we can write the current of an intrinsic device as

$$I_{DLIN} = \frac{W}{L} \mu_n C_{ox} (V_{GS} - V_T) V_{DS} = V_{DS}/R_{ch} \, . \tag{5.11}$$

When source and drain series resistors are present, the linear region current becomes

$$I_{DLIN} = V_{DS}/R_{tot} \, , \tag{5.12}$$

where

$$R_{tot} = R_{ch} + R_S + R_D = R_{ch} + R_{DS} \, . \tag{5.13}$$

(It is common to label the sum of R_S and R_D as R_{SD}). So the effect of series resistance in the linear region is to simply lower the slope of the IV characteristic as shown in Fig. 5.3.

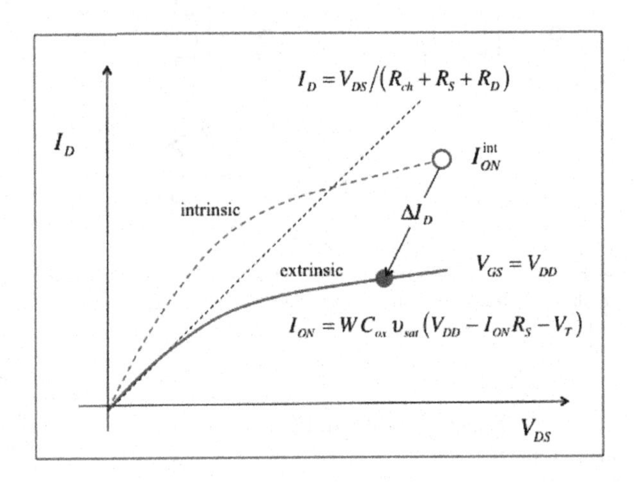

Fig. 5.3 Illustration of the effect of series resistance on the IV characteristics of a MOSFET. The dashed curve is an intrinsic MOSFET for which $R_S = R_D = 0$. As indicated by the solid line, series resistance increases the channel resistance and lowers the on-current.

Figure 5.3 also shows that series resistance decreases the value of the saturation region current. In an ideal MOSFET with no output conductance, the drain series resistance has no effect in the saturation region where $V_D > V_{DSAT}$, but the source resistance reduces the intrinsic V_{GS}, so eqn. (5.1) becomes

$$I_{DSAT} = W C_{ox} \upsilon_{sat} (V'_{GS} - I_{DSAT} R_S - V_T) \, . \tag{5.14}$$

Series resistance lowers the internal gate to source voltage of a MOSFET, and therefore lowers the saturation current. The maximum voltage applied between the gate and source is the power supply voltage, V_{DD}. Series resistance will have a small effect if $I_{DSAT}R_S \ll V_{DD}$. For high performance, we require

$$R_S \ll \frac{V_{DD}}{I_{DSAT}} . \tag{5.15}$$

Modern Si MOSFETs deliver about 1 mA/μm of on-current at $V_{DD} =$ 1 V. Accordingly, R_S must be much less than 1000 $\Omega - \mu$m; series resistances of about 100 $\Omega - \mu$m are needed. Although we will primarily be concerned with understanding the physics of the intrinsic MOSFET, we should be aware of the significance of series resistance when analyzing measured data. As channel lengths continue to scale down, keeping the series resistance to a manageable level is increasingly difficult.

Exercise 5.1: Analysis of experimental data.

Use eqn. (5.14) and the *IV* characteristic of Fig. 4.3, to deduce the "saturation velocity" for the on-current. Note that we'll regard the saturation velocity as an empirical parameter used to fit the data of Fig. 4.3 and will compare it to the high-field saturation velocity for electrons in bulk Si.

Assume the following parameters:

$$\begin{aligned} I_{ON} &= 1180\ \mu\text{A}/\mu\text{m} \\ C_{ox} &= 1.55 \times 10^{-6}\ \text{F/cm}^2 \\ R_{DS} &= 220\ \Omega \\ V_T &= 0.25\ \text{V} \\ V_{DD} &= 1.2\ \text{V} \\ W &= 1\ \mu\text{m}. \end{aligned}$$

Solving eqn. (5.14) for υ_{sat}, we find

$$\upsilon_{sat} \equiv \upsilon_{inj} = \frac{I_{DSAT}}{WC_{ox}\left(V_{GS} - V_T\right)} .$$

$$V_{GS} = V_{DD} - I_{DSAT}R_{SD}/2 .$$

Putting in numbers, we find

$$\upsilon_{sat} = 0.92 \times 10^7\ \text{cm/s} .$$

It is interesting to note that the velocity we deduce is close to the high-field, bulk saturation velocity of Si (1×10^7 cm/s), but the physics of velocity saturation in a nanoscale MOSFET is actually quite different from the physics of velocity saturation in bulk Si under high electric fields. Accordingly, from now on, we will give υ_{sat} a different name, the *injection velocity*, υ_{inj}.

5.5 Discussion

One might have expected the traditional model that we have developed to be applicable only to long channel MOSFETs because it is based on assumptions such as diffusive transport in the linear region and high-field velocity saturation in the saturated region. Surprisingly, we find that it accurately describes the IV characteristics of MOSFETs with channel lengths less than 100 nm as shown in Fig. 5.4. To achieve such fits, we view two of the physical parameters in our VS model as empirical parameters that are fit to measured data, and we find that with relatively small adjustments in these parameters, excellent fits to most transistors can be achieved. The two adjusted parameters are the injection velocity, υ_{inj}, (which is the saturation velocity in the traditional model) and the apparent mobility μ_{app}, (which is the real mobility in the traditional model). The fact that this simple traditional model describes modern transistors so well, tells us that it captures something essential about the physics of MOSFETs.

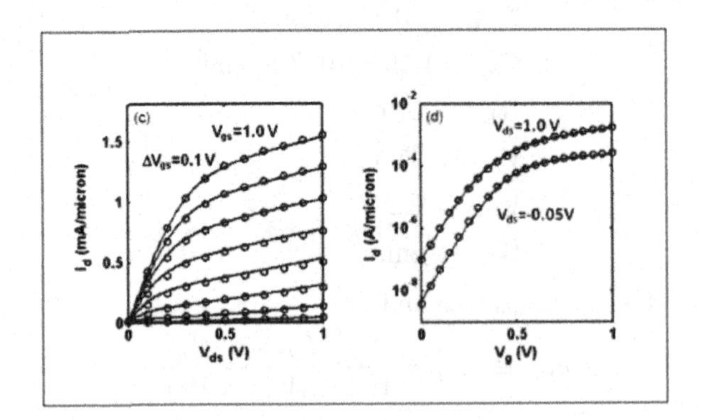

Fig. 5.4 Measured and fitted VS model data for 32 nm n-MOSFET technology. Left: Common source output characteristic. Right: Transfer characteristic. The VS model used for these fits is an extension of the model described by eqns. (7.9) that uses an improved description of MOS electrostatics to treat the subthreshold as well as above threshold conduction. (©IEEE 2009. Reprinted, with permission, from: [3].)

5.6 Summary

In this lecture we recast traditional MOSFET theory in the form of a simple virtual source model. Application of this simple model to modern transistors shows that it describes them remarkably well. This is a consequence of the fact that it describes the essential features of the barrier controlled transistor (i.e. MOS electrostatics). The weakest part of the model is the transport model, which is based on the use of a mobility and saturated velocity. Because of the simplified transport model, we need to regard the mobility and saturation velocity in the model as fitting parameters that can be adjusted to fit experimental data.

In the next few lectures (Part 2 of this volume), we will review MOS electrostatics and learn how to describe subthreshold as well as above-threshold conduction. The result will be an improved VS model, but mobility and saturation velocity will still be viewed as fitting parameters. Beginning in Part 3, we'll discuss transport and learn how to formulate the VS model so that transport is described physically.

5.7 References

For a thorough treatment of classical MOSFET theory, see:

[1] Y. Tsividis and C. McAndrew, *Operation and Modeling of the MOS Transistor*, 3rd Ed., Oxford Univ. Press, New York, 2011.

[2] Y. Taur and T. Ning, *Fundamentals of Modern VLSI Devices*, 2nd Ed., Oxford Univ. Press, New York, 2013.

The MIT Virtual Source Model, which provides a framework for these lectures, is described in:

[3] A. Khakifirooz, O.M. Nayfeh, and D.A. Antoniadis, "A Simple Semiempirical Short-Channel MOSFET Current Voltage Model Continuous Across All Regions of Operation and Employing Only Physical Parameters," *IEEE Trans. Electron. Dev.*, **56**, pp. 1674-1680, 2009.

PART 2
MOS Electrostatics

Lecture 6

Poisson Equation and the Depletion Approximation

6.1 Introduction

In Lectures 1-5 we discussed some basic MOSFET concepts. By assuming that the inversion charge at the beginning of the channel is given by

$$Q_n(V_{GS}) = 0 \quad V_{GS} \leq V_T$$

$$Q_n(V_{GS}) = -C_G\,(V_{GS} - V_T) \quad V_{GS} > V_T \tag{6.1}$$

$$V_T = V_{T0} - \delta V_{DS}\,,$$

and by using simple, traditional models for the average velocity at the beginning of the channel, we derived the IV characteristics of a MOSFET. In this lecture, we begin to address some important questions. First why does Q_n increase linearly with gate voltage for $V_{GS} > V_T$, what is the gate capacitance, C_G (we'll see that it is somewhat less than C_{ox}), and how does the small charge present for $V_{GS} < V_T$ vary with gate voltage? The answers to these questions come from an understanding of one-dimensional MOS electrostatics, the subject of this lecture and the next three. Another question has to do with the physics of DIBL; what determines the value of the parameter, δ? To answer this question, we need to understand two-dimensional MOS electrostatics, the subject of Lecture 10. A sound understanding of 1D and 2D MOS electrostatics is absolutely essential for understanding transistors because electrostatics is what determines how the terminal voltages control the source to channel barrier in a MOSFET. This chapter reviews MOS electrostatics as it is discussed in most introductory semiconductor textbooks (e.g. [1], [2]).

6.2 Energy bands and band bending

We seek to understand how the terminal voltages and design of the MOSFET affect the electrostatic potential in the device, $\psi(x,y,z)$. The x-direction is from source to drain, the y-direction is into the depth of the semiconductor, and the z-direction in along the width of the MOSFET. We seek solutions of the Poisson equation,

$$\nabla \cdot \vec{D}(x,y,z) = \rho(x,y,z)$$

$$\nabla^2 \psi(x,y,z) = -\frac{\rho(x,y,z)}{\epsilon_s}\,, \tag{6.2}$$

where $\vec{D}$ is the displacement vector, ρ is the space charge density, and ϵ_s is the dielectric constant of the semiconductor, which is assumed to be spatially uniform.

In general, a three-dimensional solution is required, but we will assume a wide transistor so that the potential is uniform in the z-direction and a 2D solution suffices. We'll begin by discussing 1D electrostatics in the direction normal to the channel. As indicated in Fig. 6.1, we imagine a long channel device and consider $\psi(y)$ *vs.* y at a location in the middle of the channel where the influence of the source and drain potentials are minimal, so that 2D effects can be neglected.

Energy band diagrams provide a convenient, qualitative solution to the Poisson equation. In this section, we'll examine the influence of a gate voltage on the energy vs. position into the depth of the semiconductor channel. Figure 6.2 shows the case where the energy bands are flat — the

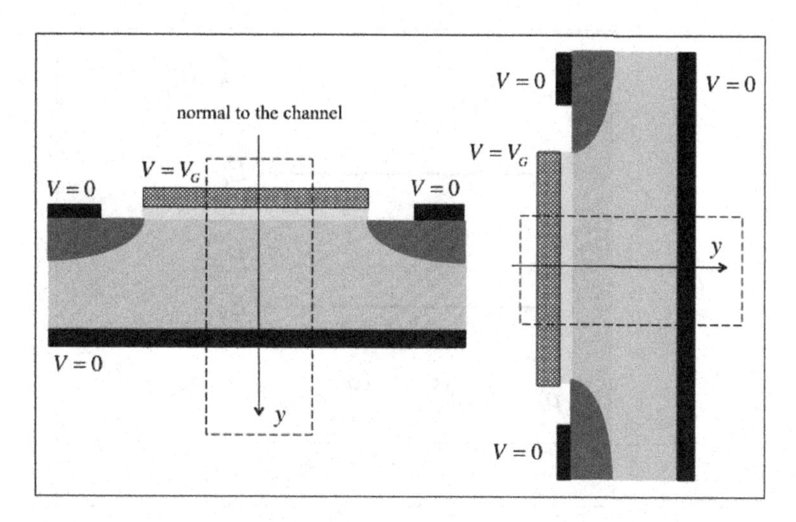

Fig. 6.1 Sketch of a long channel MOSFET for which we seek to understand 1D MOS electrostatics. Left: Illustration of how we aim to understand the potential profile vs. position into the depth of the channel. Right: Orientation that we will use when we plot energy band diagrams in this lecture.

potential is zero (or uniform in the y-direction), and the energy bands are independent of position. Note also that the electron and hole densities are exponentially related to the difference between the band edge and the Fermi level when Boltzmann carrier statistics are assumed.

Figure 6.3 shows the expected electrostatic potential vs. position when a positive voltage is applied to the gate. Some voltage will be dropped across the oxide, and the potential at the surface of the semiconductor, ψ_S, will be positive with $0 < \psi_S < V_G$. If the back of the semiconductor is grounded ($\psi(y \to \infty) = 0$), then we expect the potential in the semiconductor to decay to zero as y increases.

A positive electrostatic potential lowers the potential energy of an electron, so the bands will bend when the electrostatic potential varies with position,

$$E_C(y) = constant - q\psi(y)\,. \tag{6.3}$$

If the electrostatic potential increases from the bulk of the Si to the surface, then the energy bands will bend down, as shown on the right of Fig. 6.3.

Before we examine how the bands bend as a function of gate voltage, we define a few terms in Fig. 6.4. First, we assume for now an ideal, hypothetical gate electrode for which the Fermi level in the metal just happens to line up with the Fermi level in the semiconductor. We call this

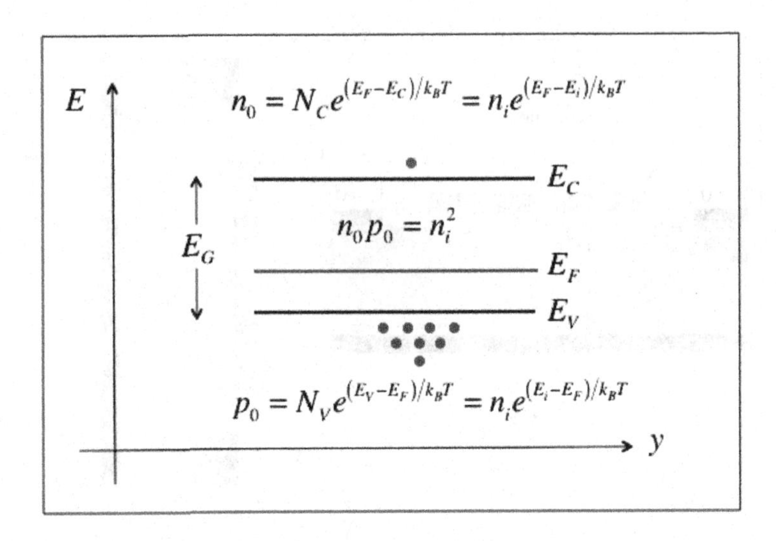

Fig. 6.2 Equilibrium energy band diagram for a uniform electrostatic potential. Also noted is the exponential relation between the electron and hole densities and the separation of the band edge and the Fermi level.

the *flatband* condition — the bands in the semiconductor and oxide are flat. Flat bands occur at $V_G' = 0$ V for this hypothetical metal gate. (The prime indicates that we're talking about a hypothetical material.) In practice, there will always be a work function difference, Φ_{MS}, between the gate electrode and the semiconductor. The flatband condition will not occurs at $V_G = 0$ but at $V_G = V_{FB} = \Phi_{MS}/q$, which is the voltage needed to "undo" the work function difference.

Recall that when a voltage is applied to a contact, it lowers the Fermi level. As shown on the right of Fig. 6.4, the Fermi level in the gate electrode is lowered from E_{FM}^0 at $V_G' = 0$ to $E_{FM}^0 - qV_G'$. The positive potential on the gate electrode lowers the electrostatic potential in the oxide and semiconductor as determined by solutions to the Laplace and Poisson equations, which will be discussed later. If we define the reference for the electrostatic potential to be in the bulk of the semiconductor, $\psi(y \to \infty) = 0$, then the electrostatic potential at any location in the semiconductor is simply related to the band bending according to

$$\psi(y) = \frac{E_C(\infty) - E_C(y)}{q} . \tag{6.4}$$

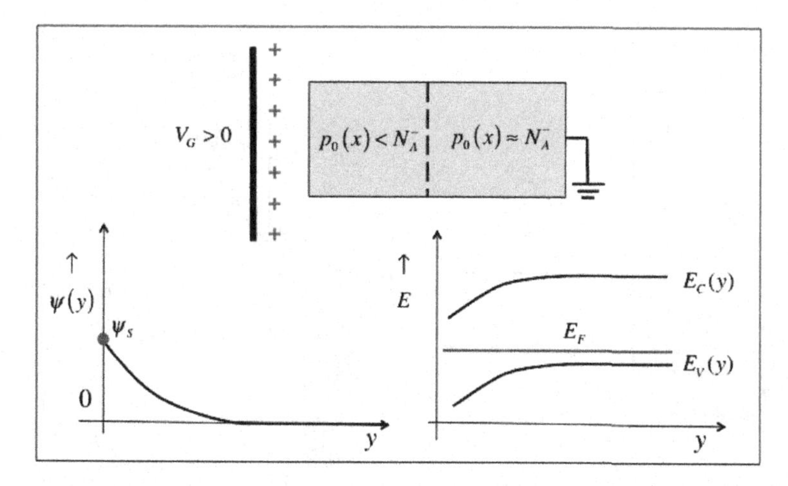

Fig. 6.3 Illustration of how the application of a positive gate voltage affects the electrostatic potential and energy bands in a semiconductor. Bottom left: Expected electrostatic potential vs. position in the semiconductor when a positive potential is applied to the gate electrode. Bottom right: Expected energy band diagram.

Note in Fig. 6.4 that the Fermi level is flat in the semiconductor — even when a gate bias is applied. This occurs because the insulator prevents current flow, so the metal and the semiconductor are separately in equilibrium with different Fermi levels.

We are now ready to discuss band bending vs. gate voltage as summarized by the energy band diagrams in Fig. 6.5. When a negative gate voltage is applied, a negative electrostatic potential is induced in the oxide, and the semiconductor, and the bands bend up. The surface potential is negative, $\psi(y=0) = \psi_S < 0$. The hole concentration increases near the oxide-semiconductor interface because the valence band bends up toward the Fermi level. The net charge near the surface is positive. This *accumulation charge* resides very close to the surface of the semiconductor and is sometimes approximated as a δ-function.

When a positive gate voltage is applied, a positive electrostatic potential is induced in the oxide and semiconductor, and the bands bend down. The surface potential is positive, $\psi(y=0) = \psi_S > 0$. Because the valence band moves aways from the Fermi level, the hole concentration decreases (we can think of the positive gate potential (charge) as pushing the positively charged holes away from the surface). The result is a *depletion layer*, a layer of thickness, W_D, in which the hole concentration is negligible, $p_0 \ll N_A^-$. If the bandbending is not too large, then the electron concentration is also

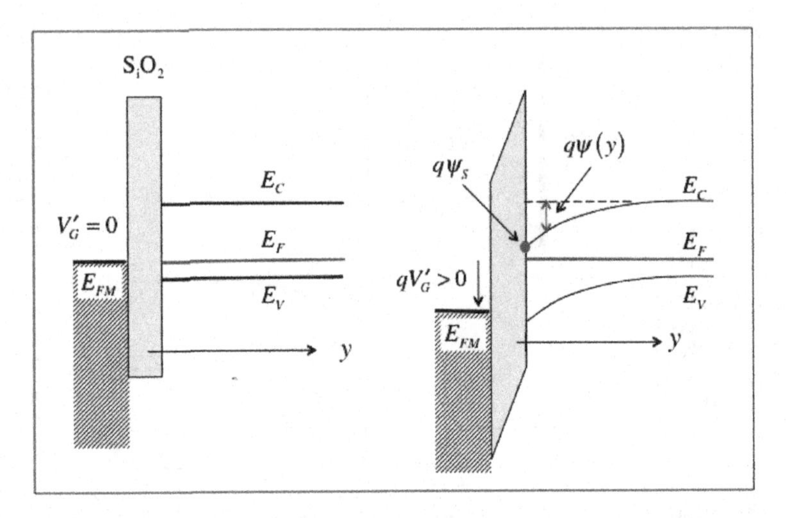

Fig. 6.4 Left: Illustration of flatband conditions in an ideal MOS structure. Right: Illustration of band bending when a positive gate voltage is applied.

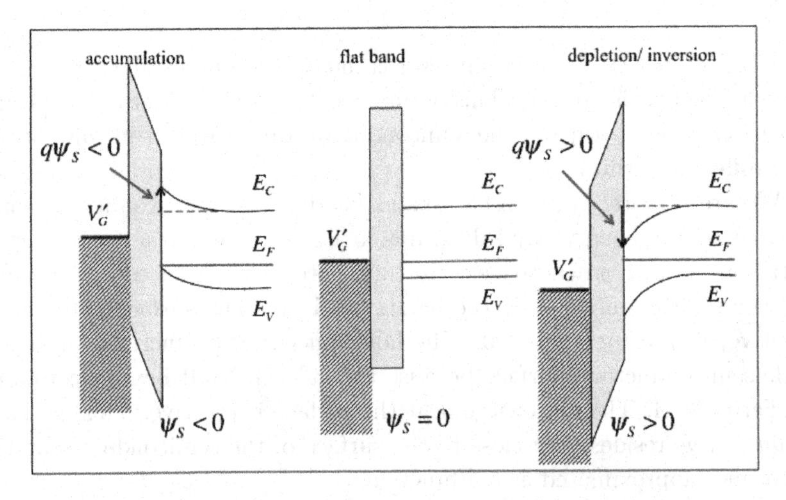

Fig. 6.5 Illustration of bandbending for three different gate voltages. Left: Accumulation of majority carriers, Center: Flatband, and Right: Depletion/Inversion.

small, and the only charge near the surface is due to the ionized acceptors in the depletion region, If the bandbending is large enough, then the electron concentration begins to build up near the surface. This *inversion layer* of mobile carriers is responsible for the current in a MOSFET. Inversion will be discussed in Sec. 6.5.

Finally, note that the band diagrams in Fig. 6.5 are for a p-type semiconductor. To test your understanding, draw corresponding energy band diagrams for an n-type semiconductor in accumulation, at flatband, and in depletion. The term, accumulation, always describes the accumulation of majority carriers, depletion always refers to the depletion of majority carriers, and inversion always refers to the build up of minority carriers.

6.3 Poisson-Boltzmann equation

Our goal is to understand how the total charge in the semiconductor,

$$Q_S = \int_0^\infty \rho(y)dy = q\int_0^\infty \left(p_0(y) - n_0(y) + N_D^+ - N_A^-\right) dy \quad \text{C/m}^2 \,, \tag{6.5}$$

depends on the electrostatic potential in the semiconductor. The subscripts, "0", are a reminder that the semiconductor is in equilibrium. We are also interested in the charge due to mobile electrons,

$$Q_n = -q\int_0^\infty n_0(y)dy \quad \text{C/m}^2 \,, \tag{6.6}$$

because the mobile electrons carry the current in a MOSFET.

Energy band diagrams provide a qualitative solution for the potential and charge in the semiconductor. To actually solve for the potential vs. position, we need to solve the Poisson equation. In this section, we'll formulate the Poisson equation for 1D semiconductors,

$$\frac{d^2\psi}{dy^2} = \frac{-q}{\varepsilon_s}\left[p_0(y) - n_0(y) + N_D^+ - N_A^-\right] . \tag{6.7}$$

To be specific, we'll assume a p-type semiconductor for which $N_D = 0$. Complete ionization of dopants will be assumed ($N_A^- = N_A$). In the bulk, we have space charge neutrality, $p_B - n_B - N_A = 0$, so $N_A = p_B - n_B$, and Poisson's equation becomes

$$\frac{d^2\psi}{dy^2} = \frac{-q}{\varepsilon_s}\left[p_0(y) - n_0(y) + n_B - p_B\right] . \tag{6.8}$$

where

$$\begin{aligned} p_B &\cong N_A \\ n_B &\cong n_i^2/N_A \end{aligned} . \tag{6.9}$$

The subscripts, "B", refer to the equilibrium concentrations in the bulk. Using eqn. (6.9), we can express (6.8) as

$$\frac{d^2\psi}{dy^2} = \frac{-q}{\varepsilon_s}\left[p_0(y) - N_A - n_0(y) + n_i^2/N_A\,\right] . \tag{6.10}$$

Equation (6.10) is one equation with three unknowns, $\psi(y)$, $n_0(y)$, and $p_0(y)$. To solve this equation, we need to find two more equations.

Recall that the MOS structure is in equilibrium for any gate bias, because the oxide prevents current from flowing. In equilibrium, the carrier densities are related to the location of the Fermi level (a constant in equilibrium) and the band edges (which follow the electrostatic potential). Accordingly, we can write,

$$\begin{aligned} p_0(y) &= p_B e^{-q\psi(y)/k_BT} = N_A e^{-q\psi(y)/k_BT} \\ n_0(y) &= n_B e^{+q\psi(y)/k_BT} = \frac{n_i^2}{N_A} e^{+q\psi(y)/k_BT} \,, \end{aligned} \tag{6.11}$$

which we can use to write eqn. (6.10) as

$$\boxed{\frac{d^2\psi}{dy^2} = \frac{-q}{\varepsilon_s}\left[N_A\left(e^{-q\psi(y)/k_BT} - 1\right) - \frac{{n_i}^2}{N_A}\left(e^{q\psi(y)/k_BT} - 1\right)\right] \,.} \tag{6.12}$$

Equation (6.12) is known as the *Poisson-Boltzmann equation*; it describes a 1D, p-type semiconductor in equilibrium with the dopants fully ionized. To complete the problem specification, we need to specify boundary equations. Assuming a semi-infinite semiconductor, we have

$$\begin{aligned} \psi(y = 0) &= \psi_S \\ \psi(y \to \infty) &= 0 \,. \end{aligned} \tag{6.13}$$

In practice, ψ_S is set by the gate voltage.

The Poisson-Boltzmann equation is a nonlinear differential equation that is a bit difficult to solve in general. Some progress can be made analytically, but a numerical integration is also needed. Those interested in seeing how this is done should refer to [3–5]. It also turns out that we can solve the Poisson-Boltzmann equation approximately when the semiconductor is in strong accumulation, in depletion, or in strong inversion. We'll make use of these approximate solutions later. In the next section, we described the approximate solution for the depletion condition.

6.4 Depletion approximation

A very good approximate solution for the electrostatic potential and electric field versus position is readily obtained when the device is biased in depletion. In depletion, the bands bend down for a p-type semiconductor, and the concentration of holes is negligibly small for $y \lesssim W_D$. In depletion, the conduction band is still far above the Fermi level (that changes in inversion), so the electron concentration is also small. As a result, the space charge density,

$$\rho(y) = q\left[p_0(y) - n_0(y) + N_D^+ - N_A^-\right], \tag{6.14}$$

simplifies considerably. By ignoring the small number of mobile carriers, assuming only p-type dopants, and assuming complete ionization of dopants, eqn. (6.14) becomes

$$\begin{aligned} \rho(y) &= -qN_A \quad y < W_D \\ \rho(y) &= 0 \qquad\quad\; y \geq W_D \,. \end{aligned} \tag{6.15}$$

The depletion approximation is typically quite good, and it is simple enough to permit analytical solutions.

Figure 6.6 shows the energy band diagram in depletion and the corresponding electric field vs. position. We find the electric field by solving the Poisson equation,

$$\begin{aligned} \frac{dD}{dx} &= \frac{d(\epsilon_s \mathcal{E})}{dx} = \epsilon_s \frac{d\mathcal{E}}{dx} = \rho(y) = -qN_A \\ \frac{d\mathcal{E}}{dx} &= \frac{-qN_A}{\epsilon_s} \,. \end{aligned} \tag{6.16}$$

If the doping density is uniform, then the electric field is a straight line with a negative slope, as indicted on the right in Fig. 6.6. Accordingly, we can write the electric field in the depletion approximation as

$$\mathcal{E}(y) = \frac{qN_A}{\epsilon_s}(W_D - y) \,. \tag{6.17}$$

The electric field at the surface of the semiconductor is an important quantity that we find from eqn. (6.17) as

$$\mathcal{E}(y=0) = \mathcal{E}_S = \frac{qN_A W_D}{\epsilon_s} \,. \tag{6.18}$$

To find the electrostatic potential versus position, $\psi(y)$, recall that

$$\psi(y) = -\int_{\infty}^{y} \mathcal{E}(y')dy' \,. \tag{6.19}$$

Accordingly, the total potential drop across the depletion region, which is ψ_S, is the area under the $\mathcal{E}(y)$ vs. y curve, or

$$\psi_S = \frac{1}{2}\mathcal{E}_S W_D \,, \tag{6.20}$$

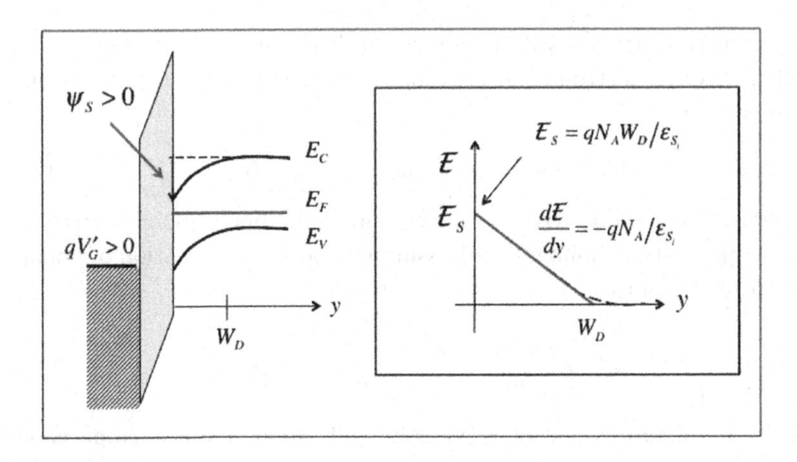

Fig. 6.6 Illustration of depletion in an MOS structure. Left: The energy band diagram. Right: The electric field vs. position. The solid line is the depletion approximation, and the dashed line is the actual electric field.

from which we find with the aid of eqn. (6.18)

$$\boxed{W_D = \sqrt{\frac{2\epsilon_s \psi_S}{qN_A}}}, \tag{6.21}$$

which is an important result.

The total charge in the semiconductor is

$$Q_S = \int_0^{\infty} \rho(y)dy \approx Q_D = -qN_A W_D = \epsilon_s \mathcal{E}_S \quad \mathrm{C/m^2}\,; \tag{6.22}$$

from eqns. (6.21) and (6.22), we find another important result

$$\boxed{Q_D \approx -\sqrt{2qN_A\epsilon_s\psi_S}}\,. \tag{6.23}$$

Note that in depletion, the total charge in the semiconductor, Q_S, is to a very good approximation, the charge in the depletion layer, Q_D, which consists of ionized acceptors.

When the semiconductor is biased in depletion, the depletion approximation provides accurate solutions for the electric field and electrostatic potential. It cannot be used, however, in the accumulation or inversion regions. Finally, to test your understanding, repeat the derivations in this section for an n-type semiconductor in depletion.

6.5 Onset of inversion

Figure 6.7 shows the band diagram and space charge profile for inversion conditions. In inversion, a large surface potential brings the conduction band at the surface very close to the Fermi level, so the concentration of electrons becomes large. The concentration of electrons at the surface can be related to the concentration of electrons in the bulk by using eqn. (6.11). Now we can ask the question: How large does the bandbending (ψ_S) need to be to make the surface as n-type and the bulk is p-type? From eqn. (6.11),

$$n_0(y=0) = \frac{n_i^2}{N_A} e^{q\psi_S/k_BT} = N_A\,, \tag{6.24}$$

we find the answer

$$\boxed{\begin{aligned} \psi_S &= 2\psi_B \\ \psi_B &= \frac{k_BT}{q} \ln\left(\frac{N_A}{n_i}\right) \end{aligned}}\,. \tag{6.25}$$

Equation (6.25) defines the onset of *inversion*; for surface potentials greater than about $2\psi_B$, there is an n-type layer at the surface of the p-type semiconductor. From Fig. 6.1, we see that for a gate voltage that produces a surface potential greater than about $2\psi_B$, there will be an n-type channel connecting the n-type source and drain regions, and the transistor will be on. The gate voltage needed to produce the required surface potential is the *threshold voltage.*

Under inversion conditions, the depletion region reaches a depth of W_T, where

$$\boxed{W_T = W_D\left(2\psi_B\right) = \sqrt{\frac{2\epsilon_s(2\psi_B)}{qN_A}}\,,} \tag{6.26}$$

which is an important result.

The total charge per unit area in the depletion region is

$$Q_D = -qN_AW_T \quad \text{C/m}^2\,. \tag{6.27}$$

There is also a significant charge due to the inversion layer electrons that pile up near the oxide-Si interface,

$$Q_n = q\int_0^\infty n_0(y)dy \quad \text{C/m}^2\,. \tag{6.28}$$

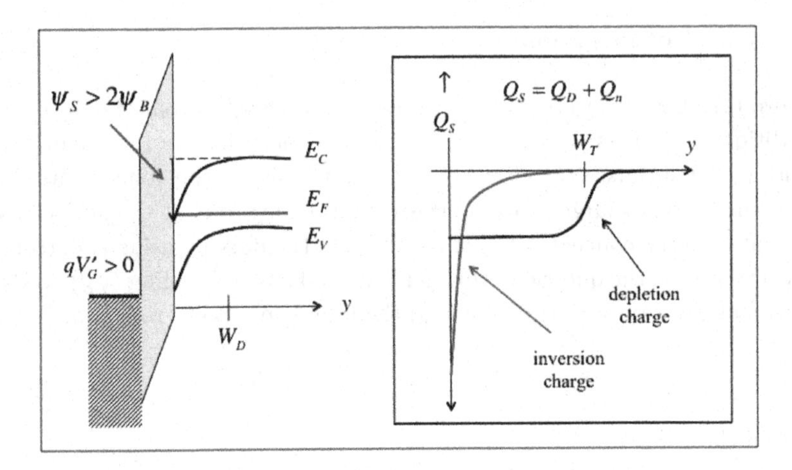

Fig. 6.7 Inversion condition in a semiconductor. Left: the energy band diagram. Right: the corresponding charge density. Note than in the depletion approximation, the depletion charge would got to zero abruptly at $y = W_D$.

The total charge in the semiconductor under inversion conditions is

$$Q_S = Q_D + Q_n \,. \tag{6.29}$$

Only the inversion layer charge carries the current in a MOSFET, so in subsequent lectures, we will relate the inversion layer charge to the gate voltage.

6.6 The body effect

In the previous section, we discussed MOS electrostatics in the middle of a long channel device in which the lateral electric fields due to the PN junctions were small, so a 1D analysis sufficed. But even in this case, PN junctions can have a strong influence. To see why, consider Fig. 6.8, which shows the case for zero voltage on the N and P regions (i.e. the case that we have been considering). The solid line shows the energy bands at the oxide-semiconductor interface from the source, across the channel, and to the drain under flatband conditions. The height of the energy barrier is just qV_{bi}, where the built in potential of the PN is given by standard semiconductor theory [1, 2] as

$$V_{bi} = \frac{k_B T}{q} \ln \frac{N_A N_D}{n_i^2} \,. \tag{6.30}$$

This energy barrier is large, so very few electrons can enter the channel from the source or drain by surmounting the energy barrier.

The dashed line in Fig. 6.8 shows the energy band diagram for a surface potential of $\psi_S = 2\psi_B$. In this case the energy barrier between source and channel is

$$E_b = q\left(V_{bi} - 2\psi_B\right) = k_B T \ln\left(N_D/N_A\right) . \tag{6.31}$$

For typical numbers ($N_D = 10^{20}$ cm^{-3} and $N_A = 10^{18}$ cm^{-3}), $E_b \approx 0.1$ eV. Electrons from the source can surmount this rather small energy barrier, and the result is an inversion layer in the channel.

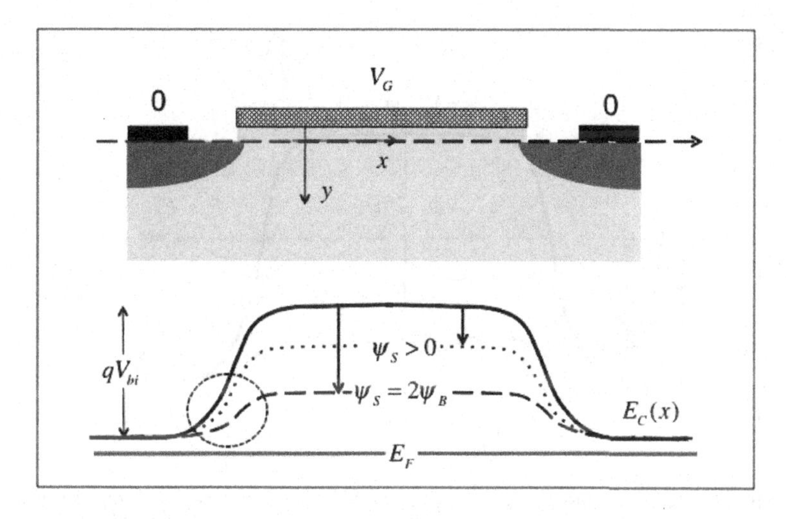

Fig. 6.8 Conduction band energy vs. position at the surface ($y = 0$) of the semiconductor along the channel from the source to the drain. Solid line: flatband condition (flat into the semiconductor). Dotted line: For a surface potential of $\psi_S > 0$ Dashed line: For a surface potential of $\psi_S = 2\psi_B$.

Now consider the situation in Fig. 6.9, which shows the energy band diagrams when a positive voltage (a reverse bias, V_R) has been applied to the source and drain. Under flatband conditions (solid line), the height of the energy barrier has increased to $q(V_{bi} + V_R)$. The dotted line shows the energy band diagram for $\psi_S = 2\psi_B$, the onset of inversion for the case of Fig. 6.8. In this case, however, the energy barrier is still very large, so electrons cannot enter from the source or drain. To achieve the same energy barrier as for the case of Fig. 6.8, the surface potential must be $\psi_S = 2\psi_B + V_R$.

We can also plot the energy band diagram into the depth of the semiconductor (the y-direction) rather than along the channel (the x-direction). The result is shown in Fig. 6.10 for a surface potential of $\psi_S = 2\psi_B + V_R$. (Figure 6.10 shows the energy band diagram normal to the channel.) Note that the hole quasi-Fermi level, F_p — the dashed line, is where the Fermi level was for zero bias on the source and drain, but the positive voltage on the source and drain lowers the electron quasi-Fermi level by qV_R. The electron quasi-Fermi level controls the electron density in the semiconductor. To achieve the same electron density at the onset of inversion as in the case for $V_R = 0$, the bands must be bent down by an additional amount of qV_R.

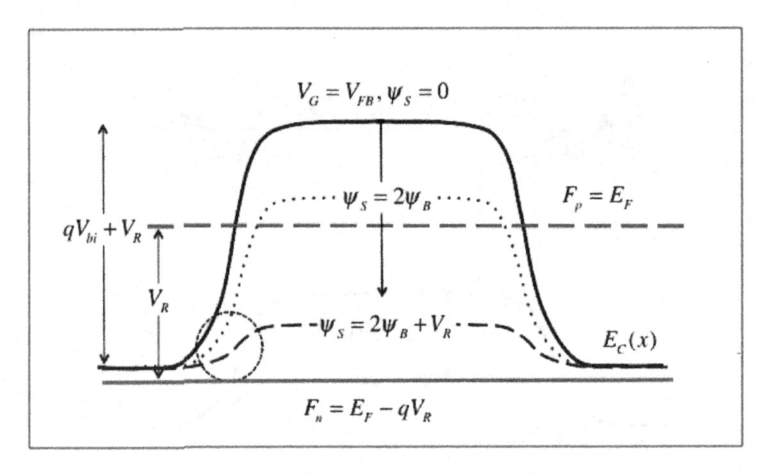

Fig. 6.9 Conduction band energy vs. position at the surface of the semiconductor along the channel from the source to the drain for the case of a reverse bias, V_R, between the source and drain and the semiconductor bulk. Solid line: Flatband condition. Dotted line: Surface potential of $\psi_S = 2\psi_B$. Dashed line: Surface potential of $\psi_S = 2\psi_B + V_R$.

In MOSFET circuits, the voltage on the source may be positive, which means that the source to substrate junction may be reverse biased. To create an inversion layer at the source end of the channel, the bands must be bent by $2\psi_B + V_R$. We'll see in the next lecture, that this increases the threshold voltage. The reason for the increased threshold voltage is an increase in the depletion charge. From eqn. (6.25), we find

$$Q_D \approx -\sqrt{2qN_A\epsilon_s\psi_S} = -\sqrt{2qN_A\epsilon_s(2\psi_B + V_R)}\,. \tag{6.32}$$

A reverse bias on the source can significantly increase the charge in the depletion layer at the onset of inversion.

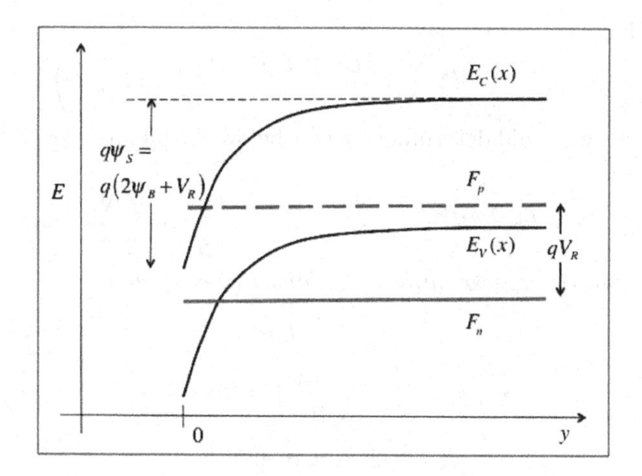

Fig. 6.10 Conduction band energy vs. position into the depth of the semiconductor at the midpoint of the channel for the case of a reverse bias, V_R, between the source and drain and the semiconductor bulk. This figure corresponds to the $\psi_S = 2\psi_B + V_R$ case in Fig. 6.9. Note the splitting of the quasi-Fermi levels under bias. The electron quasi-Fermi level has been lowered by an amount, V_R.

Finally, you may wonder how we can continue to assume that the semiconductor is in equilibrium when the PN junctions are biased. The answer is that in reverse bias (or even for small forward bias), the current is so small that we can continue to assume that the semiconductor is in equilibrium without introducing significant errors.

Exercise 6.1: Some typical numbers.

To get a feel for some of the numbers, consider an example for silicon:

$$\begin{aligned} N_A &= 1 \times 10^{18}\ \text{cm}^{-3} \\ N_V &= 1.81 \times 10^{19}\ \text{cm}^{-3} \\ n_i &= 1.00 \times 10^{10}\ \text{cm}^{-3} \\ \kappa_s &= 11.7 \\ T &= 300\ \text{K}\,. \end{aligned}$$

and consider the following questions:

1) *What is the position of the Fermi level in the bulk?*
We can answer this question by determining how far above the valence band

the Fermi level is.

$$p_{0B} = N_A = N_V e^{(E_V - E_F)/k_B T} \rightarrow \frac{E_F - E_V}{q} = \frac{k_B T}{q} \ln\left(\frac{N_V}{N_A}\right) = 0.075 \text{ eV}.$$

Alternatively, we could determine how far below the intrinsic level the Fermi level is.

$$p_{0B} = N_A = n_i e^{(E_i - E_F)/k_B T} \rightarrow \frac{E_i - E_F}{q} = \frac{k_B T}{q} \ln\left(\frac{N_A}{n_i}\right) = 0.48 \text{ eV}.$$

2) *What is the surface potential at the onset of inversion?*

$$\psi_S = 2\psi_B,$$

$$\psi_B = \frac{k_B T}{q} \ln\left(\frac{N_A}{n_i}\right) = 0.48 \text{ V},$$

$$\psi_S = 2\psi_B = 0.96 \text{ V}.$$

In the bulk, the Fermi level is very close to the valence band. To make the surface as n-type as the bulk is p-type, we need to bend the conduction band down to very close to the Fermi level, which means that it must be bent down by about the band gap, about 1 V.

3) *What is the width of the depletion layer at the onset of inversion?*

$$W_T = \sqrt{2\epsilon_s(2\psi_B)/qN_A}.$$

Inserting numbers, we find

$$W_T = 36 \text{ nm}.$$

4) *What is total charge per unit area in the depletion region?*

$$Q_D = -qN_A W_T = -\sqrt{2q\epsilon_s N_A(2\psi_B)}.$$

Inserting numbers, we find

$$Q_D = -5.8 \times 10^{-6} \text{ C/cm}^2,$$

or, in terms of the number of charges per unit area:

$$|Q_D|/q = 3.6 \times 10^{12} \text{ cm}^{-2}.$$

5) *What is electric field at the surface of the semiconductor?*
From Gauss's Law:

$$\mathcal{E}_S = -\frac{Q_D}{\epsilon_s}.$$

Inserting numbers, we find

$$\mathcal{E}_S = 5.6 \times 10^6 \text{ V/cm},$$

which is a strong electric field.

This example gives a feel for the numbers we expect to encounter for typical MOS calculations.

6.7 Discussion

i) charge in the semiconductor vs. band bending

Our goal is in this lecture was to understand how the band bending (surface potential) controls the charge in the semiconductor. Figure 6.5 illustrated the bandbending under accumulation, and depletion/inversion. Figure 6.11 shows that in accumulation, the majority carrier hole charge builds up exponentially for increasingly negative ψ_S. In the depletion region, we find from eqn. (6.23) that $|Q_S| \propto \sqrt{\psi_S}$. In inversion, the minority carrier electron charge builds up exponentially for increasingly positive $\psi_S > 2\psi_B$. In Lectures 8 and 9, we will derive approximate solutions to the Poisson-Boltzmann equation in accumulation and inversion, but the general shape of the $Q_S(\psi_S)$ characteristic is easy to understand.

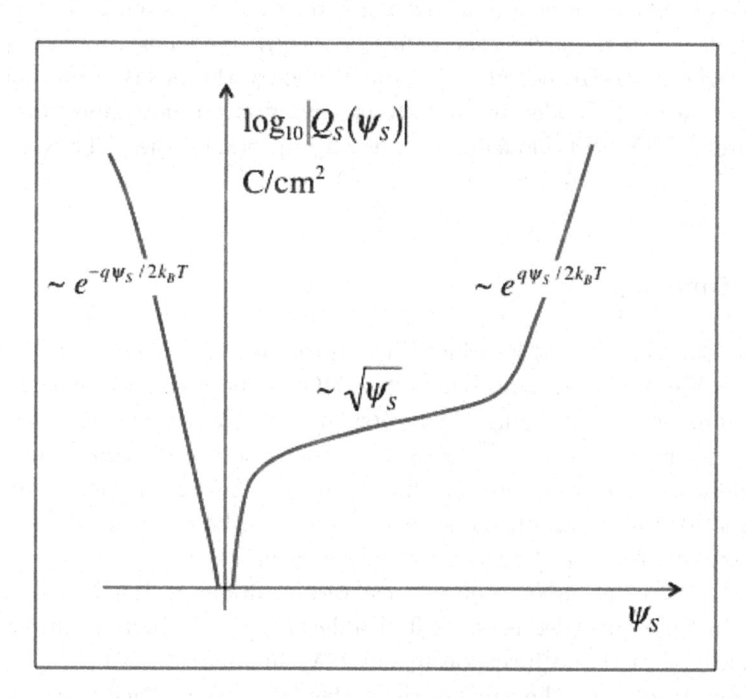

Fig. 6.11 Charge in a p-type semiconductor as a function of the surface potential. The sheet charge, Q_S in C/m^2 is the volume charge density, ρ in C/m^3 integrated into the depth of the semiconductor.

ii) criterion for weak inversion, moderate inversion, and strong inversion

In Sec. 6.5, we asserted that inversion occurs for

$$\psi_S > 2\psi_B ,$$

but inversion is a gradual process. Note that when $\psi_S = \psi_B$, the semiconductor is intrinsic at the surface, $n_0(0) = p_0(0)$. For $\psi_S > \psi_B$, there is a small, net concentration of electrons at the surface. We will see that this small concentration of electrons leads to a small "leakage" current. We say that $\psi_S = \psi_B$ is the beginning of *weak inversion.* At $\psi_S = 2\psi_B$, the surface is as n-type as the bulk is p-type, but the total number of electrons in this layer near the surface is still small. We say that $\psi_S = 2\psi_B$ is the end of the weak inversion region and the beginning of the moderate inversion region. We will see in Lecture 8 that for on-current conditions, the surface potential can be a few k_BT/q greater than $2\psi_B$. When the surface potential is a little larger than $2\psi_B$ $Q_n \gg Q_D$; moderate inversion ends and *strong inversion* begins. For our purposes, the precise values of ψ_S that define weak, moderate, and strong inversion are not important, but for careful MOSFET modeling, this is an important issue. The reader is referred to [3] for a discussion of these effects.

6.8 Summary

This lecture has been a short introduction to some very basic MOS electrostatics. We discussed band bending in MOS structures and the concepts of accumulation, depletion, and inversion. We established a criterion for the onset of inversion ($\psi_S = 2\psi_B$). We formulated the Poisson-Boltzmann equation, which can be solved to find $\psi(y)$ vs. y for any surface potential. From $\psi(y)$, the charge in the semiconductor can be determined. Finally, we discussed the depletion approximation, which assumes a space charge profile and can be used to obtain accurate solutions in the depletion region. In subsequent lectures, we'll also develop approximate solutions for the accumulation and inversion regions. As summarized in Fig. 6.11, everything depends on the value of the surface potential, which is set by the gate voltage. In the next lecture, we relate the surface potential to the gate voltage.

6.9 References

The concepts introduced in this lecture are covered in introductory semiconductor textbooks such as

[1] Robert F. Pierret *Semiconductor Device Fundamentals, 2nd Ed.*, Addison-Wesley Publishing Co, 1996.

[2] Ben Streetman and Sanjay Banerjee, *Solid State Electronic Devices, 6th Ed.*, Prentice Hall, 2005.

The exact solution of the Poisson-Boltzamnn equation is discussed in the two books and the notes listed below.

[3] Y. Tsividis and C. McAndrew, *Operation and Modeling of the MOS Transistor*, 3rd Ed., Oxford Univ. Press, New York, 2011. (See Sec. 2.3.2.1)

[4] Y. Taur and T. Ning, *Fundamentals of Modern VLSI Devices*, 2nd Ed., Oxford Univ. Press, New York, 2013. (See Sec. 2.4.4)

[5] Mark Lundstrom, "Notes on the Solution of the Poisson-Boltzmann Equation for MOS Capacitors and MOSFETs, 2nd Ed.," `https://nanohub.org/resources/5338`, 2012.

Lecture 7

Gate Voltage and Surface Potential

7.1 Introduction

Figure 7.1 summarizes the questions that we'll address in this lecture. Note that there is an electrostatic potential drop of ψ_S across the semiconductor. The first question is: "What gate voltage produced this surface potential?" or what is $\psi_S(V_G)$? We'll first explain how to do this exactly, and later discuss an approximate solution. The gate voltage needed to put the semiconductor at the onset of inversion is known as the *threshold voltage*, V_T, and is the voltage needed to make $\psi_S = 2\psi_B$. This is the gate voltage needed to turn a MOSFET on. Finally, measurements of the small signal gate capacitance as a function of the D.C. bias on the gate, V_G, which are often used to characterize MOS structures, will be discussed. The topics discussed in this lecture are covered in introductory textbooks, such as [1, 2].

7.2 Gate voltage and surface potential

To relate the gate voltage to the surface potential, note that the gate voltage is the sum of the voltage drop across the oxide and the voltage drop across

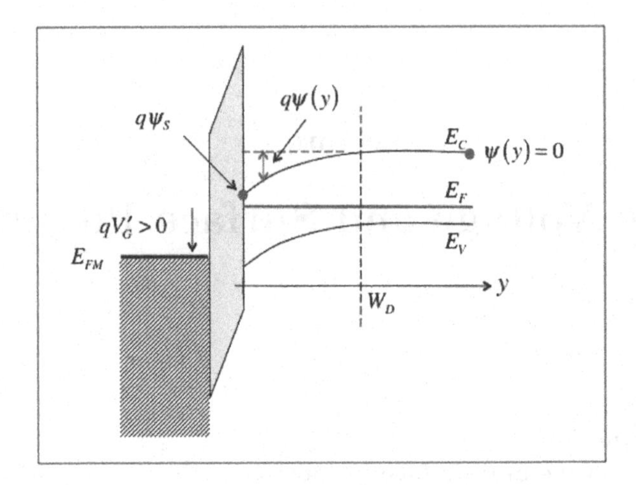

Fig. 7.1 An MOS energy band diagram for a positive gate voltage, V_G', which produces a positive surface potential in the semiconductor and a voltage drop across the oxide. We seek to relate the gate voltage, V_G' to the surface potential, ψ_S.

the semiconductor,

$$V_G' = \Delta V_{ox} + \Delta V_{semi} = \Delta V_{ox} + \psi_S \,. \tag{7.1}$$

The voltage drop across the oxide is the electric field times the thickness of the oxide,

$$\Delta V_{ox} = \mathcal{E}_{ox} t_{ox} \,. \tag{7.2}$$

To find the electric field in the oxide, we use Gauss's Law, which tells us that the normal displacement field at the oxide — Si interface is equal to the charge per unit area in the semiconductor (ignoring for now any possible charge at the oxide-Si interface). Accordingly, we find

$$\epsilon_{ox}\mathcal{E}_{ox} = -Q_S(\psi_S) \,, \tag{7.3}$$

where Q_S in C/m^2 is the charge in the semiconductor, which is a function of the surface potential, ψ_S. From eqns. (7.2) and (7.3), we find

$$\Delta V_{ox} = -\frac{Q_S(\psi_S)}{C_{ox}} \,, \tag{7.4}$$

where

$$C_{ox} = \frac{\epsilon_{ox}}{t_{ox}} \ \text{F/m}^2 \,, \tag{7.5}$$

is the gate insulator capacitance per unit area. Finally, using eqns. (7.4) and (7.1), we find the desired relation,

$$V'_G = -\frac{Q_S(\psi_S)}{C_{ox}} + \psi_S \,. \tag{7.6}$$

Equation (7.6) assumes an ideal gate electrode (and the absence of charge at the oxide-semiconductor interface) so that at $V'_G = 0$, the bands are flat and $\psi_S = Q_S = 0$.

Consider the case illustrated in Fig. 7.2 for which the work function of the gate electrode, Φ_M, is less than the work function of the semiconductor, Φ_S. The equilibrium energy band diagram shows that there is a built-in potential across the structure. For zero applied voltage at the gate electrode, the electrostatic potential at the gate is $-(\Phi_M - \Phi_S)/q$. It is apparent that if we apply a gate voltage equal to the metal — semiconductor work function difference, then the effect of the difference in work functions will be undone, and the bands will be flat. Accordingly, the flatband voltage won't be at $V_G = 0$, but at $V_G = V_{FB}$ where

$$qV_{FB} = (\Phi_M - \Phi_S) = \Phi_{MS} \,. \tag{7.7}$$

Alternatively consider the case where there is no work function difference, but there is a fixed charge, Q_F, in C/m^2 at the oxide-semiconductor interface. In this case, Gauss's Law for the electric field in the oxide, eqn. (7.3), becomes

$$\epsilon_{ox}\mathcal{E}_{ox} = -Q_S\,(\psi_S) - Q_F \,, \tag{7.8}$$

so eqn. (7.4) becomes

$$\Delta V_{ox} = -\frac{Q_S(\psi_S)}{C_{ox}} - \frac{Q_F}{C_{ox}} \,. \tag{7.9}$$

When $\psi_S = 0$, $Q_S = 0$, and the bands in the semiconductor are flat. According to eqn. (7.1), this flatband condition occurs at $V_G = V_{FB} = -Q_F/C_{ox}$.

In general, there is both a work function difference and charge at the oxide-semiconductor interface, so the flatband condition occurs at a gate voltage of

$$\boxed{V_{FB} = \frac{\Phi_{MS}}{q} - \frac{Q_F}{C_{ox}} \,,} \tag{7.10}$$

and the general relation between the gate voltage and the surface potential is

$$\boxed{V_G' = V_G - V_{FB} = -\frac{Q_S(\psi_S)}{C_{ox}} + \psi_S}\,. \tag{7.11}$$

It is also possible that the charge at the oxide-semiconductor interface is not fixed but depends on the surface potential and that there is charge distributed throughout the oxide layer. See [1] for a discussion of these effects.

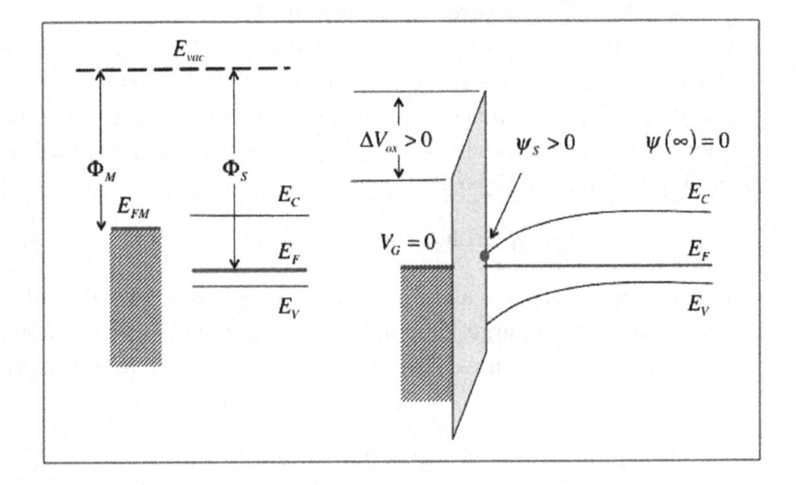

Fig. 7.2 Illustration of how a metal-semiconductor work function difference affects a 1D MOS structure. Left: The isolated components with separate Fermi levels in the gate electrode and in the semiconductor. Right: The resulting equilibrium energy band diagram for $V_G = 0$. Note that the built-in potential for this structure is analogous to the built-in potential of a PN junction, and just as for a PN junction, it cannot be measured directly.

Equation (7.11) is our desired relation between the gate voltage and the surface potential in the semiconductor. In general, eqn. (7.11) cannot be analytically solved for ψ_S as a function of V_G. In practice, however, we can assume a ψ_S and then compute the V_G that produced it. We saw in Lecture 6 how to calculate $Q_S(\psi_S)$ in depletion; in Lectures 8 and 9 we'll discuss how to calculate $Q_S(\psi_S)$ more generally and will examine the $\psi_S(V_G)$ relation in detail.

Although the computation of the ψ_S vs. V_G characteristic takes a bit of work (see [3, 4, 5]), the qualitative shape of the characteristic, which is shown in Fig. 7.3, is easy to understand. Recall the $Q_S(\psi_S)$ characteristic sketched in Fig. 6.11. As ψ_S increases from zero to a positive value, the charge in the depletion layer builds up slowly (as $\sqrt{\psi_S}$); the charge in the semiconductor is modest, so from eqn. (7.11), we see that most of the gate voltage is dropped across the semiconductor. Once the surface potential exceeds $2\psi_B$, the inversion charge becomes significant; it builds up exponentially, and the voltage drop across the oxide becomes very large. Most of the gate voltage in excess of the amount needed to bend the bands by $2\psi_B$ is dropped across the oxide, so it is very hard to increase the surface potential beyond $2\psi_B$. When the gate voltage is negative, a strong accumulation layer of charge quickly builds up. In accumulation, most of the gate voltage is dropped across the oxide and very little across the semiconductor.

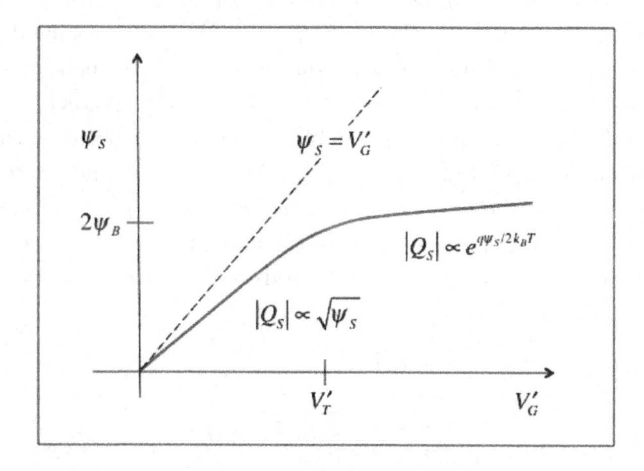

Fig. 7.3 Sketch of the expected ψ_S vs. V_G characteristics. Below threshold, the surface potential varies directly with V_{GS} according to $\psi_S = V_{GS}/m$, where $m \approx 1$, but above threshold, $\psi_S \approx 2\psi_B$, and the surface potential varies slowly with V_{GS} because $m \gg 1$.

7.3 Threshold voltage

One way to define the threshold voltage is as the gate to source voltage needed to bend the bands so that $\psi_S = 2\psi_B$, which is the point at which a significant inversion charge begins to build up (see [3], pp. 732, 733). From eqn. (7.11) we can write

$$V_T = V_{FB} - \frac{Q_S(2\psi_B)}{C_{ox}} + 2\psi_B\,. \tag{7.12}$$

At the onset of inversion, $Q_S = Q_D + Q_n$ consists mostly of depletion charge; the charge in the inversion layer is still small. By assuming $Q_S(2\psi_B) \approx Q_D(2\psi_B)$, we find

$$V_T = V_{FB} - \frac{Q_D(2\psi_B)}{C_{ox}} + 2\psi_B\,.$$

$$\boxed{V_T = V_{FB} + \frac{\sqrt{2qN_A\epsilon_s(2\psi_B)}}{C_{ox}} + 2\psi_B}\,. \tag{7.13}$$

Equation (7.13) is a key result that allows us to compute the threshold voltage if we know the channel doping and oxide thickness. Higher channel doping densities lead to higher threshold voltages, and thinner gate oxide thicknesses lead to lower threshold voltages. We have assumed uniform channel doping, but non-uniform channel doping profiles, such as *retrograde* or *ground plane* profiles are also used (see [6] for a discussion).

As discussed in relation to eqn. (6.32), a reverse bias between the source and channel lowers the quasi-Fermi level for electrons and increases the surface potential for the onset of inversion from $2\psi_B$ to $2\psi_B + V_{SB}$, where V_{SB} is the reverse bias between the source and the body. Accordingly, the gate voltage between the gate and the body needed to bend the bands to the onset of inversion increases to

$$\begin{aligned} V_{GB} &= V_{FB} - \frac{Q_D(2\psi_B + V_{SB})}{C_{ox}} + 2\psi_B + V_{SB} \\ &= V_{FB} + \frac{\sqrt{2qN_A\epsilon_s(2\psi_B + V_{SB})}}{C_{ox}} + 2\psi_B + V_{SB}\,. \end{aligned} \tag{7.14}$$

The voltage between the source and the body is V_{SB}, so the gate to source voltage, V_{GS}, at the onset of inversion is $V_{GS} = V_T$, where

$$V_T = V_{GB} - V_{SB} = V_{FB} - \frac{Q_D(2\psi_B + V_{SB})}{C_{ox}} + 2\psi_B\,.$$

$$\boxed{V_T = V_{FB} + \frac{\sqrt{2qN_A\epsilon_s(2\psi_B + V_{SB})}}{C_{ox}} + 2\psi_B}\,. \tag{7.15}$$

We see that a heavily doped channel not only increases V_T, it also makes the threshold voltage more sensitive to the reverse bias between the source and the body. The dependence of the threshold voltage on the source to body voltage is known as the *body effect.*

Finally, note that threshold voltage usually refers to the onset of strong inversion. As discussed in Sec. 6.7, $\psi_S > 2\psi_B$ for strong inversion, so $2\psi_B$ should be replaced by a potential that is a few k_BT/q larger. Nevertheless, it is common practice to use $\psi_S = 2\psi_B$ in the V_T equation, except in careful MOSFET modeling where this issue becomes important. See Chapter 2 of [1] for a discussion.

7.4 Gate capacitance

A common way to characterize MOS structures is to measure the small signal, A.C. capacitance between the gate electrode and the bottom of the substrate as a function of the D.C. bias on the gate. Figure 7.4 reviews some basic concepts.

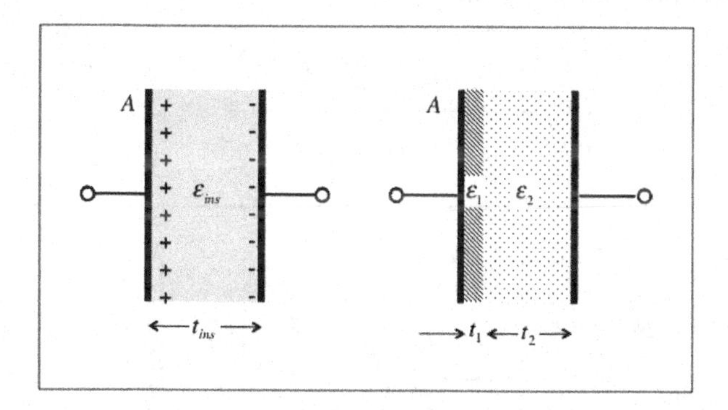

Fig. 7.4 Left: A simple parallel plate capacitor with a single dielectric between the two plates. Right: A parallel plate capacitor with two different dielectrics between the plates. The cross-sectional area of the plates is A.

For a simple parallel plate capacitor (shown on the left of Fig. 7.4) the capacitance per unit area is readily shown to be

$$\frac{C}{A} = \frac{\epsilon_{\text{ins}}}{t_{ins}} \ \text{F/m}^2 . \tag{7.16}$$

Consider next the parallel plate capacitor shown on the right of Fig. 7.4. In this case, there are two different dielectrics between the two plates with two different dielectric constants and two different thicknesses. The capacitance per unit area is readily shown to be

$$\frac{1}{C/A} = \frac{1}{C_1/A} + \frac{1}{C_2/A} = \frac{1}{\epsilon_1/t_1} + \frac{1}{\epsilon_2/t_2} \text{ F/m}^2 . \tag{7.17}$$

With this background, let's consider the MOS capacitance at three different D.C. biases. Figure 7.5 shows the band diagrams in depletion, inversion, and accumulation. In the first case, the gate electrode is the first metal plate, the gate insulator the first dielectric, the depleted semiconductor the second dielectric, and the undepleted p-layer the second "metal" plate. Accordingly, we expect to measure a gate capacitance of

$$\frac{1}{C_G(\text{depl})} = \frac{1}{C_{ox}} + \frac{1}{C_D} \text{ F/m}^2 , \tag{7.18}$$

where C_G is the gate capacitance per unit area, and

$$\boxed{C_{ox} = \frac{\epsilon_{ox}}{t_{ox}} \text{ F/m}^2} , \tag{7.19}$$

is the *oxide capacitance* per unit area and

$$\boxed{C_D = \frac{\epsilon_s}{W_D(\psi_S)} \text{ F/m}^2} , \tag{7.20}$$

is the *depletion capacitance* per unit area.

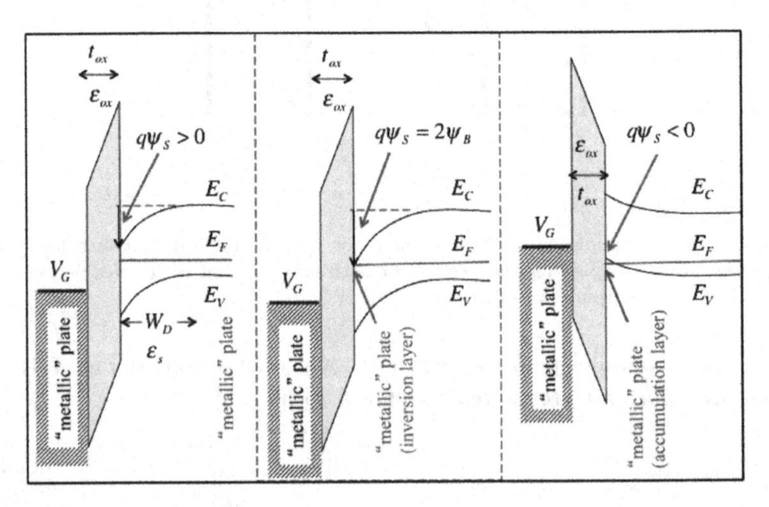

Fig. 7.5 Left: Energy band diagrams under three different D.C. biases. Left: depletion, Center: inversion, Right: accumulation.

Consider next the inversion capacitance illustrated in the middle of Fig. 7.5. In this case, the first dielectric is still the oxide layer, but the second "metal" plate is the highly conductive inversion layer of electrons at the oxide-semiconductor interface. Accordingly, we expect the gate capacitance in inversion to be

$$C_G(\mathrm{inv}) \approx C_{ox} \,. \tag{7.21}$$

Finally, consider the accumulation capacitance on the right of Fig. 7.5. The first dielectric is still the oxide layer and the second "metal" plate is the highly conductive accumulation layer of holes at the oxide-semiconductor interface. Accordingly, we expect the capacitance in accumulation to be

$$C_G(\mathrm{acc}) \approx C_{ox} \,. \tag{7.22}$$

These examples show that the gate capacitance is the series combination of two capacitors,

$$\frac{1}{C_G} = \frac{1}{C_{ox}} + \frac{1}{C_S(\psi_S)} \,, \tag{7.23}$$

where $C_S(\psi_S)$ is the *semiconductor capacitance*, which depends strongly on the value of the D.C. surface potential.

These qualitative ideas about how the gate capacitance varies with the D.C. bias on the gate can be made more quantitative. The gate capacitance is defined as

$$C_G \equiv \frac{dQ_G}{dV_G} \ \mathrm{F/m^2} \,, \tag{7.24}$$

where Q_G is the charge per unit area on the gate electrode. Because the charge must balance, $Q_G = -Q_S$, where Q_S is the charge per unit area in the semiconductor. By differentiating eqn. (7.11), we find

$$\frac{dV_G}{d(-Q_S)} = \frac{d\psi_S}{d(-Q_S)} + \frac{1}{C_{ox}} \,, \tag{7.25}$$

which can be written as

$$\boxed{\frac{1}{C_G} = \frac{1}{C_{ox}} + \frac{1}{C_S}} \,, \tag{7.26}$$

where

$$\boxed{C_S \equiv \frac{d(-Q_S)}{d\psi_S}} \,, \tag{7.27}$$

is the *semiconductor capacitance.* (Note that an increase in surface potential increases the magnitude of the *negative* charge in the depletion and inversion layers, so the semiconductor capacitance is a positive quantity.)

Figure 7.6 shows the equivalent circuit that represents the gate capacitance. To compute C_G vs. V_G. we need to understand how $C_S = d(-Q_S)/d\psi_S$ varies with bias. Figure 6.11 gives the qualitative answer. In accumulation and inversion, the semiconductor capacitance is very large, so the total capacitance is close to the oxide capacitance, as sketched in Fig. 7.7. In depletion, the semiconductor capacitance is moderate, so that total capacitance is lowered, as shown in Fig. 7.7.

For the solid line in Fig. 7.7, the A.C. signal used to measure the small signal capacitance is assumed to be at a low enough frequency such that electrons in the inversion layer can respond to the A.C. signal. That is

$$C_S = \frac{d(-Q_S)}{d\psi_S} \approx \frac{d(-Q_n)}{d\psi_S} \gg C_{ox} ,$$

At high frequencies, a low small signal capacitance is measured when the D.C. bias is in inversion. This occurs when the frequency is so high that the inversion layer cannot respond to the A.C. signal. Rather slow recombination-generation processes are needed to increase or decrease the inversion layer density. Accordingly, when the small signal frequency is high and the D.C. bias is in inversion, we find

$$C_S = \frac{d(-Q_S)}{d\psi_S} \approx \frac{d(-Q_D)}{d\psi_S} = \frac{\epsilon_s}{W_T} ,$$

which is the dashed line in Fig. 7.7. (For a typical silicon MOSFET, the high frequency limit is well below 1 MHz.) When the capacitor is part of a MOSFET, however, electrons can quickly enter and leave the semiconductor through the source and drain contacts, so the high frequency characteristic is observed. See [1] for a discussion.

7.5 Approximate gate voltage — surface potential relation

Equation (7.11) relates the gate voltage to the surface potential of the semiconductor. It can be solved numerically for a general surface potential, and in depletion, it can be solved analytically. Assuming that the semiconductor charge is only the depletion layer charge, eqn. (7.11) becomes

$$V_G = V_{FB} + \frac{\sqrt{2q\epsilon_s N_A \psi_S}}{C_{ox}} + \psi_S , \tag{7.28}$$

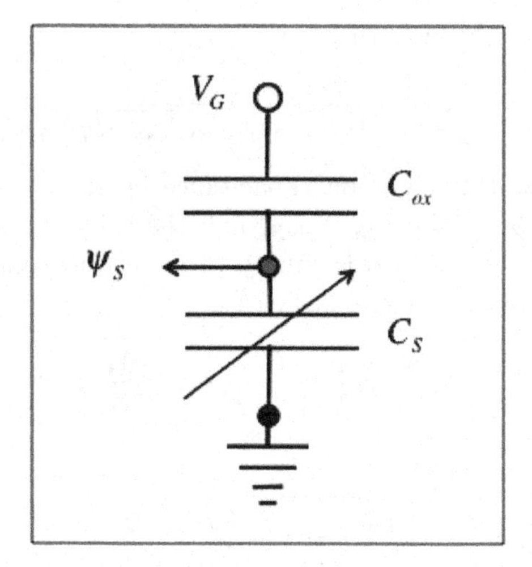

Fig. 7.6 Equivalent circuit illustrating how the gate capacitance is the series combination of the oxide capacitance, C_{ox} and the semiconductor capacitance, C_S.

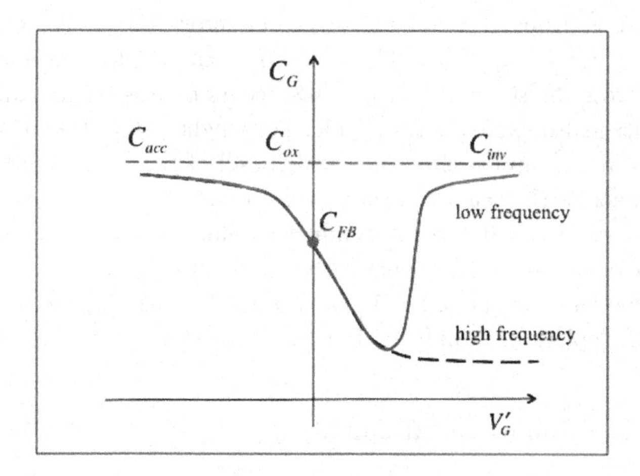

Fig. 7.7 Sketch of how the small signal gate capacitance is expected to vary with D.C. bias. Solid line — low frequency characteristic. Dashed line — high frequency characteristic.

which is a quadratic equation for $\sqrt{\psi_S}$ (See [1] for the solution.) For many applications, a simpler relation is needed, and the equivalent circuit of Fig. 7.6 provides an approach.

The semiconductor capacitance is a function of the surface potential, but in depletion the semiconductor capacitance is the depletion capacitance, which varies rather slowly with ψ_S,

$$C_s \approx C_D = \frac{\epsilon_s}{W_D(\psi_S)} = \frac{\epsilon_s}{\sqrt{2\epsilon_s \psi_S / qN_A}} . \tag{7.29}$$

If we approximate the depletion capacitance by its average value in depletion (perhaps by setting $\psi_S = \psi_B$, half the value needed to invert the semiconductor), then Fig. 7.6 is simply two constant capacitors in series, and voltage division in this circuit gives

$$\psi_S = V_G \left(\frac{C_{ox}}{C_{ox} + C_D} \right) = \frac{V_G}{m} , \tag{7.30}$$

where

$$\boxed{m = 1 + \frac{C_D}{C_{ox}}} \tag{7.31}$$

is known as the *body effect coefficient in depletion.*

The body effect coefficient, m, tells us what fraction of the applied gate voltage is dropped across the semiconductor. For a very thin oxide, $C_{ox} \gg C_D$, and m approaches one — all of the applied gate voltage is dropped across the semiconductor. This occurs because there can only be a small voltage drop across a thin oxide. For a lightly doped semiconductor, $C_D \ll C_{ox}$ and m approaches one — again all of the applied gate voltage is dropped across the semiconductor. This occurs because the light doping leads to a small charge in the semiconductor, which produces a small electric field in the oxide and a correspondingly small voltage drop in the oxide. A typical value for m is about 1.1–1.3, so a plot of ψ_S vs. V_G has a slope less than one in depletion, as indicated in Fig. 7.3.

Exercise 7.1: Some typical numbers.

To get a feel for some of the numbers that result from the formulas developed in this lecture, consider the silicon example of Exercise 6.1:

$$\begin{aligned} N_A &= 4 \times 10^{18}\ \text{cm}^{-3} \\ N_V &= 1.81 \times 10^{19}\ \text{cm}^{-3} \\ n_i &= 1.00 \times 10^{10}\ \text{cm}^{-3} \\ \kappa_s &= 11.7 \\ T &= 300\ \text{K} . \end{aligned}$$

Now, also assume

$$t_{ox} = 1.8 \text{ nm}$$
$$\kappa_{ox} = 4.0$$
an n$^+$ polysilicon gate
no charge at the oxide – semiconductor interface.

and consider some questions:

1) *What is the metal-semiconductor workfunction difference and the flat-band voltage?*
In Exercise 6.1, we found that the Fermi level in the semiconductor was 0.075 eV above the valence band in the bulk. Assume that the polysilicon gate electrode is doped heavily so that $E_F = E_C$, a reasonable assumption. The difference between the Fermi level in the metal-like gate and the p-type semiconductor is just a little less than the semiconductor band gap:

$$\Phi_{MS} = -(1.1 - 0.075) = -1.03 \text{ eV},$$

and the flatband voltage is

$$V_{FB} = \frac{\Phi_{MS}}{q} = -1.03 \text{ V}.$$

2) *What is the threshold voltage?*
In Exercise 6.1, we found that $\psi_B = k_B T/q \ln(N_A/n_i) = 0.48$ V so at the onset of inversion $\psi_S = 2\psi_B = 0.96$ V.

At the onset of inversion, the charge in the semiconductor is mostly charge in the depletion layer. We found in Exercise 6.1 that $Q_D = \sqrt{2qN_A\epsilon_s(2\psi_B)} = 1.2 \times 10^{-6}$ C/cm^2. The oxide capacitance is

$$C_{ox} = \frac{\epsilon_{ox}}{t_{ox}} = 2.0 \times 10^{-6} \text{F/cm}^2.$$

Finally, from eqn. (7.13), we find the threshold voltage as

$$V_T = V_{FB} - \frac{Q_D(2\psi_B)}{C_{ox}} + 2\psi_B = 0.19 \text{ V}.$$

3) *What is the value of the body effect coefficient?*
First, we need the depletion layer capacitance. Let's evaluate it at $\psi_S = \psi_B = 0.48$:

$$C_D = \frac{\epsilon_s}{W_D} = \frac{\epsilon_s}{\sqrt{2\epsilon_s\psi_S/qN_A}} = 8.3 \times 10^{-7} \text{ F/cm}^2.$$

From eqn. (7.31) we find

$$m = 1 + \frac{C_D}{C_{ox}} = 1.4.$$

In depletion, 70% of the gate voltage is dropped across the semiconductor.

7.6 Discussion

Figure 7.8 shows a typical MOS *gate stack.* Note that the gate electrode is not a metal but, rather, a heavily doped layer of polycrystalline (so-called "poly") silicon. If doped heavily enough, it acts more or less like a metal. (Note that manufacturers are currently replacing SiO_2 with higher dielectric constant materials (so-called "hi-k" dielectrics) to increase the gate capacitance. The poly silicon gate is also being replaced with a metal gate, but poly Si gate stacks are still common.)

As shown in Fig. 7.6, the gate capacitance consists of a series combination of the oxide and semiconductor capacitance, so the total gate capacitance is less than C_{ox}. In depletion, the total capacitance is significantly less than C_{ox}, but in inversion, the semiconductor capacitance becomes very large. Ideally, we'd like C_S to be much larger than C_{ox} in inversion so that $C_G \approx C_{ox}$. As gate oxides have scaled down in thickness over the past few decades, the lowering of the gate capacitance in inversion by the semiconductor capacitance has become a significant factor. To treat this problem quantitatively, numerical calculations are needed; we'll discuss this issue briefly in the next two lectures.

For polysilicon gates there is one more factor that lowers the overall gate capacitance, so-called *poly depletion.* As shown in Fig. 7.8, under inversion conditions, there is a strong electric field in the $+y$-direction pointing from the positive charge on the gate to the negative charge in the semiconductor. This electric field depletes and then inverts the semiconductor substrate. But this electric field can also deplete (a little) the heavily doped n^+ polysilicon gate. The gate capacitance now consists of three capacitors in series, the oxide capacitance, the semiconductor capacitance, and the depletion capacitance of the polysilicon gate:

$$\frac{1}{C_G} = \frac{1}{C_{poly}} + \frac{1}{C_{ox}} + \frac{1}{C_S} .$$

Device engineers often describe these effects in terms of an *capacitance equivalent thickness* (or capacitance extracted thickness), CET, which is defined as the thickness of SiO_2 that produces the measured gate capacitance in strong inversion — including the effects of the semiconductor capacitance and poly depletion as well as the dielectric itself. The CET is defined by

$$C_G \equiv \frac{\epsilon_{ox}}{CET} . \tag{7.32}$$

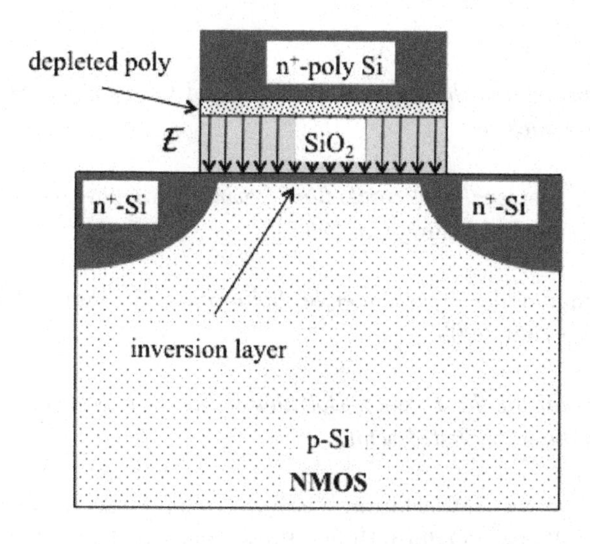

Fig. 7.8 Sketch of a typical "gate stack." Traditionally, the "metal" gate electrode is a heavily doped layer of poly crystalline (so-called *poly* silicon).

In Exercise 7.1, $t_{ox} = 1.8$ nm should really have been given as $CET = 1.8$ nm.

7.7 Summary

What happens in the semiconductor is determined by the bandbending in the semiconductor, ψ_S, but the "knob" we have to control the surface potential is the gate voltage. In this lecture, we developed a relation between the gate voltage and the surface potential, eqn. (7.11). We also showed that in depletion, there is a simple relation between V_G and ψ_S, eqn. (7.30), which we will use frequently.

Before we proceed, let's re-cap. MOS electrostatics can be qualitatively described in terms of energy band diagrams, as we did in the previous lecture. The charge in the semiconductors is a function of the bandbending, as described by ψ_S. In Lecture 6 we discussed qualitatively how the total charge in the semiconductor, Q_S, varies with ψ_S and developed an approximate expression valid in depletion. In this lecture, we related ψ_S to the gate voltage that produces it. In the following two lectures, we'll examine the mobile electron charge, $Q_n(\psi_S)$ and $Q_n(V_G)$.

7.8 References

The concepts introduced in this lecture are covered in introductory semiconductor textbooks such as

[1] Robert F. Pierret, *Semiconductor Device Fundamentals, 2nd Ed.*, Addison-Wesley Publishing Co, 1996.

[2] Ben Streetman and Sanjay Banerjee, *Solid State Electronic Devices, 6th Ed.*, Prentice Hall, 2005.

The exact solution of the Poisson-Boltzmann equation is discussed in the two books and the notes listed below.

[3] Y. Tsividis and C. McAndrew, *Operation and Modeling of the MOS Transistor*, 3rd Ed., Oxford Univ. Press, New York, 2011. (See Sec. 2.3.2.1)

[4] Y. Taur and T. Ning, *Fundamentals of Modern VLSI Devices*, 2nd Ed., Oxford Univ. Press, New York, 2013. (See Sec. 2.4.4)

[5] Mark Lundstrom and Xingshu Sun, "Notes on the Solution of the Poisson-Boltzmann Equation for MOS Capacitors and MOSFETs, 2nd Edition," `https://nanohub.org/resources/5338`, 2012.

Various channel doping profiles that can be used are discussed in:

[6] Mark Lundstrom, "ECE 612 Lecture 14: V_T Engineering," `https://nanohub.org/resources/5670`, 2008.

Lecture 8

Mobile Charge: Bulk MOS

8.1 Introduction

In Lecture 6 we discussed how the charge in the semiconductor varies with the bandbending as measured by the surface potential. The goal was to understand $Q_S(\psi_S)$ qualitatively, but we also showed how the Poisson equation can be solved in depletion for $Q_D(\psi_S)$, which is due to the ionized acceptors (or donors in an n-type semiconductor). In this lecture, our focus is on understanding the part of the charge due to mobile electrons, $Q_n(\psi_S)$, the inversion charge. For a P-MOSFET, the corresponding quantity is the hole inversion layer in the n-type channel, $Q_p(\psi_S)$. Solving the Poisson-Boltzmann equation, (6.12), as discussed in [1–3] provides a way to compute the inversion charge vs. surface potential.

The solution to the Poisson-Boltzmann equation is often referred to as the "exact" solution of the MOS problem, although it is far from exact. It assumes, for example, Maxwell-Boltzmann statistics, whereas in strong inversion or accumulation, Fermi-Dirac statistics should be used. It also ignores the quantum confinement due to the potential well formed at the oxide-semiconductor interface in a bulk MOS structure. Quantum confinement has become important in modern MOS structures. Nevertheless, the solution to the Poisson-Boltzmann equation provides a reasonable (and widely-used) approximate solution. In this lecture, we'll develop approximate analytical solutions to the Poisson-Boltzmann equation for a bulk MOS structure in weak and strong inversion. By the term "bulk MOS structure" we mean that the semiconductor begins at $y = 0$ and extends to infinity. In practice, a Si wafer can be considered to be infinitely thick for our purposes. In addition to $Q_n(\psi_S)$, we will also develop approximate solutions for $Q_n(V_G)$ in weak and strong inversion.

8.2 The mobile charge

The mobile electron charge is

$$Q_n = -q\int_0^{\infty} n_0(y)dy = -qn_S \ \text{C/m}^2 . \tag{8.1}$$

Because the electron density depends exponentially on the separation between the conduction band edge and the Fermi level, it increases near the surface where the electrostatic potential increases and E_c bends down. The result is

$$n_0(y) = \left(\frac{n_i^2}{N_A}\right) e^{q\psi(y)/k_BT} . \tag{8.2}$$

(We are assuming a structure like that of Fig. 6.1. with $V_S = V_D = 0$.) Equation (8.2) can be used in (8.1) to write

$$\begin{aligned} Q_n &= -q\left(\frac{n_i^2}{N_A}\right)\int_0^{\infty} e^{q\psi(y)/k_BT}dy \\ &= -q\left(\frac{n_i^2}{N_A}\right)\int_{\psi_S}^{0} e^{q\psi(y)/k_BT}\frac{dy}{d\psi}d\psi \end{aligned} . \tag{8.3}$$

In general, a numerical simulation is needed to solve for $\psi(y)$ and perform the integral of eqn. (8.3), but because most electrons reside very near the surface, it's reasonable to assume that the electric field, $\mathcal{E} = -d\psi/dy$, is approximately constant over the important range of the integral. The average value of the electric field in the electron layer, is $\mathcal{E}_{\text{ave}}$. Accordingly, eqn. (8.3) can be approximated as

$$Q_n = -q\left(\frac{n_i^2}{N_A}\right)\frac{1}{\mathcal{E}_{\text{ave}}}\int_{\psi_S}^{0} e^{q\psi/k_BT}d\psi , \tag{8.4}$$

which is an integral that can be performed to find

$$Q_n(\psi_S) = -q\left[\left(\frac{n_i^2}{N_A}\right)e^{q\psi_S/k_BT}\right]\left(\frac{k_BT/q}{\mathcal{E}_{\text{ave}}}\right). \tag{8.5}$$

Recognizing the quantity in brackets as the electron density at the surface and defining a thickness for the electron layer, we can write (8.5) as

$$\boxed{\begin{aligned} Q_n &= -q\,n(0)\,t_{inv} \\ n(0) &= \frac{n_i^2}{N_A}e^{q\psi_S/k_BT} \\ t_{inv} &= \left(\frac{k_BT/q}{\mathcal{E}_{\text{ave}}}\right) \end{aligned}}, \tag{8.6}$$

According to eqn. (8.6), the electron sheet charge is just $-q$ times the electron concentration at the surface, $n(0)$, times the thickness of the electron layer, t_{inv}. Equation (8.6) applies below and above threshold. We'll begin by considering the subthreshold case.

8.3 The mobile charge below threshold

Equation (8.6) is an equation for $Q_n(\psi_S)$ below threshold when we can use the depletion approximation to determine $\mathcal{E}_{\text{ave}}$. Because the electron layer is thin compared to the depletion layer thickness, we can assume $\mathcal{E}_{\text{ave}} \approx \mathcal{E}_S$. Equation (8.6) is expressed in terms of the surface potential, ψ_S, but it will be more convenient, to express Q_n in terms of the gate voltage, V_G, and the body effect coefficient, m, so eqn. (8.5) for $Q_n(\psi_S)$ needs to be converted to an expression for $Q_n(V_G)$. We begin with the surface electric field.

According to Gauss's Law, the normal component of the displacement field at the surface of the semiconductor is equal to the charge in the semiconductor. From this, we find

$$\mathcal{E}_{\text{ave}} \approx \mathcal{E}_S = \frac{qN_AW_D}{\epsilon_s} = \frac{qN_A}{C_D}, \tag{8.7}$$

where W_D is the thickness of the depletion layer, and $C_D = \epsilon_s/W_D$ is the depletion layer capacitance. Next, according to eqn. (7.31), the depletion layer capacitance is related to the body effect coefficient by $m = 1 + C_D/C_{ox}$, so $C_D = (m-1)C_{ox}$, and we can re-express (8.7) as

$$\mathcal{E}_S = \frac{qN_A}{(m-1)C_{ox}}. \tag{8.8}$$

Equation (8.8) can be used to re-express (8.5) as

$$Q_n(\psi_S) = -(m-1)C_{ox}\left(\frac{n_i}{N_A}\right)^2 e^{q\psi_S/k_BT}\left(\frac{k_BT}{q}\right) . \tag{8.9}$$

According to eqn. (6.27), the quantity, n_i^2/N_A, is related to ψ_B, so we can write

$$\left(\frac{n_i}{N_A}\right)^2 = e^{-q2\psi_B/k_BT} , \tag{8.10}$$

which can be used to write eqn. (8.9) as

$$Q_n(\psi_S) = -(m-1)C_{ox}\left(\frac{k_BT}{q}\right) e^{q(\psi_S-2\psi_B)/k_BT} . \tag{8.11}$$

Finally, we can use eqn. (7.30) to express the result in terms of gate voltage rather than surface potential,

$$\boxed{Q_n(V_G) = -(m-1)C_{ox}\left(\frac{k_BT}{q}\right) e^{q(V_G-V_T)/mk_BT}} . \tag{8.12}$$

Equation (8.12) is an important result; it expresses the small subthreshold mobile charge in terms of the gate voltage. Below threshold the small charge indicated in Fig. 8.1 increases exponentially with gate voltage. This occurs because as the bands bend down with increasing gate voltage, the electron concentration increases exponentially. The exponential increase of Q_n with gate voltage below threshold leads to an exponentially increasing subthreshold current. Our next task is to understand how Q_n varies with gate voltage above threshold, but first, we discuss a couple of points about eqn. (8.12).

We should point out that the threshold voltage in eqn. (8.12) actually depends on the gate voltage (see [1], p. 232). Near the onset of inversion, V_T is given by eqn. (7.12), but the bands don't completely stop bending when $\psi_S = 2\psi_B$, they continue to bend slowly with increasing gate voltage, which causes V_T to increase in strong inversion (see [1], p. 232).

The $(m-1)$ factor in eqn. (8.12) also deserves a comment. For a well-designed transistor, $m \to 1$, and the $(m-1)$ factor in eqn. (8.12) causes Q_n to approach zero. This clearly non-physical result can be traced to our simple treatment of the ψ_S *vs.* V_G relation as given by eqn. (7.30). Although this equation is widely-used (e.g. see eqn. (3.40) in [2]), it is only valid for m greater than about 1.1 or so. A more careful treatment of this problem resolves the issue (e.g. see eqn. (4.84) in [1]). For the fully-depleted, double gate, SOI MOSFET to be treated in Lecture 9, $m = 1$. Equation (9.42), which is developed in Lecture 9 is analogous to eqn. (8.12) but is valid for $m = 1$.

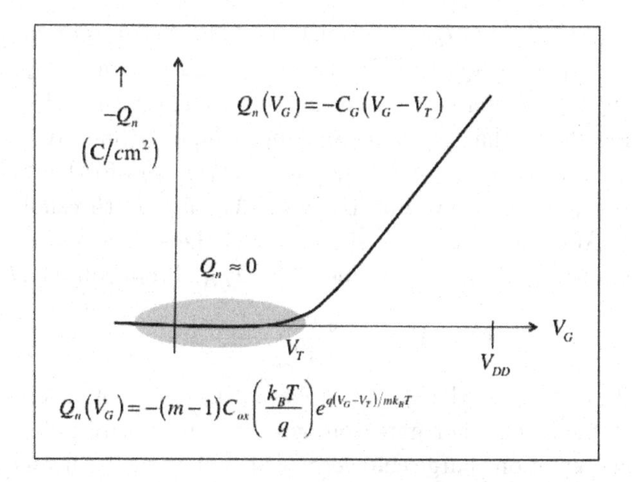

Fig. 8.1 The electron charge, Q_n vs. gate voltage for an n-channel device. The linear vertical scale used here does not show the exponential increase of Q_n with V_G below $V_G = V_T$, but it does show that Q_n varies linearly with V_T for $V_G > V_T$. A more careful treatment of V_T show that it is not a constant. It is given by eqn. (7.12) at the onset of inversion and is larger in strong inversion where $\psi_S > 2\psi_B$ (see p. 232 in [1]).

8.4 The mobile charge above threshold

Equation (8.6) applies for surface potentials below or above threshold. Below threshold, we used the depletion approximation for $\mathcal{E}_S$. In strong inversion, $Q_S \approx Q_n \gg Q_D$. Instead of eqn. (8.7), Gauss's Law gives

$$\mathcal{E}_S = -\frac{Q_n}{\epsilon_s}\,. \tag{8.13}$$

The electric field varies rapidly within the inversion layer, going from $\mathcal{E}_S$ at the surface to approximately zero at the bottom of the inversion layer. Accordingly, we assume that $\mathcal{E}_{\text{ave}} \approx \mathcal{E}_S/2$. With this assumption, we can use eqn. (8.13) in (8.6) to write the electron charge in strong inversion as

$$Q_n = -\sqrt{2\epsilon_s k_B T n(0)}\,, \tag{8.14}$$

or, using eqn. (8.6) for $n(0)$,

$$Q_n(\psi_S) = -\sqrt{2\epsilon_s k_B T (n_i^2/N_A)}\, e^{q\psi_S/2k_BT}\,. \tag{8.15}$$

Equation (8.15) shows that in strong inversion, $Q_n \propto e^{\psi_S/2k_BT}$, as was indicated in Fig. 6.11. Similar arguments for the accumulation regime show that $Q_p \propto e^{-\psi_S/2k_BT}$ in accumulation.

Equation (8.15) gives Q_n as a function of the surface potential, ψ_S; we need an expression for Q_n as a function of the gate voltage, V_G. We could compute $Q_n(V_G)$ numerically by using eqn. (7.11) with (8.15), but such a calculation shows that Q_n increases approximately linearly with V_G for $V_G > V_T$, as indicated in Fig. 8.1; i.e. $Q_n \propto (V_G - V_T)$ for $V_G > V_T$.

To see why Q_n varies linearly with V_G above threshold consider eqn. (7.11). At the onset of inversion, most of the semiconductor charge is the charge in the depletion layer, and $\psi_S = 2\psi_B$. From eqn. (7.11), we find

$$V_T = V_{FB} - \frac{Q_D(2\psi_B)}{C_{ox}} + 2\psi_B \,, \tag{8.16}$$

where we have labeled the gate voltage at the onset of inversion as the threshold voltage, V_T. For gate voltages well above threshold, the band-bending and depletion charge change very little, but a large inversion charge builds up. From eqn. (7.11), we find

$$V_G \approx V_{FB} - \frac{Q_D(2\psi_B) + Q_n}{C_{ox}} + 2\psi_B \,, \tag{8.17}$$

By subtracting eqn. (8.16) from (8.17), we find

$$Q_n \approx -C_{ox}\,(V_G - V_T) \,. \tag{8.18}$$

In practice, $d(-Q_n)/dV_G$ is a little less than C_{ox} because ψ_S is not clamped at $2\psi_B$ as assumed in eqn. (8.17). We can find the slope from

$$\frac{d(-Q_n)}{dV_G} \approx \frac{d(-Q_S)}{dV_G} = \frac{dQ_M}{dV_G} = C_G \,, \tag{8.19}$$

so above threshold, we write the inversion charge as

$$Q_n(V_G) \approx -C_G\,(V_G - V_T) \,, \tag{8.20}$$

where $C_G < C_{ox}$ is approximately constant. We saw in Sec. 7.4 that

$$\frac{1}{C_G} = \frac{1}{C_{ox}} + \frac{1}{C_S} \,, \tag{8.21}$$

where C_S is the semiconductor capacitance, which is the depletion capacitance in depletion or the inversion layer capacitance in inversion,

$$C_S(inv) = \frac{d(-Q_n)}{d\psi_S} = \frac{-Q_n}{2k_BT/q} \,. \tag{8.22}$$

(The last expression follows from eqn. (8.15)). Alternatively, we can define the semiconductor capacitance in inversion to be

$$C_S(inv) \equiv \frac{\epsilon_s}{t_{inv}} \,, \tag{8.23}$$

where the inversion layer thickness is

$$t_{inv} = \frac{2(k_B T/q)\epsilon_s}{-Q_n} . \tag{8.24}$$

To summarize, in strong inversion (i.e. for gate voltages well above the threshold voltage), the inversion layer charge is given by

$$\boxed{\begin{aligned} & Q_n = -C_G\,(V_G - V_T) \quad (V_G > V_T) \\ & \frac{1}{C_G} = \frac{1}{C_{ox}} + \frac{1}{C_S(inv)} \\ & C_S(inv) = \frac{\epsilon_s}{t_{inv}} \\ & t_{inv} = \frac{2(k_B T/q)\epsilon_s}{-Q_n} \end{aligned}} . \tag{8.25}$$

These results show that when $C_S \gg C_{ox}$, then $C_G \approx C_{ox}$. This was the case for MOS technology for a long time, but as gate oxides got thinner and thinner, the assumption began to break down. In addition, Fermi-Dirac statistics and quantum confinement effects (neglected here) also lower C_S. The result is that C_S significantly lowers the gate capacitance below C_{ox} in modern MOSFETs.

Exercise 8.1: Inversion layer capacitance and thickness.

To get a feel for some of the numbers that result from the formulas developed in this lecture, consider the silicon example of Exercises 6.1 and 7.1:

$$\begin{aligned} & N_A = 1.00 \times 10^{18}\ \text{cm}^{-3} \\ & N_V = 1.81 \times 10^{19}\ \text{cm}^{-3} \\ & n_i = 1.00 \times 10^{10}\ \text{cm}^{-3} \\ & \kappa_s = 11.7 \\ & \kappa_{ox} = 4.0 \\ & T = 300\ \text{K} \\ & t_{ox} = 1.8\ \text{nm} \\ & \text{an n}^+ \text{ polysilicon gate} \\ & \text{no charge at the oxide — semiconductor interface} . \end{aligned}$$

1) *What is the semiconductor capacitance when* $n_S = 1 \times 10^{13}\mathrm{cm}^{-2}$*?*
The sheet carrier density here is typical for the on-state of a modern MOSFET. (Note that it is expressed in units of cm^{-2}, not in m^{-2} as it should be for MKS units. This is common practice in semiconductor work, but we have to be careful to convert to MKS units when evaluating formulas.)

From eqn. (8.22) we find

$$C_S(inv) = \frac{-Q_n}{2k_BT/q} = \frac{qn_S}{2k_BT/q} = 30.8 \times 10^{-6}\ \mathrm{F/cm^2}\,.$$

In comparison to the $C_{ox} = 2.0 \times 10^{-6}\mathrm{F/cm^2}$ that we found Exercise 9.1, this is a very large value, but we should mention that it is unrealistically large. As we will see in Lecture 9, Fermi-Dirac carrier statistics and quantum confinement will lower C_S significantly.

2) *What is the gate capacitance?*
According to eqn. (8.21)

$$C_G = \frac{C_{ox}C_S}{C_{ox}+C_S} = \frac{C_{ox}}{1+C_{ox}/C_S}\,.$$

Putting in numbers, we find

$$C_G = \frac{C_{ox}}{1+2.0/30.8} = 0.94\,C_{ox} = 1.9 \times 10^{-6}\ \mathrm{F/cm^2}\,.$$

As expected, $C_G < C_{ox}$. When Fermi-Dirac statistics and quantum confinement are considered, the C_G/C_{ox} ratio is even smaller.

3) *What is the Capacitance Equivalent Thickness, CET?*
First, recall the definition of CET from eqn. (7.32):

$$C_G \equiv \frac{\epsilon_{ox}}{CET} \rightarrow CET = \frac{\epsilon_{ox}}{C_G}\,.$$

Inserting numbers, we find

$$CET = \frac{4.0 \times 8.854 \times 10^{-14}}{1.9 \times 10^{-6}} = 1.86\ \mathrm{nm}\,.$$

Note that the CET is a little thicker than the actual oxide thickness of 1.8 nm. In Lecture 9, we'll show that the effect is even larger when Fermi-Dirac statistics and quantum confinement are considered. For polysilicon gates, poly depletion also increases CET.

4) *What is the semiconductor surface potential when* $n_S = 1 \times 10^{13}$ cm^{-2}*?* From eqn. (8.15), we find

$$\psi_S = 2\left(\frac{k_BT}{q}\right)\ln\left(\frac{qn_S}{\sqrt{\epsilon_s k_BT(n_i^2/N_A)}}\right).$$

Inserting numbers, we find

$$\psi_S = 1.12 \ V .$$

Recall from Exercise 6.1 that $2\psi_B = 0.96$ V, so ψ_S is a little bigger than $2\psi_B$ in strong inversion. For this example, ψ_S is about $6k_BT/q$ larger than $2\psi_B$. Again, when Fermi-Dirac statistics and quantum confinement are treated, the effect is larger.

8.5 Surface potential vs. gate voltage

It is often said that the bandbending in an MOS structure is limited to $\psi_S \approx 2\psi_B$. We saw in Exercise 8.1 that the surface potential in strong inversion is a few k_BT/q larger than $2\psi_B$, but the point is that it is hard to bend the bands very much beyond $2\psi_B$. To see why, consider a simple example.

According to eqn. (8.15), Q_n varies exponentially with surface potential in strong inversion. Assume that the gate voltage produces a band bending that results in $n_S = 5 \times 10^{12}$ cm^{-2}. How much additional bandbending is required to double to inversion layer charge to $n_S = 1 \times 10^{13}$ cm^{-2}? From eqn. (8.15), we see that the answer is

$$\Delta\psi_S = 2\left(\frac{k_BT}{q}\right)\ln(2) = 0.036 \ V ,$$

so a very small change in surface potential doubles the strong inversion charge. How much does the voltage drop across the oxide increase? The answer is

$$\Delta V_{ox} = -\frac{\Delta Q_n}{C_{ox}} = \frac{1.6 \times 10^{-19} \times (5 \times 10^{12})}{2 \times 10^{-6}} = 0.4 \ V ,$$

where we have assumed the same oxide capacitance as in Exercise 8.1. We see that the increase in the voltage drop across the oxide is more than 10 times the increase in the surface potential.

This example shows that because a small change in surface potential produces a large increase in the charge, a large increase in the voltage drop across the oxide results. For this example, the gate voltage must increase by 0.44 V to increase the surface potential by 0.04 V. So above threshold, most of an increase in gate voltage is dropped across the oxide and very little is dropped across the semiconductor. This explains why ψ_S varies slowly with V_G for $V_G > V_T$, as sketched in Fig. 7.3.

Equation (7.30) gives another view of this problem. We find

$$\psi_S = \frac{V_G}{m},$$

where from eqn. (7.31)

$$m = 1 + \frac{C_S}{C_{ox}}.$$

Below threshold, $C_S < C_{ox}$ ($C_S = C_D$ below threshold), and m is close to one, but above threshold, the semiconductor capacitance becomes very large and $C_S \gg C_{ox}$ and $m \gg 1$. Using the numbers from Exercise 8.1, we find $m \approx 9$, so the two capacitor voltage divider in Fig. 7.8 shields ψ_S from V_G.

8.6 Discussion

In this section we have shown that the electron charge, $Q_n(\psi_S)$, varies exponentially with ψ_S both below and above threshold. The dependence below threshold, eqn. (8.11), is as $\exp(\psi_S/k_BT)$, while the dependence above threshold, eqn. (8.15), is as $\exp(\psi_S/2k_BT)$, but the exponential dependence on ψ_S is there in both cases.

Below threshold, $Q_n(V_{GS})$ varies exponentially with V_{GS} because $\psi_S \propto V_{GS}$ (see eqn. (8.12)). Above threshold, however, things are different. Above threshold, $Q_n(V_{GS})$ varies linearly with V_{GS} as given by eqn. (8.25). This occurs because above threshold, $\psi_S \propto \ln(V_{GS})$.

To summarize, we have derived eqns. (8.12) and (8.18) to describe the bulk MOS structure below and above threshold:

$$\boxed{\begin{aligned} & V_G \ll V_T : \\ & Q_n(V_G) = -(m-1)C_{ox}\left(\frac{k_BT}{q}\right)e^{q(V_G-V_T)/mk_BT} \\ & V_G \gg V_T \\ & Q_n(V_G) = -C_G\left(V_G - V_T\right). \end{aligned}} \tag{8.26}$$

From this equation and eqn. (5.3),

$$I_{DS}/W = \left|Q_n\left(x=0\right)\right|\left\langle \upsilon\left(x=0\right)\right\rangle, \tag{8.27}$$

we can compute the drain current below and above threshold. It would be useful, however, to have a single expression that works below and above threshold. The general $Q_n(V_{GS})$ relation can be evaluated numerically, but as we'll discuss in Lecture 11, an empirical expression that reduces to the correct result below and above threshold can also be used.

8.7 Summary

In this lecture, we have discussed how Q_n varies with surface potential and with gate voltage, considering both the subthreshold and above threshold regions. The correct results in subthreshold and in strong inversion are readily obtained without numerically solving the Poisson-Boltzmann equation, but the numerical solution of the Poisson-Boltzmann equation gives the results from subthreshold to strong inversion and in between.

In the next lecture, we will consider $Q_n(\psi_S)$ and $Q_n(V_G)$ for a different MOS structure, an extremely thin layer of silicon. This structure is more typical of the channel structures now being used to scale devices to their limit. We will find, however, that the basic considerations for this *extremely thin silicon on insulator* (ETSOI) structure are quite similar to the bulk MOS structure discussed in this lecture.

8.8 References

The exact solution of the Poisson-Boltzamnn equation is discussed in the two books and the notes listed below.

[1] Y. Tsividis and C. McAndrew, *Operation and Modeling of the MOS Transistor*, 3rd Ed., Oxford Univ. Press, New York, 2011. (See Sec. 2.3.2.1)

[2] Y. Taur and T. Ning, *Fundamentals of Modern VLSI Devices*, 2nd Ed., Oxford Univ. Press, New York, 2013. (See Sec. 2.4.4)

[3] Mark Lundstrom and Xingshu Sun, "Notes on the Solution of the Poisson-Boltzmann Equation for MOS Capacitors and MOSFETs, 2nd Ed.," `https://nanohub.org/resources/5338`, 2012.

Lecture 9

Mobile Charge: Extremely Thin SOI

9.1 Introduction

In Lecture 8 we discussed MOS electrostatics for a bulk semiconductor substrate. Modern MOS structures often make use of extremely thin silicon layers. An example is shown in Fig. 9.1. An electron in such a thin layer behaves as quantum mechanical "particle in a box." Because of the confinement in one direction, electrons are quasi-two-dimensional particles, and we must use a 2D density-of-states when evaluating carrier densities. In a bulk MOSFET, the electrostatic potential well at the oxide-Si interface produces a quantum well, so quantum confinement occurs in all MOS structures and should be considered in the bulk MOSFETs that were discussed in the previous lecture. This lecture examines structures more like those used in state-of-the-art MOSFETs, and it is also an opportunity to examine quantum confinement in MOS structures.

In Lecture 8, our goal was to understand how the electron charge, Q_n, varied with surface potential, ψ_S, and with gate voltage, V_G, in a bulk

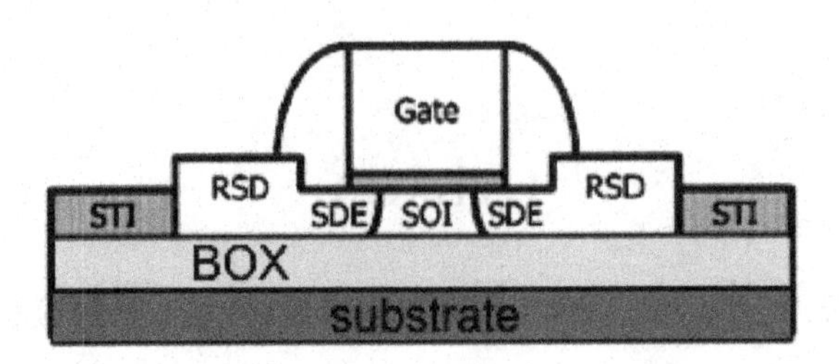

Fig. 9.1 Illustration of a single gate extremely thin silicon on insulator (ETSOI) MOSFET. (©IEEE 2009. Reprinted, with permission, from: A. Majumdar, Z. Ren, S. J. Koester, and W. Haensch, "Undoped-Body Extremely Thin SOI MOSFETs With Back Gates," *IEEE Trans. Electron Dev.*, **56**, pp. 2270-2276, 2009.) The model structure discussed in this lecture is a double gate version of this device.

MOS structure. Our goal in this lecture is to do the same for the ETSOI structure. We will treat the electrons as quantum-confined, 2D particles. In Lecture 8, we assumed classical, 3D particles. Had we included quantum confinement in the bulk MOS case, the numerical values of the results would have changed (enough to be important in modern MOSFETs), but the qualitative features would be similar to those obtained from the classical analysis. For the ETSOI structure, we will treat quantum confinement from the outset (because it is easy to do so), but we will see that the results are qualitatively similar to those obtained from the classical analysis of the bulk case.

9.2 A primer on quantum confinement

This section is a very brief introduction to quantum confinement in MOS structures. We'll also discuss the role of band structure by examining how quantum confinement affects the six constant energy ellipsoids in the conduction band of silicon.

Quantum confinement

Quantum mechanics tells us that electrons behave both as particles and as waves and that the wave aspects become important when the potential energy changes spatially on the scale of the electron's wavelength (the so-called de Broglie wavelength, λ_B). We can estimate the average electron wavelength from

$$p = \hbar k = \hbar \frac{2\pi}{\lambda_B}, \tag{9.1}$$

where p is the crystal momentum, k the electron's wave vector, and λ_B the electron's wavelength. The energy of the electron is $E = p^2/2m^*$, and the thermal equilibrium average electron energy is $3k_BT/2$. Using these relations, we obtain a rough estimate of the thermal average de Broglie wavelength as

$$\langle \lambda_B \rangle \approx \frac{h}{\sqrt{3m^* k_B T}} \approx 6 \text{ nm}, \tag{9.2}$$

where we have assumed for a rough estimate that $m^* = m_0$. Electrostatic potential wells to confine electrons to dimensions less than 10 nm are readily produced with a gate voltage, and semiconductor layers less than 10 nm thick are also readily achieved. The behavior of electrons confined in these *quantum wells* is different from the behavior of electrons in the bulk, and it is important to understand the differences.

Figure 9.2 sketches two quantum wells; the one on the left is a rectangular quantum well with infinitely high barriers on the sides, and the one on the right is a triangular quantum well. The direction of confinement is the y-direction, but we assume that electrons are free to move in the x-z plane. Just as the Coulomb potential of the nucleus of a hydrogen atom confines the electron to the vicinity of the nucleus, which leads to the occurrence discrete energy levels of the hydrogen atom, we find that the energies of electrons in these quantum wells consists of discrete *subbands* associated with confinement in the y-direction.

The time independent Schrödinger equation for electrons is

$$\left[-\frac{\hbar^2}{2m^*} \nabla^2 - E_c(x, y, z) \right] \psi(x, y, z) = E\psi(x, y, z). \tag{9.3}$$

If E_c is a constant, then the solutions are plane waves,

$$\psi(x, y, z) = \frac{1}{\sqrt{\Omega}} e^{i\vec{k}\cdot\vec{r}}, \tag{9.4}$$

where Ω is an arbitrary normalization volume, and $1/\sqrt{\Omega}$ insures that the the integral of $\psi\psi^*$ over the volume is one. The magnitude of the wavevector, $\vec{k}$, is obtained from

$$\frac{\hbar^2 k^2}{2m^*} = (E - E_c). \tag{9.5}$$

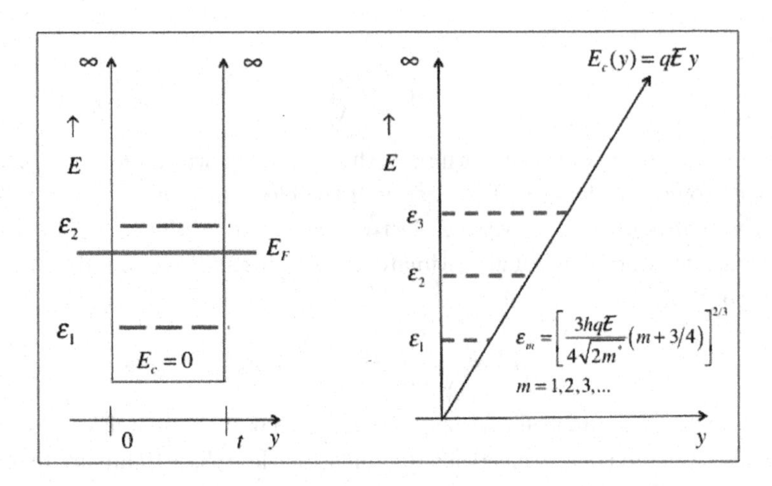

Fig. 9.2 Illustration of two simple quantum wells. The direction of confinement is the y-direction and electrons are free to move in the x-z plane. Left: Rectangular quantum well with infinitely high barriers. Right: Triangular quantum well with infinitely high barriers.

The solution to the wave equation for a quantum well is the product of a plane wave in the x-z plane times a function in the y-direction that depends on the shape of the quantum well in the y-direction,

$$\psi(x, y, z) = \frac{1}{\sqrt{A}} e^{i(\vec{k_{||}}\cdot\vec{\rho})} \times \phi(y)\,, \tag{9.6}$$

where A is an area in x-z plane used to normalize the wavefunction in the $x - z$ plane. To find $\phi(y)$, we solve

$$\left[-\frac{\hbar^2}{2m^*}\frac{d^2}{dy^2} - E_c(y)\right]\phi(y) = E\phi(y)\,. \tag{9.7}$$

Consider the rectangular quantum well on the left of Fig. 9.2 and take $E_c = 0$. The solutions to eqn. (9.7) are $\phi(y) = \sin(k_y y)$ and $\cos(k_y y)$, where

$$\frac{\hbar^2 k_y^2}{2m^*} = \epsilon = E - \frac{\hbar^2 k_{||}^2}{2m^*}\,. \tag{9.8}$$

The boundary conditions are $\phi(0) = \phi(t) = 0$ because the infinitely high barriers force the wave function to zero at the boundaries. Only $\sin(k_y y)$ satisfies the boundary condition at $z = 0$, and to satisfy the boundary condition at $y = t$, k_y must take on discrete values of

$$k_y t = m\pi\,, \tag{9.9}$$

where $m = 1, 2, ...$ The result is that the energy in eqn. (9.8) becomes quantized; only the energies

$$\epsilon_m = \frac{\hbar^2 m^2 \pi^2}{2m^* t^2}, \tag{9.10}$$

are allowed. The total energy is

$$E = \frac{\hbar^2 \left(k_{||}^2 + k_m^2\right)}{2m^*} = \epsilon_m + \frac{\hbar^2 k_{||}^2}{2m^*}. \tag{9.11}$$

Quantum confinement produces a set of *subbands* in the conduction band (and a corresponding set in the valence band). For $m = 1$, the lowest energy is ϵ_1, but there is an additional kinetic energy of $\hbar^2 k_{||}^2/2m^*$ associated with the electron's velocity in the x-z plane. Quantum confinement effectively raises the bottom of the conduction band. The number of subbands that are occupied depends on the location of the Fermi level. The subband energies are determined by the shape of the potential well and by the height of the barriers. For the triangular quantum well shown on the right of Fig. 9.2, we also expect subbands, but the values of ϵ_m are different, and the wavefunctions are Airy functions rather than sine functions. In general, light effective masses and thin quantum wells give high subband energies, as illustrated by Eq. (9.10) for the rectangular quantum well with infinite barriers.

In addition to changing the energies of electrons in the conduction and valence bands, quantum confinement also changes the spatial distribution of electrons. For the rectangular quantum well, $n(y) \propto \sin^2(k_m y)$ for electrons in the mth subband. The contributions from all of the occupied subbands should be added to get the total electron density. The electrons in a quantum well are free to move in the x-z plane, but can move very little in the y-direction. They are called *quasi-two-dimensional* electrons.

The two-dimensional nature of the electrons changes the density of states. Instead of the bulk density-of-states for 3D (unconfined) electrons [1], we have for each subband, m [1, 2]

$$D_{2D}^m = g_v^m \frac{m_m^*}{\pi \hbar^2} \Theta(E - \epsilon_m). \tag{9.12}$$

Instead of $n = N_{3D} \mathcal{F}_{1/2}(\eta_F)$ m^{-3} for the 3D carrier density, we have for the 2D sheet carrier density,

$$n_S^m = N_{2D}^m \mathcal{F}_0(\eta_F^m) \text{ m}^{-2}, \tag{9.13}$$

where

$$N_{2D}^m \equiv g_v^m \frac{m_m^* k_B T}{\pi \hbar^2}\,, \tag{9.14}$$

is the effective density-of-states in 2D, and

$$\eta_F^m = (E_F - \epsilon_m)/k_B T\,, \tag{9.15}$$

$$\mathcal{F}_0(\eta_F^m) \equiv \int_0^\infty \frac{\eta^0\, d\eta}{1 + e^{\eta - \eta_F^m}} = \ln\left(1 + e^{\eta_F}\right)\,. \tag{9.16}$$

To get the total electron density, we add the contributions of all of the occupied subbands.

To relate this discussion to MOSFETs, note that the quantum well on the left of Fig. 9.2 is similar to what happens in an ETSOI MOSFET where the quantum confinement is produced by the ultra-thin Si film. The quantum well on the right of Fig. 9.2 is like the quantum well produced electrostatically in a bulk MOSFET when the gate voltage strongly inverts the semiconductor. The direction of confinement (the y-direction) is normal to the channel of the MOSFET and electrons are free to move in the $x - z$ plane, which is the plane of the channel.

Bandstructure effects

Some interesting effects occur when electrons in the conduction band of Si are confined in a quantum well. Figure 9.3 shows the constant energy surfaces for electrons in the conduction band of Si. The lowest energies occur at six different locations in the Brillouin zone along the three coordinate axes (the valley degeneracy is $g_v = 6$). The constant energy surfaces are ellipsoids of revolution described by

$$E = \frac{\hbar^2 k_x^2}{2m_{xx}^*} + \frac{\hbar^2 k_y^2}{2m_{yy}^*} + \frac{\hbar^2 k_z^2}{2m_{zz}^*}\,. \tag{9.17}$$

There are two different effective masses, a heavy, *longitudinal effective mass*, m_l^* and a light, *transverse effective mass*, m_t^*. For Si, $m_l^* = 0.90m_0$ and $m_t^* = 0.19m_0$. For example, for the ellipsoids oriented along the x-axis, $m_{xx}^* = m_l^*$ and $m_{yy}^* = m_{zz}^* = m_t^*$.

According to eqn. (9.10), the subband energies are determined by the effective mass, but which effective mass should we use? The answer is to use the effective mass in the direction of confinement, the y-direction in this case. From Fig. 9.3, we see that for (100) Si with confinement in the y-direction two of the six ellipsoids have the heavy, longitudinal effective mass in the y-direction and four of the six have the light, transverse effective mass in the y-direction. The result is two different series of subbands — an *unprimed ladder* of subbands with the energies determined by m_l^* and a degeneracy of $g_v = 2$, and a *primed ladder* of subbands with the energies determined by m_t^* and a valley degeneracy of 4. The lowest subband is the $m = 1$ unprimed subband. In the x-z plane, electrons in these two degenerate subbands respond with the light, transverse effective mass.

In the simple examples considered in this lecture, we will assume that only the bottom, unprimed subband (for which the mass in the confinement direction is $m^* = m_l^*$ and the mass in the $x - z$ plane is $m^* = m_t^*$) is occupied. If higher subbands are occupied, the different subband energies and the different effective masses in the x and z directions must be accounted for, and the total sheet carrier density is the sum of the contribution from each occupied subband.

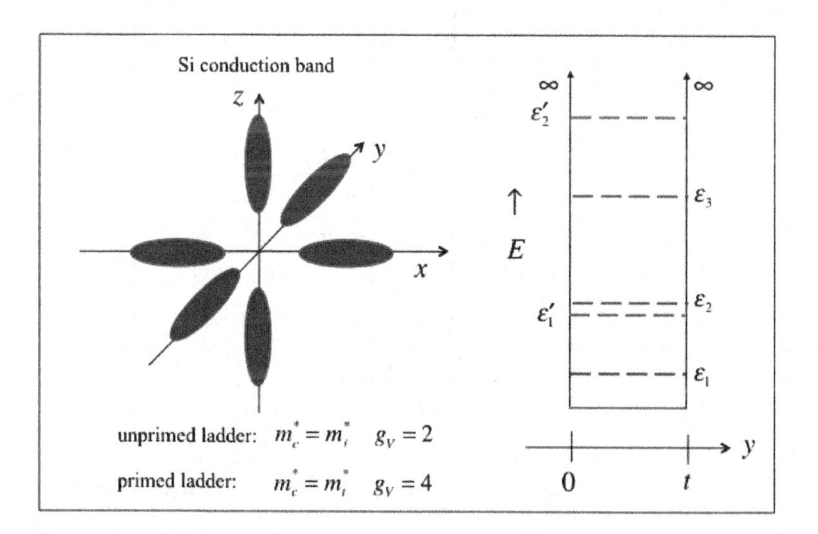

Fig. 9.3 Left: A sketch of the constant energy surfaces for electrons in silicon. Right: the corresponding unprimed and primed "ladders" of subbands for (100) Si. The effective masses listed here are the confinement masses in the y-direction.

9.3 The mobile charge

The mobile electron charge is

$$Q_n = -q \int_0^{t_{Si}} n(y)dy = -qn_S \ \text{C/m}^2 , \tag{9.18}$$

where t_{Si} is the thickness of the silicon layer. Consider the quantum well shown in Fig. 9.4 for which two subbands in the conduction and valence bands and the Fermi level are shown. If we were to treat the electrons as classical particles, the electron density would be uniform in the well with a value of

$$\begin{aligned} n_0 &= N_{3D}^c \mathcal{F}_{1/2}\left[(E_F - E_C)/k_B T\right] \ \text{m}^{-3} \\ n_S &= n_0 t_{Si} \ \text{m}^{-2} , \end{aligned} \tag{9.19}$$

where N_{3D}^c is the effective density of states for three-dimensional electrons, n_0 the volume density of electrons, and n_S the sheet density.

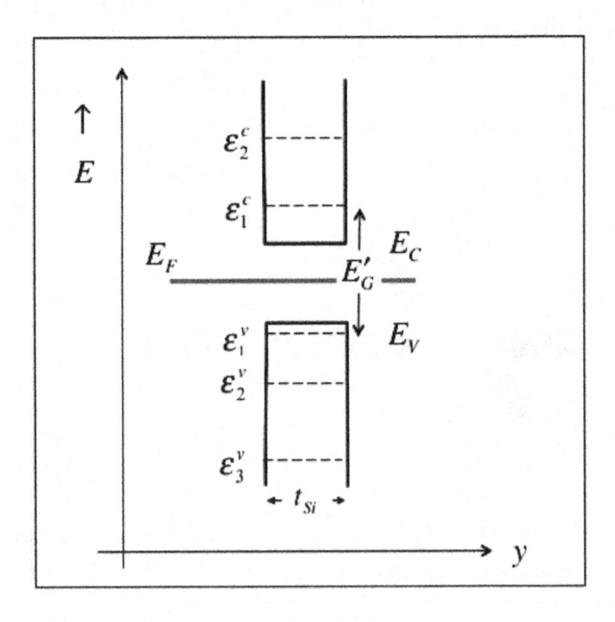

Fig. 9.4 Extremely thin silicon on insulator energy band diagram. Only the silicon layer is shown. Two subbands are shown in the conduction band and in the valence band.

Quantum confinement creates a series of conduction bands (and valence bands; these are the so-called subbands); the bottom of each one is at an energy, $E_C + \epsilon_m$. Quantum confinement also changes the electron's wave function; for a infinite potential well $\psi(y) \propto \sin(m\pi y/t_{Si})$. The density of electrons per m^3 in each subband is $n(y)$, which is proportional to $\psi^*(y)\psi(y) = \sin^2(m\pi y/t_{Si})$. The spatial distribution of electrons inside the quantum well is given by $n(y)$. The integrated total electron density per m^2, eqn. (9.18), is found by integrating the 2D density of states times the Fermi function. The result for subband m is, from eqn. (9.13),

$$
\begin{aligned}
n_S^m &= \int_0^{t_{Si}} n(y)dy = \int_0^\infty D_{2D}(E) f_0(E) dE \\
&= N_{2Dm}^c \ln\left(1 + e^{(E_F - E_C - \epsilon_m^c)/k_B T}\right) ,
\end{aligned}
\tag{9.20}
$$

where N_{2Dm}^c is given by eqn. (9.14). The total sheet electron density is found by summing the contributions from each subband.

In this lecture, we will assume that only the lowest subband is occupied, which is a reasonable assumption when the well is very thin and the subband energies are widely spaced. Accordingly, the electron charge per m^2 is

$$
Q_n(\psi_S) = -qn_S = -qN_{2D1}^c \ln\left(1 + e^{(E_F - E_C - \epsilon_1^c)/k_B T}\right) . \tag{9.21}
$$

In an ETSOI MOS structure, a gate is used to change the potential, ψ_S, in the quantum well. The geometry of the ETSOI MOS capacitor is shown in Fig. 9.5. A symmetrical, double gate structure is assumed. The same voltage is applied to the top and bottom gates, and the Fermi level is grounded. We also assume that the Si layer is thin enough and the electron density small enough so that the bottom of the well is nearly flat, which means that the electrostatic potential in the well, ψ_S, is not a function of y. (More generally, we would need to solve the Schrödinger equation self-consistently with the Poisson equation to solve for $\psi(y)$ [3].) With these assumptions, eqn. (9.21) becomes

$$
Q_n(\psi_S) = -qn_S = -qN_{2D1}^c \ln\left(1 + e^{(E_F - E_{C0} + q\psi_S - \epsilon_1^c)/k_B T}\right) , \tag{9.22}
$$

where $E_C = E_{C0} - q\psi_S$ and ψ_S is controlled by the potential of the two gates. Here E_{C0} is the location of E_C when $\psi_S = 0$, which is determined by the gate workfunction. Finally, we will assume non-degenerate carrier statistics, so that the ETSOI results can be compared directly with the bulk MOS results discussed in Lecture 8, which also used non-degenerate carrier statistics. Our final expression for $Q_n(\psi_S)$ is

$$
Q_n(\psi_S) = -qn_S = -qN_{2D1}^c e^{(E_F - E_{C0} + q\psi_S - \epsilon_1^c)/k_B T} , \tag{9.23}
$$

Equation (9.23) can be written as

$$Q_n(\psi_S) = -qn_{S0}\, e^{q\psi_S/k_BT}\,, \tag{9.24}$$

where

$$n_{S0} = N_{2D1}^{c}\, e^{(E_F - E_{C0} - \epsilon_1^c)/k_BT}\,. \tag{9.25}$$

Similarly, the hole charge can be written as

$$Q_p(\psi_S) = qp_{S0}\, e^{-q\psi_S/k_BT}\,, \tag{9.26}$$

where

$$p_{S0} = N_{2D1}^{v} e^{(E_{V0} - \epsilon_1^v - E_F)/k_BT}\,. \tag{9.27}$$

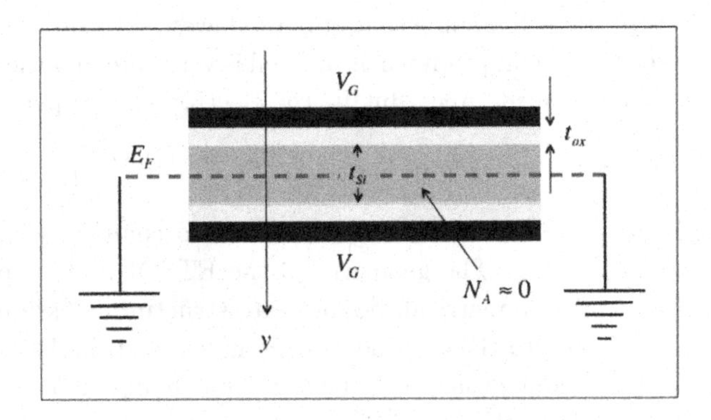

Fig. 9.5 Extremely thin silicon on insulator channel structure. A symmetrical double gate structure will be examined. The top and bottom gate insulators are identical and the same voltage is applied to both the top and bottom gates. The Fermi level is grounded, so E_F is the equilibrium Fermi level. The electric field in the y-direction is symmetrical about the dashed line.

Our first objective is to understand how the charge in the semiconductor varies with the potential in the semiconductor — that is, we want to compare $Q_S(\psi_S)$ for the ETSOI structure with the corresponding result for the bulk MOS structure, as summarized in Fig. 6.11.

If the ETSOI structure is undoped, then the charge in the silicon is due only to mobile holes and electrons:

$$\begin{aligned} Q_S(\psi_S) &= q(p_S - n_S) \\ &= q\left(p_{S0}\, e^{-q\psi_S/k_BT} - n_{S0}\, e^{q\psi_S/k_BT}\right)\ \mathrm{C/m^2}\,. \end{aligned} \tag{9.28}$$

If we assume that the reference for the potential has been chosen so that for $\psi_S = 0$, $n_{S0} = p_{S0} = n_{Si}$, then

$$Q_S(\psi_S = 0) = 0\,, \tag{9.29}$$

and we can write eqn. (9.28) as

$$Q_S = qn_{Si}\left(e^{-q\psi_S/k_BT} - e^{q\psi_S/k_BT}\right)\ \mathrm{C/m^2}\,. \tag{9.30}$$

If we are dealing with an n-channel MOSFET, then we are interested in the sheet density of mobile electrons,

$$n_S(\psi_S) = n_{Si}\,e^{q\psi_S/k_BT}\ \mathrm{m^{-2}}\,. \tag{9.31}$$

Figure 9.6 illustrates how the gate voltage affects the energy bands. For positive gate voltage, the potential in the semiconductor increases, the conduction band decreases in energy and moves closer to the Fermi level, and the electron concentration increases exponentially. For a negative gate voltage, the valence band moves up, and the hole concentration increases exponentially. Figure 9.7 shows the resulting $Q_S(\psi_S)$ characteristic, which should be compared to Fig. 6.11 for the bulk MOSFET. For the ETSOI structure, we have assumed undoped silicon, so there is no $\sqrt{\psi_S}$ region due to depletion. As soon as ψ_S is positive or negative enough, a large electron or hole density builds up. In strong inversion or in accumulation, the charge in the bulk MOS structure increased exponentially with surface potential. The same happens for the ETSOI structure, but note that the inversion or accumulation charge varied as $\exp(q\psi_S/2k_BT)$ for the bulk case and that it varies as $\exp(q\psi_S/1k_BT)$ for the ETSOI case. This difference can be traced to the fact that the potential well in the bulk case is related to the surface electric field while in the ETSOI case, the thin Si film itself creates the potential well.

Exercise 9.1: Intrinsic electron sheet density.

Equation (9.30) is analogous to eqns. (8.14) and (8.15) for the bulk MOS structure, but to evaluate (9.30), we need to compute n_{Si}. In general, n_{S0} is the electron sheet density when $\psi_S = 0$; it depends on where the Fermi level is located, which in turn depends on the workfunction of the gate electrode. In this exercise, we will assume that the semiconductor is intrinsic when $\psi_S = 0$, so $n_{S0} = p_{S0} = n_{Si}$. We'll evaluate n_{Si} assuming

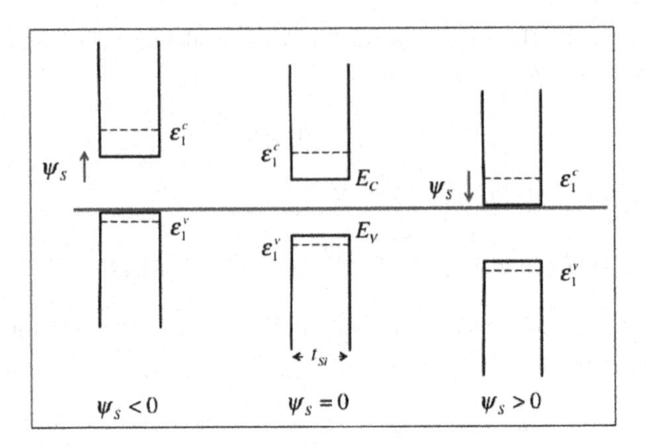

Fig. 9.6 Illustration of how a negative, zero, and positive electrostatic potential, ψ_S affect the ETSOI energy band diagram.

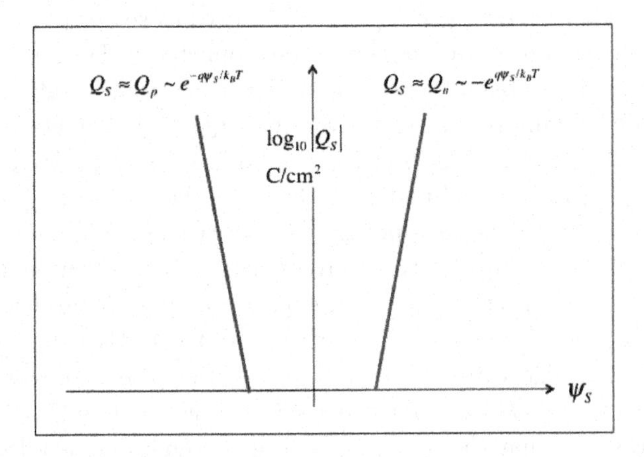

Fig. 9.7 Sketch of the net charge in the semiconductor vs. the potential in the semiconductor. This sketch should be compared to the corresponding sketch for the bulk MOS structure, Fig. 8.11.

some typical numbers for a Si ETSOI structure:

$$t_{Si} = 5 \text{ nm}$$
$$m_l^* = 0.92 m_0$$
$$m_t^* = 0.19 m_0$$
$$m_{hh}^* = 0.54 m_0$$
$$E_G = 1.125 \text{ eV}$$
$$T = 300 \text{ K}.$$

First, let's compute some quantities we'll need. The lowest $n = 1$ subband for the conduction band comes from the two ellipsoids oriented along the confinement direction for which $m^* = m_l^*$. For holes, the first subband comes from the heavy hole valence band for which $m^* = m_{hh}^*$. Using eqn. (9.10), we find

$$\epsilon_1^c = 0.016 \text{ eV}$$
$$\epsilon_1^v = 0.028 \text{ eV}.$$

Quantum confinement increases the effective bandgap because the conduction begins at $E_C + \epsilon_1^c$ and the top of the valence band is at $E_V - \epsilon_1^v$. The effective bandgap for ETSOI structure is

$$E_G' = E_G + \epsilon_1^c + \epsilon_1^v = 1.169 \ \text{eV}.$$

The 2D effective density-of-states is given by eqn. (9.14). For the first subband in the conduction band, the valley degeneracy is two, and the effective mass in the $x - z$ plane is m_t^*, so the density-of-states effective mass is $m_D^* = 2m_t^*$. For the valence band, there is one heavy hole band, so $g_v = 1$ and $m_D^* = m_{hh}^*$. Putting numbers in eqn. (9.14), we find

$$N_{2D}^c = 4.11 \times 10^{12} \text{ cm}^{-2}$$
$$N_{2D}^v = 5.84 \times 10^{12} \text{ cm}^{-2}.$$

From eqns. (9.25) and (9.27), we find the sheet carrier densities as

$$n_S = N_{2D}^c e^{(E_F - E_C - \epsilon_1^c)/k_B T}$$
$$p_S = N_{2D}^v e^{(E_V - E_F - \epsilon_1^v)/k_B T}.$$

By multiplying the two equations, we see that

$$n_S p_S = N_{2D}^c N_{2D}^v e^{-E_G'/k_B T} = n_{Si}^2$$

is independent of the location of the Fermi level. When the quantum well is intrinsic, $n_S = p_S = \sqrt{n_S p_S}$. We call this concentration the intrinsic sheet carrier concentration, n_{Si}, and find it as

$$n_{Si} = \sqrt{n_S p_S} = \sqrt{N_{2D}^c N_{2D}^v}\, e^{-E_G'/2k_B T} \quad \text{m}^{-2}.$$

Note the similarity of this expression to the standard expression for the intrinsic carrier density in a bulk semiconductor [1]. Putting in the numbers we've computed we find

$$n_{Si} = 8.5 \times 10^2 \quad \text{cm}^{-2},$$

which is a very small number. It is likely that this small number would be overwhelmed by charges at the oxide-Si interface or by unintentional dopants in the Si film.

9.4 The mobile charge below threshold

In Lecture 8, we developed one expression for $Q_n(\psi_S)$, eqn. (8.6), but the average electric field in the electron layer was different below and above threshold, so we needed to develop separate expressions for $Q_n(\psi_S)$ below and above threshold. For the ETSOI structure,

$$\boxed{Q_n(\psi_S) = -qn_{Si}\, e^{q\psi_S/k_BT}\,,} \tag{9.32}$$

where we have assumed that the Fermi level is set so that $n_{S0} = n_{Si}$ when $\psi_S = 0$. This expression is valid both below and above threshold.

Our next step is to relate $Q_n(\psi_S)$ below threshold to the gate voltage. From Fig. 9.5 we see that there is a line of symmetry along the middle of the channel (dashed line). Half of the charge in the semiconductor images on the top gate and the other half on the bottom gate. Because of this symmetry, we only need to relate the voltage on the top gate to the charge in the top half of the channel. Our starting point is eqn. (7.1), which is

$$V_G' = \Delta V_{ox} + \psi_S\,. \tag{9.33}$$

According to Gauss's Law, the electric field at the top oxide-Si interface is obtained from

$$\epsilon_s \mathcal{E}_{\text{ox}} = -\frac{Q_S(\psi_S)}{2} = -\frac{Q_n(\psi_S)}{2}\,. \tag{9.34}$$

The potential drop across the oxide is

$$\Delta V_{\text{ox}} = \mathcal{E}_{\text{ox}} t_{\text{ox}}\,. \tag{9.35}$$

Using eqns. (9.33), (9.34), and (9.35), we find

$$V_G' = -\frac{Q_n(\psi_S)}{2C_{\text{ox}}} + \psi_S\,. \tag{9.36}$$

Below threshold, the charge in the semiconductor is very small, so the volt drop across the oxide is very small and eqn. (9.36) simplifies to

$$V_G' = \psi_S\,. \tag{9.37}$$

For the bulk MOS structure, we saw that $\psi_S = V_G'/m$, where $m > 1$, but in the double gate ETSOI case $m = 1$. The fact that the gate has complete control of ψ_S is an advantage of the ETSOI double gate structure.

It is now easy to convert eqn. (9.32) to an expression for $Q_n(V_G)$ below threshold,

$$Q_n(V_G) = -qn_{Si}\, e^{qV_G/k_BT}\,. \tag{9.38}$$

Equation (9.38) describes the electron charge in the subthreshold region. For the bulk MOS structure, $\psi_S = 2\psi_B$ produced an electron density at the surface that was equal to the hole density in the bulk, and we used this potential to define the end of weak inversion as $\psi_S = 2\psi_B$. We cannot use this definition in this case, because the ETSOI structure is not doped. In this case, we might argue that when the conduction band has been pulled down so that $E_F = E_C + \epsilon_1^c$, then the electron concentration will become significant. Equation (9.4) shows that when this condition holds and when non-degenerate statistics are used, then $n_S = N_{2D}^c$. Accordingly, we can find the semiconductor potential at the onset of inversion from

$$n_S(\psi_S) = n_{Si}\, e^{q\psi_S/k_BT} = N_{2D}^c\,, \tag{9.39}$$

which gives the potential at the onset of inversion as

$$\boxed{\psi_S^{\text{inv}} = \frac{k_BT}{q} \ln\left(\frac{N_{2D}^c}{n_{Si}}\right)}\,. \tag{9.40}$$

From eqn. (9.36) and recognizing that at the onset of inversion, the charge in the semiconductor is very small and, therefore, the voltage drop across the oxide is negligible, we find

$$V_T' = \psi_S^{\text{inv}} = \frac{k_BT}{q} \ln\left(\frac{N_{2D}^c}{n_{Si}}\right)\,. \tag{9.41}$$

Using eqn. (9.41) in (9.38), we can eliminate n_{Si} and write the result as

$$\boxed{Q_n(V_G) = -C_Q \frac{k_BT}{q}\, e^{q(V_G - V_T)/k_BT}\,,} \tag{9.42}$$

where

$$C_Q = q^2 D_{2D}\,, \tag{9.43}$$

is called the *quantum capacitance.* (Here D_{2D} is the two-dimensional density-of-states.) We will discuss the quantum capacitance later; at this point, it is simply a convenient way to express the various constants and material parameters in eqn. (9.42).

Equation (9.42) is the key result, it should be compared to eqn. (8.12) for the bulk MOS case. In both cases, we see that the subthreshold charge increases exponentially with gate voltage. The difference is that the double gate device has an ideal subthreshold slope, $m = 1$, while the bulk MOS structure typically has $m \approx 1.1 - 1.3$. This means that for a given increase in gate voltage, the electron charge in a double gate structure will increase more than in a bulk MOS structure.

Exercise 9.2: Semiconductor potential at the beginning of inversion.

For the bulk MOS structure, the surface potential at the onset of inversion was $\psi_S = 2\psi_B$. For the ETSOI structure, we have defined the onset of inversion with eqn. (9.40). How does ψ_S^{inv} for the ETSOI structure compare to $\psi_S^{inv} = 2\psi_B$ for the bulk structure?

Using numbers from Exercise 9.1 in eqn. (9.40), we find

$$\psi_S^{\text{inv}} = \frac{k_B T}{q} \ln \left(\frac{N_{2D}^c}{n_{Si}} \right) = 0026 \ln \left(\frac{4.11 \times 10^{12}}{8.5 \times 10^2} \right) = 0.58 \text{ V} .$$

The result that $q\psi_S$ is about one-half of the effective band gap is expected. The Fermi level for $\psi_S = 0$ was positioned near the middle of the band gap, so than $n_{S0} = p_{S0} = n_{Si}$. The potential at the onset of inversion is such that it lowers $E_{C0} + \epsilon_1^c$ to E_F, which is our criterion for inversion.

In this example, the required bandbending for inversion is about one-half of the result for the bulk MOS structure because in the bulk structure, the Fermi level in the p-type bulk is positioned near the valence band, so the bandbending must be almost the bandgap to bring the Fermi level in alignment with the conduction band.

9.5 The mobile charge above threshold

Equation (9.32) applies for surface potentials below and above threshold. Below threshold, we could assume that the voltage drop across the oxide was small and relate the subthreshold electron charge to the gate voltage according to eqn. (9.42). Above threshold, the voltage drop across the oxide becomes very large, and $Q_n(V_G)$ changes.

Equation (9.32) gives $Q_n(\psi_S)$, and eqn. (9.36) relates the surface potential to the gate voltage. We could compute $Q_n(V_G)$ numerically by solving these two equation, but such a calculation shows that Q_n increases approximately linearly with V_G for $V_G > V_T$, i.e. $Q_n \propto (V_G - V_T)$ for $V_G > V_T$ — just as it was for the bulk MOS case. We find the slope of this line from

$$C_G = \frac{dQ_M}{dV_G} = \frac{d(-Q_S)}{dV_G} = \frac{d(-Q_n)}{dV_G} . \tag{9.44}$$

Differentiating eqn. (9.36) with respect to $(-Q_n)$, we find

$$\frac{1}{C_G} = \frac{1}{2C_{ox}} + \frac{1}{C_S} , \tag{9.45}$$

and write the inversion charge above threshold as

$$Q_n(V_G) = -C_G\,(V_{GS} - V_T)\,, \tag{9.46}$$

where C_G is approximately constant. For $C_S \gg 2C_{ox}$, $C_G \approx 2C_{ox}$. The factor of two comes from the two gates in Fig. 9.5.

Because the semiconductor capacitance is finite, C_G is a little less than $2C_{ox}$, just as it was for the bulk MOS case. Using eqn. (9.32) we find the semiconductor capacitance for the ETSOI structure to be

$$C_S(inv) = \frac{d(-Q_n)}{d\psi_S} = \frac{-Q_n}{k_BT/q}\,. \tag{9.47}$$

Equation (9.47) should be compared to eqn. (8.22) for the bulk MOS case. We see that the semiconductor capacitance for the ETSOI structure is twice the corresponding value for the bulk structure.

Equation (9.47) and the corresponding result for the bulk case, eqn. (8.22), assume non-degenerate carrier statistics. What happens in the degenerate limit? In the degenerate limit, $E_F >> E_{C0} + \epsilon_1^c$, and eqn. (9.20) becomes

$$Q_n = -qn_S = -qN_{2D}^c\,(E_F - E_{C0} - \epsilon_1^c + q\psi_S)/k_BT)\,, \tag{9.48}$$

so the semiconductor capacitance becomes

$$\begin{aligned} C_S = C_{inv} &= \frac{d(-Q_n)}{d\psi_S} = \frac{q^2 N_{2D}^c}{k_BT} = q^2\left(\frac{m_D^*}{\pi\hbar^2}\right) \\ &= q^2 D_{2D} = C_Q\,, \end{aligned} \tag{9.49}$$

where D_{2D} is the two-dimensional density-of-states and C_Q is known as the *density-of-states capacitance* or the *quantum capacitance* that we saw in eqn. (9.43).

For more general conditions (i.e. between the non-degenerate and fully degenerate cases, multiple subbands occupied, thicker Si layers for which bandbending in the Si layer is important, etc.), the semiconductor capacitance must be computed numerically. But the general point is that the semiconductor capacitance is related to the density-of-states. For MOS structures that use semiconductors with a light effective mass (e.g. III-V semiconductors), we should expect the semiconductor capacitance and overall gate capacitance to be reduced in comparison to silicon.

Finally, let's re-write eqn. (9.45) as

$$C_G = \frac{(2C_{ox})C_S}{(2C_{ox}) + C_S}\,. \tag{9.50}$$

We might have expected the total gate capacitance of the double gate structure to be twice the gate capacitance of the corresponding single gate ETSOI structure, but it is actually a little less than twice the corresponding single gate result. To see why, we can re-write eqn. (9.50) as

$$C_G = 2 \times \left[\frac{C_{ox}(C_S/2)}{C_{ox} + (C_S)/2} \right] . \tag{9.51}$$

The quantity in brackets is the series combination of C_{ox} and $C_S/2$. The semiconductor capacitance is shared between the two gates, so each of the two gates has a capacitance that is a little less than the capacitance of a single gate SOI structure. For a discussion of these effects in single and double gate ETSOI structures, see Majumdar [3].

Exercise 9.3: Inversion layer capacitance and capacitance equivalent thickness.

To get a feel for some of the numbers that result from the formulas developed in this lecture, consider the silicon example of Exercises 9.1 and 9.2 with the additional information:

$$\kappa_{ox} = 4.0$$
$$t_{ox} = 1.8 \text{ nm} .$$

1) *What is the semiconductor capacitance when* $n_S = 1 \times 10^{13}$ cm^{-2}*?*
The sheet carrier density here is typical for the on-state of a modern MOSFET. (Note that it is expressed in units of cm^{-2}, not in m^{-2} as it should be for MKS units. This is common practice in semiconductor work, but we have to be careful to convert to MKS units when evaluating formulas.)

From eqn. (9.47) we find

$$C_S(inv) = \frac{-Q_n}{k_BT/q} = \frac{qn_S}{k_BT/q} = 61.6 \times 10^{-6} \text{ F/cm}^2 ,$$

which is twice the value found in Ex. 10.1 for the bulk MOS structure. In comparison to the $C_{ox} = 2.0 \times 10^{-6}$F/cm^2, this is a very large value, but it is unrealistically large because Fermi-Dirac carrier statistics and quantum confinement will lower C_S significantly. Assuming complete degeneracy, we can evaluate C_S from eqn. (9.49). We find

$$C_S = C_Q = 25.4 \times 10^{-6} \text{ F/cm}^2 ,$$

which is less than one-half the value obtained assuming non-degenerate statistics.

2) *What is the gate capacitance?*

According to eqn. (9.50)

$$C_G = \frac{(2C_{ox})C_S}{(2C_{ox}) + C_S} = \frac{2C_{ox}}{1 + 2C_{ox}/C_S} .$$

Putting in numbers, we find

$$C_G = \frac{2C_{ox}}{1 + 4.0/25.4} = 0.86\,(2C_{ox}) = 3.44 \times 10^{-6} \ \text{F/cm}^2 .$$

As expected, $C_G < 2C_{ox}$.

3) *What is the Capacitance Equivalent Thickness, CET?*
First, recall the definition of CET from eqn. (7.32) and adjust for the two gates:

$$C_G/2 \equiv \frac{\epsilon_{ox}}{CET} \rightarrow CET = \frac{\epsilon_{ox}}{C_G/2} .$$

Inserting numbers, we find

$$CET = \frac{4.0 \times 8.854 \times 10^{-14}}{1.72 \times 10^{-6}} = 2.06 \ \ \text{nm} .$$

Note that the CET is a little thicker than the actual oxide thickness of 1.8 nm and that the effect is greater than in Exercise 8.1 because of our use of Fermi-Dirac statistics and because C_S is shared between the two gates.

4) *What is the semiconductor surface potential when* $n_S = 1 \times 10^{13}$ cm^{-2}*?*
From eqn. (9.31), we have

$$\psi_S = \frac{k_B T}{q} \ln\left(\frac{n_S}{n_{Si}}\right) .$$

Inserting numbers, we find

$$\psi_S(n_S = 1 \times 10^{13} \ \text{cm}^{-2}) = 0.60 \ \ V .$$

Recall that we defined the onset of inversion to be at $\psi_S^{inv} = 0.58$, so this value is a little higher. For the bulk MOS example, the surface potential in strong inversion was several k_BT/q larger than $2\psi_B$. In this case, the surface potential in strong inversion is about one k_BT/q larger than ψ_S^{inv}, The difference is partially due to the fact that for the ETSOI structure, Q_n varies as $\exp(q\psi_S/k_BT)$ and for the bulk structure, Q_n varies as $\exp(q\psi_S/2k_BT)$, so it takes more bandbending in the bulk case to increase the inversion layer charge.

5) *What is the semiconductor surface potential when Fermi-Dirac statistics are used?*

Equation (9.22) relates n_S to ψ_S for general carrier statistics:

$$n_S = N_{2D}^c \ln\left(1 + e^{(E_F - E_{C0} + q\psi_S - \epsilon_1^c)/k_BT}\right).$$

Assuming that $\psi_S = 0$ when $n_S = n_{Si}$ and that the semiconductor is non-degenerate when $n_S = n_{Si}$, we find

$$n_{Si} = N_{2D}^c \, e^{(E_F - E_{C0} + q\psi_S - \epsilon_1^c)/k_BT},$$

which can be used in the first equation to write

$$n_S = N_{2D}^c \ln\left(1 + \frac{n_{Si}}{N_{2D}^c} e^{+q\psi_S/k_BT}\right),$$

which can be solved for

$$\psi_S = \frac{k_BT}{q} \ln\left[\frac{N_{2D}^c}{n_{Si}}\left(e^{n_S/N_{2D}^c} - 1\right)\right].$$

Using numbers from Exercise 9.1,

$$N_{2D}^c = 4.11 \times 10^{12} \text{ cm}^{-2}$$
$$n_{Si} = 8.5 \times 10^2 \quad \text{cm}^{-2}.$$

with $n_{Si} = 1 \times 10^{13} \quad \text{cm}^{-2}$, we find

$$\psi_S(n_S = 1 \times 10^{13} \text{ cm}^{-2}) = 0.64 \text{ V},$$

which is 0.04 V larger than the value obtained with Maxwell-Boltzmann statistics.

9.6 Surface potential vs. gate voltage

Figure 7.3 summarized ψ_S vs. V_G for a bulk MOS structure. Below threshold, $\psi_S = V_G/m$, where m is a little larger than 1. Above threshold, ψ_S varied slowly with V_G because m became very large so most of the increase in gate voltage went into the volt drop across the oxide, not the semiconductor. We expect qualitatively similar results for the ETSOI structure.

Figure 9.8 compares ψ_S vs. V_G for a bulk and ETSOI MOS structures. In the subthreshold region, $\psi_S = V_G$ because $m = 1$ for the double gate structure. Above threshold, ψ_S varies slowly with V_G for the same reasons as for the bulk structure. In fact, the variation is a little weaker with V_G, because the inversion charge in an ETSOI structure increases more rapidly with ψ_S.

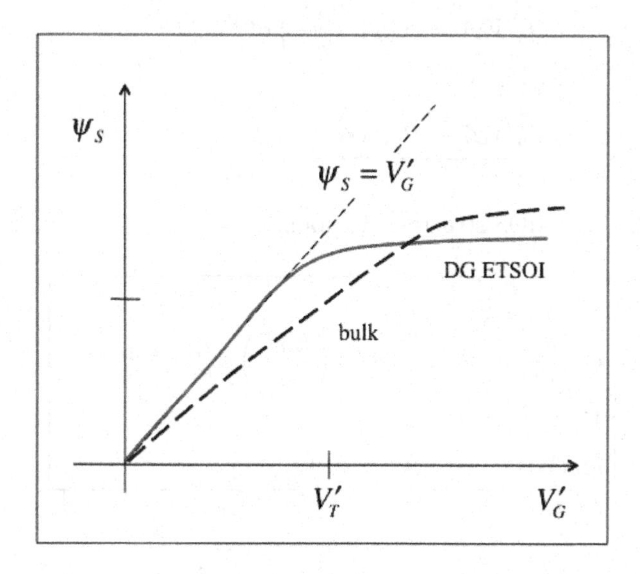

Fig. 9.8 The surface potential vs. gate voltage. The solid line sketches the ETSOI characteristic, and the dashed line shows the corresponding characteristic for a bulk MOS structure. The sketch assumes that for $\psi_S = 0$, the Fermi level is near the valence band in both cases.

9.7 Discussion

In this section we have shown that the electron charge, $Q_n(\psi_S)$, varies exponentially with ψ_S both below and above threshold. The dependence below threshold, eqn. (9.32), is as $\exp(\psi_S/k_BT)$, and the dependence above threshold, is the same. These results are very similar to those obtained for the bulk MOS case, eqns. (8.12) and (8.15).

Below threshold, $Q_n(V_G)$ varies exponentially with V_G because $\psi_S = V_G$) (see eqn. (9.42)). Above threshold, however, $Q_n(V_G)$ varies linearly with V_G as given by eqn. (9.46). Again, the results are similar to the corresponding results for the bulk MOS case.

To summarize, we have derived eqns. (9.42) and (9.46) to describe $Q_n(V_G)$ for the ETSOI MOS structure below and above threshold. For the double gate ETSOI structure, we found

$$\boxed{\begin{aligned}&V_G \ll V_T: \\ &Q_n(V_G) = -C_Q\left(\frac{k_BT}{q}\right)e^{q(V_G-V_T)/k_BT} \\ &V_G \gg V_T \\ &Q_n(V_G) = -C_G\left(V_G - V_T\right).\end{aligned}}, \quad (9.52)$$

and for the bulk MOS structure, we found

$$\boxed{\begin{aligned}&V_G \ll V_T: \\ &Q_n(V_G) = -(m-1)C_{ox}\left(\frac{k_BT}{q}\right)e^{q(V_G-V_T)/mk_BT} \\ &V_G \gg V_T \\ &Q_n(V_G) = -C_G\left(V_G - V_T\right).\end{aligned}}.$$

Specifics depend on the actual channel structure (e.g. bulk, double gate ETSOI, single gate ETSOI, etc.), but these two examples show that in general, Q_n varies exponentially with V_G below threshold and linearly with V_G above threshold. The general $Q_n(V_G)$ relation can be evaluated numerically, but as we'll discuss in Lecture 11, an empirical expression that reduces to the correct result below and above threshold can also be used.

9.8 Summary

In this lecture, we have discussed how Q_n in an ETSOI structure varies with surface potential and with gate voltage, considering both the subthreshold and above threshold regions. The correct results in subthreshold and in strong inversion are readily obtained, but a numerical solution (or an empirical one) is needed to cover the entire range. The results show that one-dimensional electrostatics is similar in bulk and ETSOI MOS structures. In the next lecture we'll consider two-dimensional electrostatics and will explain why the double gate structure is preferable for very short channel lengths.

9.9 References

For as review of concepts such as a particle in a box, 2D density-of-states, intrinsic carrier concentration, see:

[1] Robert F. Pierret, *Advanced Semiconductor Fundamentals, 2nd Ed.*, Vol. VI, Modular Series on Solid-State Devices, Prentice Hall, Upper Saddle River, N.J., USA, 2003.

Lecture 1 of the following online course discusses bandstructure fundamentals and Lecture 4 the density-of-states.

[2] Mark Lundstrom, "ECE 656: Electronic Transport in Semiconductors," Purdue University, Fall 2013, `//https://www.nanohub.org/groups/ece656_f13`.

For an example of how the Schrödinger and Poisson equations are solved self-consistently for an MOS structure, see:

[3] D. Vasileska, D.K. Schroder, and D.K. Ferry, "Scaled silicon MOSFETs: degradation of the total gate capacitance," *IEEE Trans. Electron Devices*, **44**, pp. 584-587, 1997.

The following paper contains an interesting discussion of the difference in the gate capacitance of single and double gate ETSOI structures.

[4] Amlan Majumdar, "Semiconductor Capacitance Penalty per Gate in Single- and Double-Gate FETs," *IEEE Electron Device Letters*, **35**, 609-611, 2014.

A more extensive treatment of the electrostatics of ultra-thin SOI structures is presented by Fossum and Trivedi.

[5] Jerry G. Fossum and Vishal P. Trivedi, *Fundamentals of Ultra-Thin-Body MOSFETs and FinFETs*, Cambridge Univ. Press, Cambridge, U.K., 2013.

Lecture 10

2D MOS Electrostatics

10.1 Introduction

In the previous two lectures, we discussed one-dimensional electrostatics by asking how the potential in the semiconductor varied in response to the gate voltage. In a short channel transistor, however, the source and drain potentials can produce strong electric fields along the direction of the channel. As suggested by Fig. 10.1, the electrostatic potential in the channel of a short channel MOSFET should vary strongly in both the x and y directions. Two-dimensional electrostatics have important consequences for the operation of a transistor. As shown on the left of Fig. 10.2, the application of a large drain bias shifts the $\log_{10} I_{DS}$ vs. V_{GS} characteristics to the left. In Lecture 2, the shift in the characteristics was related to the DIBL device metric; as defined in Fig. 2.12, $DIBL = -\Delta V_{GS}/\Delta V_{DS}$, where ΔV_{GS} is the change in gate voltage needed to keep the drain current constant when the drain voltage changes by ΔV_{DS}.

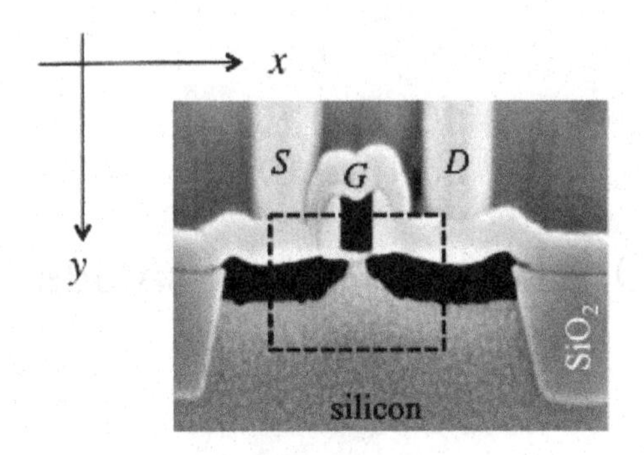

Fig. 10.1 The region of interest for 2D MOS electrostatics. The electric field in the channel is strong in the direction normal to the channel because of the gate voltage, and it is strong along the channel when there is a significant drain bias.

If we pick a small current as the definition of when a transistor is "on" (the horizontal dashed line in Fig. 10.2), then we see that the large drain bias decreases the magnitude of the threshold voltage. Note that the threshold voltage expression developed in Lecture 7 (eqn. (7.13)) has no drain bias in it because 2D electrostatics were not considered. Another manifestation of 2D electrostatics is a channel length dependence to the threshold voltage, as illustrated on the right of Fig. 10.2. The output resistance of a transistor is also due to two-dimensional electrostatics. Our goal in this lecture is to understand how 2D electrostatics affects the terminal characteristics of MOSFETs.

Two-dimensional electrostatics can be treated by numerically solving the 2D Poisson equation (in a very small transistor, the 3D equation should be solved). Numerical simulations are indispensable for designing modern transistors. Our goal in this lecture, however, is not quantitative predictions but, rather, to develop physical insight.

10.2 The 2D Poisson equation

Gauss's Law states that

$$\nabla \cdot \vec{D}(x,y) = \rho(x,y)\,, \tag{10.1}$$

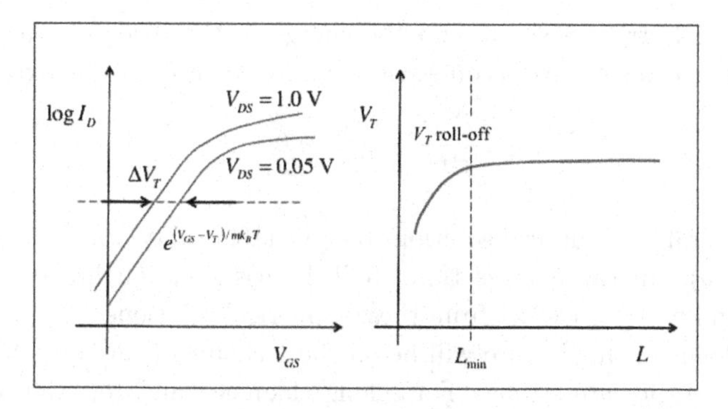

Fig. 10.2 Illustration of how 2D electrostatics affects the terminal performance of a short channel transistor. Left: DIBL, which shifts the $\log_{10} I_{DS}$ vs. V_{GS} transfer characteristic to the left. This behavior can also be interpreted as a reduction in threshold voltage with increasing drain bias. Right: V_T roll-off, which is the reduction of V_T for short channel lengths.

where $\vec{D}$ is the displacement vector and ρ the space charge density. The displacement field is related to the electric field by

$$\vec{D}(x, y) = \epsilon_s \vec{\mathcal{E}}(x, y), \tag{10.2}$$

and the dielectric constant, ϵ_s, is assumed to be a constant in the semiconductor and another constant in the oxide. The electrostatic potential is related to the electric field by

$$\vec{\mathcal{E}}(x, y) = -\vec{\nabla}\psi(x, y). \tag{10.3}$$

Putting these equations together, we obtain the 2D Poisson equation as

$$\frac{\partial^2 \psi}{\partial x^2} + \frac{\partial^2 \psi}{\partial y^2} = -\frac{\rho(x, y)}{\epsilon_s}. \tag{10.4}$$

Equation (10.4) is to be solved in the channel region of the MOSFET. We are most interested in the subthresold region, or just at the beginning of inversion where 2D electrostatics leads to DIBL and V_T roll-off. In the subthreshold region

$$\rho(x, y) \approx q\left[N_D^+(x, y) - N_A^-(x, y)\right] \approx -qN_A, \tag{10.5}$$

where the last expression comes by assuming that there are only p-type dopants in the channel, that they are fully ionized, and that their concentration is uniform.

The gate oxide and the gate electrode are also part of the channel region and must be included to evaluate $\psi(x, y)$ in the channel. The oxide has a different dielectric constant, and the charge in the oxide can usually be neglected, so eqn. (10.4) becomes the Laplace equation in the oxide:

$$\frac{\partial^2 \psi}{\partial x^2} + \frac{\partial^2 \psi}{\partial y^2} = 0 \,. \tag{10.6}$$

In general, a numerical solution to eqns. (10.4) and (10.6) is needed to find $\psi(x, y)$. In the next section, we'll discuss some qualitative ways to understand what to expect from these numerical solutions.

Our focus in this lecture will be on short channel transistors for which 2D electrostatics are strong. For a long channel transistor, the potential varies slowly along the direction of the channel, so

$$\frac{\partial^2 \psi}{\partial x^2} \ll \frac{\partial^2 \psi}{\partial y^2} \,, \tag{10.7}$$

and eqn. (10.4) reduces to the 1D Poisson equation discussed in Lectures 6-9. Most of traditional MOSFET theory is based on the assumption of eqn. (10.7), and the approach is known as the *gradual channel approximation.* The standard approach to modeling short channel MOSFETs is to develop a model for a long channel transistor and then to add the effects produced by 2D electrostatics to the model. References [1-4] discuss this approach.

10.3 Threshold voltage roll-off and DIBL

We begin we re-writing eqn. (10.4) in depletion as

$$\frac{\partial^2 \psi}{\partial y^2} = \frac{q N_A}{\epsilon_s} - \frac{\partial^2 \psi}{\partial x^2} \,. \tag{10.8}$$

In an n-channel MOSFET, the electrostatic potential increases from the source to the drain, so $d\psi/dx > 0$. In practice, we find that the electric field, $-d\psi/dx$, also increases from the source to the drain, so we conclude that the curvature, $d^2\psi/dx^2$, is positive. This can be clearly seen from numerical simulations, as shown in the computed energy band diagrams of Fig. 10.3 (which are the same as Figs. 3.5 and 3.6). It is clear that under both low and high drain bias, $E_C(x)$ has negative curvature, so $\psi(x)$ has a positive curvature.

Realizing that $\psi(x)$ has positive curvature, we can write eqn. (10.8) as

$$\frac{\partial^2 \psi}{\partial y^2} = \frac{q N_A|_{\text{eff}}}{\epsilon_s} \,. \tag{10.9}$$

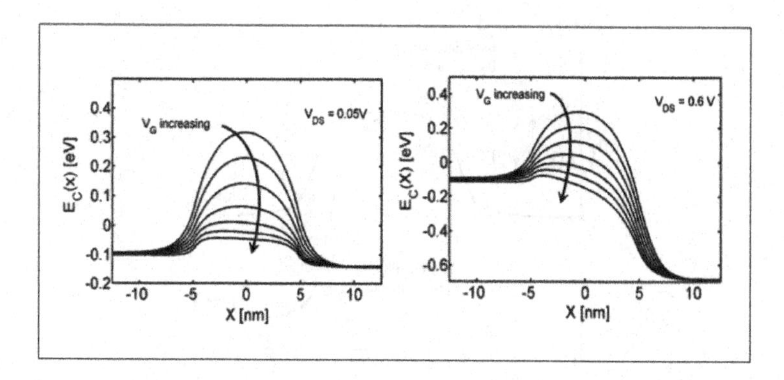

Fig. 10.3 Simulations of $E_c(x)$ vs. x for a short channel transistor. Each line corresponds to a different gate voltage, with the gate voltage increasing from the top down. Left: Low drain bias. Right: High drain bias. The simulations are the same as those in Figs. 5.5 and 5.6. (©IEEE 2002. Reprinted, with permission, from: Mark Lundstrom and Zhibin Ren, "Essential Physics of Carrier Transport in Nanoscale MOSFETs," *IEEE Trans. Electron Dev.*, **49**, pp. 133-141, 2002.)

where

$$N_A\big|_{\text{eff}} = \frac{qN_A}{\epsilon_s} - \frac{\partial^2 \psi}{\partial x^2} < N_A\,. \tag{10.10}$$

Equation (10.9) is a 1D Poisson equation for $\psi(y)$ with an "effective doping density" that is lower than the actual doping density. According to eqn. (7.13), the threshold voltage is related to the doping density as

$$V_T = V_{FB} + \frac{\sqrt{2q\epsilon_s N_A(2\psi_B)}}{C_{ox}} + 2\psi_B\,. \tag{10.11}$$

Because 2D electrostatics effectively lowers N_A, we expect it to decrease the threshold voltage. As we decrease the channel length, $d^2\psi/dx^2$ increases, which reduces the effective doping and lowers the threshold voltage. This explains why V_T decreases as the channel length decreases (so-called threshold voltage roll-off). With the same argument, we can also understand DIBL and the reduction of V_T with increasing drain voltage at a fixed channel length. As the drain voltage increases, $d^2\psi/dx^2$, increases, $N_A|_{\text{eff}}$ decreases, and V_T decreases.

Figure 10.4 presents another view of 2D electrostatics. Recall from Lecture 3 that the source to channel energy barrier plays a critical role in the operation of a transistor. In the ideal case, the height of the energy barrier is solely under the control of the gate voltage, and is not affected by the drain voltage (top of Fig. 10.4). In a real device, the drain potential

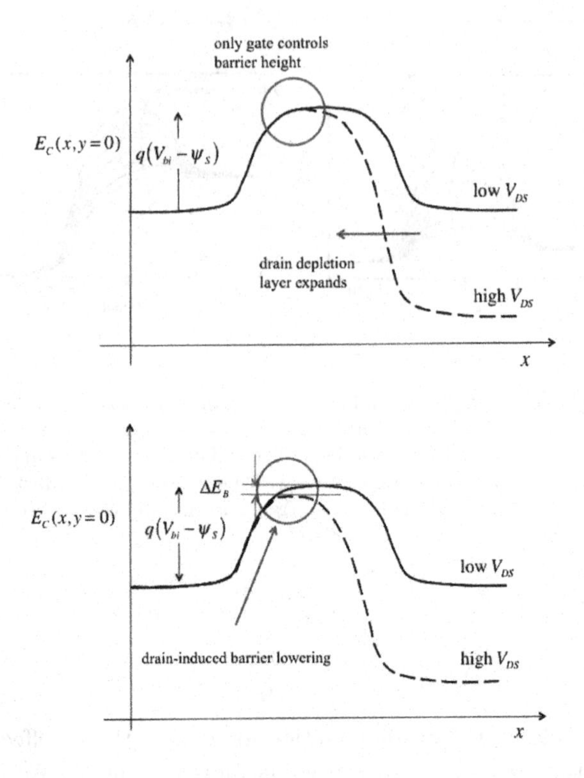

Fig. 10.4 Illustration of the effect of a drain voltage on the source to channel energy barrier. Top: No DIBL — the drain voltage has no effect on the height of the barrier. Bottom Significant DIBL. The drain voltage lowers the barrier a small amount.

reaches through and lowers the barrier (bottom of Fig. 10.4). The lower barrier allows more current to flow at the given gate voltage. Alternatively, a smaller gate voltage is needed to reach a specified current, because the barrier is being pulled down by both the gate and drain potentials. This drain-induced-barrier lowering causes the $\log_{10} I_D$ vs. V_{GS} characteristic to shift to the left.

The barrier lowering view also helps explain why 2D electrostatics reduces the effective doping. For heavy doping, the bands are hard to bend, but with 2D electrostatics, the drain helps the gate pull the barrier down. From the perspective of a 1D Poisson equation, the doping density has been effectively lowered, as in eqn. (10.9). The length of the region over which the drain potential is felt is, to first order, the depletion width of the drain-channel NP junction. In practice, the length of this region depends on the 2D geometry of the transistor as will be discussed in the next section.

10.4 Geometric screening

Screening is a general phenomena in metals and semiconductors. If a charge perturbation is produced, mobile carriers rearrange themselves to neutralize ("screen out") the charge. The characteristic distance over which the charge is screened out is the *screening length*, or *Debye length*, L_D,

$$L_D = \sqrt{\frac{\epsilon_s k_B T}{q^2 n_0}}, \tag{10.12}$$

where n_0 is the electron density. (Non-degenerate carrier statistics are assumed in eqn. (10.12).)

In a MOSFET, there is another way that electric fields can be screened out. Figure 10.5 is an illustration of what is called "geometric screening." The lines that emanate from the drain are electric field lines. Three different structures are show; a bulk MOSFET, a single gate (SG) SOI MOSFET, and a double gate (DG) MOSFET. For the DG SOI MOSFET, the field lines are seen to terminate on the two metal gates. On average, they only reach a distance, Λ, into the channel. If $\Lambda < L$, then the electric field from the drain cannot reach to the beginning of the channel and pull down the barrier. DIBL is low. The precise value of the *geometric screening length* is determined by the two-dimensional geometry, but intuitively, it is clear that the more we surround the channel with the gate electrode, the more effective this geometric screening will be. In Fig. 10.5, the DG SOI MOSEFT has the strongest geometrical screening (shortest Λ) and, therefore, suffers the least from 2D electrostatics.

While a calculation of Λ for an arbitrary geometry can get complicated ([5-6]), an heuristic derivation shows what Λ depends on. First, recall the 1D Poisson equation in the direction normal to the channel,

$$\frac{\partial^2 \psi}{\partial y^2} = \frac{q N_A}{\epsilon_s}. \tag{10.13}$$

We can also write phenomenologically

$$\frac{\partial^2 \psi}{\partial y^2} \approx \frac{V_G - \psi_S}{\Lambda^2}, \tag{10.14}$$

which simply says when $V_G > \psi_S$, then $\partial^2\psi/\partial y^2$ will be positive and $1/\Lambda^2$ is the constant of proportionality. By equating these two expressions, we

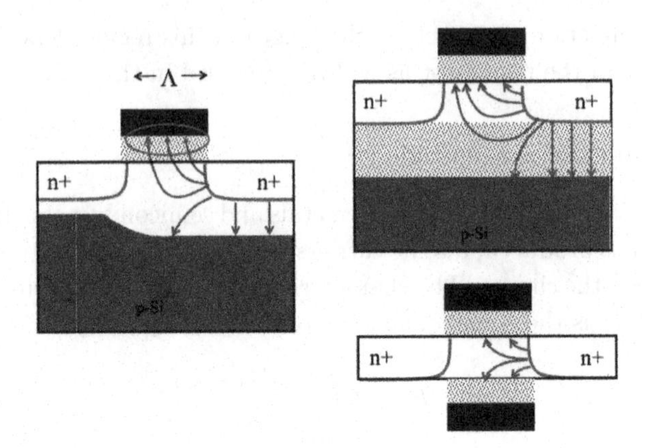

Fig. 10.5 Geometric screening in three types of MOSFETs: Left: a bulk MOSFET. Top right: A single gate SOI MOSFET. Bottom right: a double gate MOSFET. In a well-designed MOSFET, the lines representing the electric field from the drain penetrate only a distance, $\approx \Lambda$, into the channel because most terminate on the top and bottom gate electrodes. (After David Frank, Yuan Taur, and Hon-Sum Philip Wong, "Future Prospects for Si CMOS Technology," Technical Digest, IEEE Device Research Conf., pp. 18-21, 1999.)

find

$$\frac{V_G - \psi_S}{\Lambda^2} = \frac{qN_A}{\epsilon_s} . \tag{10.15}$$

We also know the solution to the 1D MOS problem in depletion,

$$V_G = -\frac{Q_D(\psi_S)}{C_{ox}} + \psi_S = \frac{qN_A W_D}{C_{ox}} + \psi_S , \tag{10.16}$$

where W_D is the width of the depletion layer at the surface. From eqns. (10.15) and (10.16), we find

$$\Lambda = \sqrt{\frac{\epsilon_s}{\epsilon_{ox}} W_D t_{ox}} . \tag{10.17}$$

Using eqn. (10.8), and (10.14), we can write the 2D Poisson equation as

$$\frac{d^2\psi_S}{dx^2} + \frac{V_G - \psi_S}{\Lambda^2} = \frac{qN_A}{\epsilon_s} , \tag{10.18}$$

where we have specified that we want the solution at the surface, $\psi_S(x) = \psi(x, y = 0)$. With a change variables to

$$\phi = \psi_S - V_G + \frac{qN_A}{\epsilon_s}\Lambda^2 \tag{10.19}$$

eqn. (10.18) becomes

$$\frac{d^2\phi}{dx^2} - \frac{\phi}{\Lambda^2} = 0\,, \tag{10.20}$$

which is a simple differential equation with solutions that vary as $\exp(\pm x/\Lambda)$, where Λ is given by eqn. (10.17).

We conclude that for a MOSFET, the characteristic length over which potential perturbations die out is the geometric scaling length, Λ. If $L > \Lambda$, then short channel effects such as DIBL will be modest. Typically, $L \approx (1.5 - 2)\Lambda$ is adequate for modest short channel effects. According to eqn. (10.17), a thin oxide is beneficial and so is a thin depletion region. As illustrated in Fig. 10.5, for these cases, the electric field lines are more likely to terminate either on the gate electrode or on the neutral semiconductor bulk rather that reaching through and pulling down the barrier.

We have developed an approximate expression for the geometrical screening length for a bulk MOSFET using heuristic arguments. The formal derivation of Λ for a variety of MOS structures is discussed in [5, 6, 12]. In general, $\Lambda_{bulk} > \Lambda_{DG-SOI} > \Lambda_{NW}$. The shorter the geometric screening length, the better the transistor. The general principle is that the more the channel is surrounded by conductive plates, especially by the gate electrode, the shorter the geometric screening length.

10.5 Capacitor model for 2D electrostatics

Figure 10.6 shows a useful way to view 2D electrostatics. As discussed in Lecture 3, MOSFETs operate by modulating the energy barrier between the source and the channel. Each capacitor in this figure represents the electrostatic coupling of a terminal to the top of the energy barrier, the virtual source (VS). The top of the barrier is near the middle of the channel for low V_{DS} and moves toward the source with increasing drain bias. As a result, the capacitors in the circuit depend on the drain bias [7]. Solutions to the 2D Poisson equation for the specific MOSFET geometry are needed to evaluate the magnitude of each capacitor, but the capacitor analysis is useful for the insight it provides. Figure 10.6 is for a bulk MOSFET with its four terminals, source, drain, bulk, and gate, but similar circuits can be drawn for other transistors, such as single and double gate SOI MOSFETs [7].

To analyze the simple circuit shown in Fig. 10.6, we will use superposition and first assume that no voltage is applied to the terminals but that

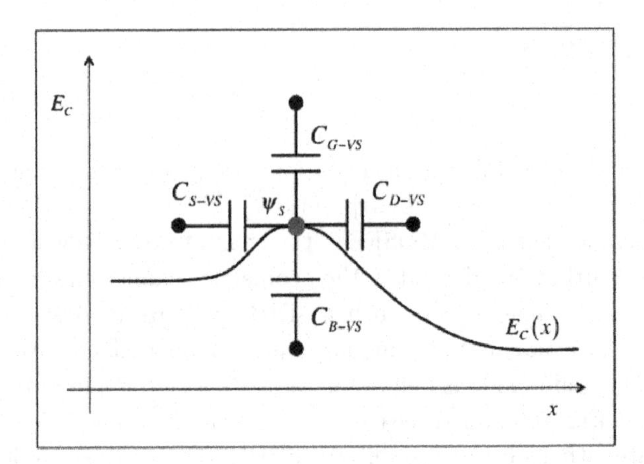

Fig. 10.6 The capacitor model of 2D electrostatics for a bulk MOSFETs. Each capacitor represents the electrostatic coupling of an electrode (source, drain, bulk, gate) to the top of the energy barrier, which is the virtual source. Note that the top of the barrier moves with drain bias, from near the middle of the channel at low drain bias to near the source at high drain bias, so the capacitors are voltage dependent, in principle.

there is a charge at the top of the barrier. The relevant circuit is shown on the left of Fig. 10.7. The total capacitance at the VS is

$$C_\Sigma = C_{G-VS} + C_{S-VS} + C_{D-VS} + C_{B-VS}\,, \tag{10.21}$$

and the corresponding potential on the VS node is

$$\psi_S = \frac{Q_S}{C_\Sigma}\,. \tag{10.22}$$

Next, we will assume that there is a voltage on the gate, but that the other three terminals are grounded. The relevant circuit is shown on the right of Fig. 10.7. Voltage division gives the potential at the VS as

$$\psi_S = \left(\frac{C_{G-VS}}{C_\Sigma}\right) V_G\,, \tag{10.23}$$

and a similar procedure can be used for each of the other electrodes - apply a voltage on the electrode of interest and ground all the others. After adding the contributions for the four voltages along with the contribution for charge present but with all of the voltages zero, the final result is

$$\psi_S = \left(\frac{C_{G-VS}}{C_\Sigma}\right) V_G + \left(\frac{C_{S-VS}}{C_\Sigma}\right) V_S + \left(\frac{C_{D-VS}}{C_\Sigma}\right) V_D + \left(\frac{C_{B-VS}}{C_\Sigma}\right) V_B + \frac{Q_S}{C_\Sigma}\,. \tag{10.24}$$

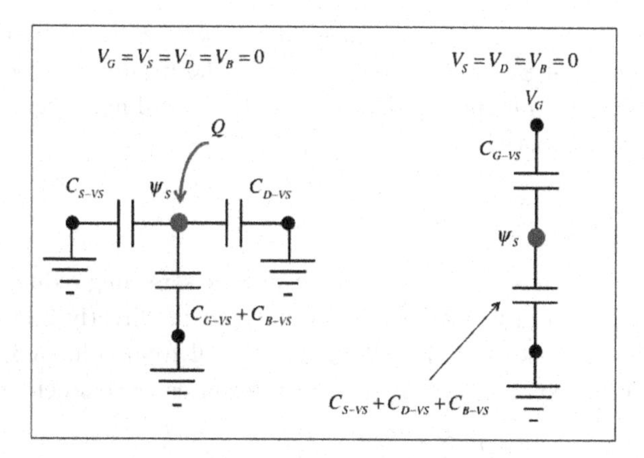

Fig. 10.7 Simplified capacitor circuits for no voltages applied but with charge, Q_S, on the virtual source (Left) and with no charge and a voltage applied to the gate (Right).

Equation (10.25) should be compared to the corresponding 1D result, eqn. (7.11),

$$\begin{aligned} V_G &= -\frac{Q_S}{C_{ox}} + \psi_S \\ \psi_S &= V_G + \frac{Q_S}{C_{ox}} \,. \end{aligned} \tag{10.25}$$

The 2D result reduces to the 1D result when the oxide capacitance is much larger that the others. In this case, the potential at the top of the barrier is totally controlled by the gate voltage, and voltages at the other terminals have no effect. This behavior is what a transistor designer works to achieve. There are two approaches — make the gate capacitance as large as possible (make t_{ox} thin or use a material with a high dielectric constant) or reduce the other capacitors by using geometric screening to electrostatically isolate the other terminals from the top of the barrier.

Consider the case where only gate and drain voltages are applied and the other terminals are grounded. Assuming subthreshold operation, where the charge is negligible, eqn. (10.24) simplifies to

$$\psi_S = \left(\frac{C_{G-VS}}{C_\Sigma}\right) V_G + \left(\frac{C_{D-VS}}{C_\Sigma}\right) V_D \,. \tag{10.26}$$

The gate and the drain voltages both affect the potential at the VS:

$$\begin{aligned} \frac{\partial \psi_S}{\partial V_G} &= \frac{C_{G-VS}}{C_\Sigma} \\ \frac{\partial \psi_S}{\partial V_D} &= \frac{C_{D-VS}}{C_\Sigma} \,. \end{aligned} \tag{10.27}$$

For a well-designed transistor, we must have $\partial\psi_S/\partial V_G \gg \partial\psi_S/\partial V_D$ so that the gate control of ψ_S is much stronger than the drain. We also want ψ_S to follow the gate voltage, i.e. $\partial\psi_S/\partial V_G \approx 1$. Accordingly, the criteria for a well-designed transistor are

$$\begin{aligned} C_{G-VS} &\gg C_{D-VS} \\ C_{G-VS} &\approx C_\Sigma \,. \end{aligned} \tag{10.28}$$

Thin gate oxides increase C_{G-VS}, and geometric screening reduces C_{D-VS}.

The capacitors in the equivalent circuit can be directly related to the terminal characteristics of the MOSFET. Recall from eqn. (3.3) that the drain current is exponentially related to the source to channel barrier,

$$I_{DS} \propto e^{-E_{SB}/k_BT} = e^{q\psi_S/k_BT} \,. \tag{10.29}$$

We can write eqn. (10.26) as

$$\psi_S = \frac{V_G}{m} + \frac{DIBL}{m} V_D \,, \tag{10.30}$$

where

$$\begin{aligned} m &\equiv \frac{C_\Sigma}{C_{G-VS}} \\ DIBL &\equiv \frac{C_{D-VS}}{C_{G-VS}} \,. \end{aligned} \tag{10.31}$$

Using eqn. (10.30), the drain current, (10.29), can be written as

$$I_{DS} \propto e^{q\psi_S/k_BT} = e^{q(V_G + DIBL \times V_D)/mk_BT} \,. \tag{10.32}$$

The subthreshold swing at a constant drain voltage is defined according to eqn. (2.1) as

$$SS = \left[\frac{\partial\left(\log_{10} I_{DS}\right)}{\partial V_G}\right]^{-1} = 2.3 m k_B T \tag{10.33}$$

and gives the change in gate voltage needed to increase the drain current by a factor of 10. The subthreshold swing is controlled by the value of m, which is ≥ 1, so $SS \geq 60$ mV/decade. Assuming that $C_{G-VS} = C_{ox}$ and $C_{B-VS} = C_D$, the depletion capacitance of the semiconductor, eqn. (10.31) gives

$$m = 1 + \frac{C_D}{C_{ox}} + \frac{C_{S-VS} + C_{D-VS}}{C_{ox}} \,. \tag{10.34}$$

Equation (10.34) should be compared to eqn. (7.31), which was derived assuming 1D electrostatics. The first term (1) gives the ideal subthreshold swing. The second term is a 1D effect that accounts for the voltage division between the gate and semiconductor depletion capacitance. This term is missing in the fully depleted ETSOI structure but present in a bulk MOSFET. The third term in (10.34) is due to 2D electrostatics. We see that 2D electrostatics increases m and therefore increases the subthreshold slope. This effect, illustrated in Fig. 10.8, is undesirable, and transistor designers work to minimize it.

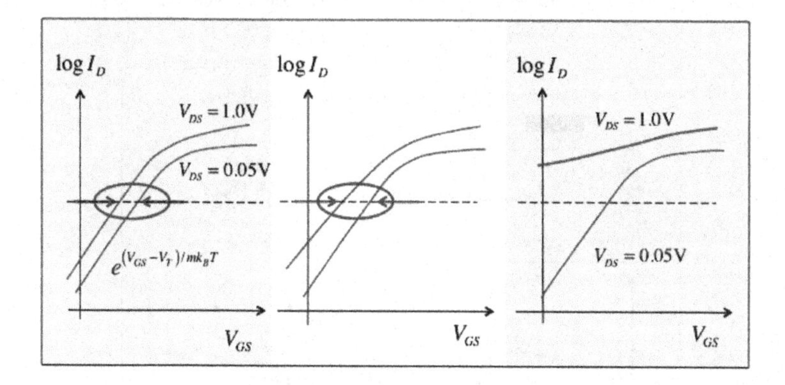

Fig. 10.8 Illustration of how the drain to virtual source capacitor not only produces DIBL (Left) but also increases the subthreshold swing (Center). Punch through (Right) occurs when 2D effects are very strong and will be discussed in Sec. 10.7.

Finally, note that the capacitor model also describes DIBL. According to eqn. (10.32), if we increase V_D by ΔV_D, then to keep the drain current constant, we must decrease V_G. The required change is V_G is

$$\Delta V_G = -DIBL \times \Delta V_D \,, \tag{10.35}$$

which is how we defined DIBL in Sec. 2.4.

To summarize, the capacitor model is a simple way to understand the results of 2D solutions to the Poisson equation. For a well-designed MOSFET, the capacitance from the gate to the virtual source should dominate. The other capacitors increase the subthreshold swing and produces DIBL.

10.6 Constant field (Dennard) scaling

Progress in semiconductor integrated circuits has been driven by *device scaling* for the past 50+ years. Each technology generation (typically 1-2 years), the size of a transistor shrinks by a factor of two, so the number of transistors on an integrated circuit chip doubles. If the downscaling of transistors is done properly, the performance of the integrated circuit improves. When downscaling transistor dimensions, the main challenge is to deal with the short channel effects caused by 2D electrostatics.

Figure 10.9 illustrates the goal of device scaling; it is to scale down the linear dimensions of a transistor by a factor of κ and to do it so that the resulting IV characteristic is a scaled version of the original IV characteristic with all the currents and voltages reduced by the factor κ.

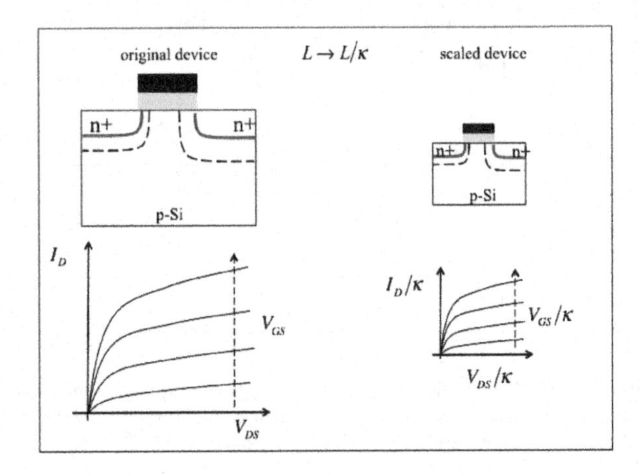

Fig. 10.9 The goal of device scaling. On the left is a transistor and its IV characteristics. On the right is a scaled version of the transistor, where the scaling factor is $\kappa > 1$. If the scaling is done properly, then a well-behaved IV characteristic results with all currents and voltages scaled by the factor κ.

Figure 10.2 sketched the expected threshold voltage vs. channel length characteristic of a MOSFET. The roll-off (decrease) of V_T at small channel lengths is due to the 2D electrostatic effects that we have discussed in previous sections. Below some minimum channel length, L_{min}, V_T is too small and too sensitive to L. Below L_{min}, the subthreshold swing and DIBL also become unacceptable, and the device may even punch through, as discussed in the next section. The goal of scaling is to reduce L_{min} by the same scaling factor, κ, so that scaled transistors with $L = L_{min}/\kappa$ do not suffer from severe short channel effects.

An approach to device scaling was first presented by R.H. Dennard and colleagues [8] and served as a guide for scaling device for decades. The basic idea is to reduce all dimensions by the scaling factor, κ, increase the doping by the same factor, and reduce the power supply voltage by the same factor. This approach maintains a constant electric field in the channel as devices scale down. Dennard scaling consists of:

1) Scaling down all dimensions:

$$\begin{aligned} L, W &\rightarrow (L, W)/\kappa \\ t_{ox} &\rightarrow t_{ox}/\kappa \\ W_D &\rightarrow W_D/\kappa \\ y_j &\rightarrow y_j/\kappa \end{aligned} \tag{10.36}$$

2) Increasing the channel doping:

$$N_A \rightarrow \kappa N_A \tag{10.37}$$

3) Scaling down the power supply voltage:

$$V_{DD} \rightarrow V_{DD}/\kappa\,. \tag{10.38}$$

Here W_D is the width of the depletion region, y_j is the source/drain junction depth, and V_{DD} is the power supply voltage.

Consider how this works using some very simple arguments. First, the average electric field is $\mathcal{E} \approx V_{DD}/L$ and since both V_{DD} and L are scaled by the same factor, the electric field in the channel of the scaled device is the same as in the original device.

The low-field velocity is mobility times electric field. Assuming that the mobility in the scaled device is the same as in the original device, the velocity in the scaled device does not change. Dennard assumed that the high field velocity was υ_{sat}, which is a material parameter that does not change with scaling. So the velocity in the scaled device is the same as in the original device.

It is important to scale depletion region thicknesses also. The drain depletion region has a strong effect on 2D electrostatics and is given by depletion theory as

$$W_D = \sqrt{\frac{2\epsilon_s}{qN_A}\left(V_{bi} + V_{DD}\right)}\,.$$

If $V_{DD} \gg V_{bi}$, then scaling the doping up by κ and V_{DD} down by κ results in W_D scaling down by κ. If t_{ox} and y_j are also scaled down, then 2D electrostatics in the scaled device will become strong at a channel length that is κ times smaller than the original device. This process scales L_{min} down by approximately a factor of κ.

The capacitances are

$$C = \frac{\epsilon A}{t} \text{ F},$$

where t is the thickness of the oxide or depletion layer. Since all thicknesses scale down by κ and area scales down by κ^2, capacitances will scale down by κ, but C_{ox}, which is a capacitance per unit area will scale up by κ.

Now let's consider the effect of Dennard scaling on some important quantities. The inversion layer charge is

$$Q_n = -C_{ox}(V_G - V_T) .$$

Since C_{ox} scales up by κ and the voltages scale down by the same factor, the inversion layer charge per unit area does not change with scaling.

Now consider the current,

$$I_{DS} = WQ_n\upsilon .$$

Since Q_n and υ do not change with scaling and W scales down by κ, the current will scale down by κ.

To summarize, constant electric field (Dennard) scaling results in the follow scaled performance parameters.

$$\begin{aligned} &Q_n \to Q_n \\ &\upsilon \to \upsilon \\ &C \to C/\kappa \\ &C_{ox} \to \kappa C_{ox} \\ &I_{DS} \to I_{DS}/\kappa . \end{aligned} \tag{10.39}$$

We can now examine the performance of scaled circuits. The circuit delay is the time required to remove the charge, CV_{DD}, stored on the circuit capacitance. The resulting delay is

$$\tau = \frac{CV_{DD}}{I_{DS}} .$$

The circuit delay is seen to scale down by the factor, κ. Power is $P_D = V_{DD}I_{DS}$, so power scales down by κ^2. The power density in W/m^2 stays the same after scaling. The density of transistors increases by κ^2, because of the size of each transistor decreases by κ^2. Finally, the power-delay product, $P_D\tau$, an important metric, scales down by κ^3. To summarize, constant field scaling results in:

$$\begin{aligned}
&\tau = CV_{DD}/I_{DS} \rightarrow \tau/\kappa \\
&P_D = V_{DD}I_{DS} \rightarrow P_D/\kappa^2 \\
&P_D/A \rightarrow P_D/A \\
&D = no./A \rightarrow D \times \kappa^2 \\
&P_D\tau = CV_{DD}^2 \rightarrow P_D\tau/\kappa^3 \,.
\end{aligned} \tag{10.40}$$

Dennard scaling is not quite as easy as it may seem because several quantities do not scale. For example, recall that V_T is given by eqn. (10.11). The flatband voltage does not scale, and ψ_B is relatively insensitive to scaling, so it is challenging to make $V_T \rightarrow V_T/\kappa$ as the scaling scenario requires. Sophisticated channel doping profiles are often used [4].

The drain depletion region varies as $\sqrt{(V_{bi} + V_{DD})/N_A}$. Since V_{bi} does not scale, it is challenging to make $W_D \rightarrow W_D/\kappa$. The subthreshold swing is insensitive to scaling. These factors make device scaling challenging, but the Dennard scaling approach provides a starting point that device designers refine to produce scaled devices with well-behaved characteristics that operate with reduced power and delay.

Device scaling is currently facing some serious challenges, and many now see an end to device scaling within a decade or so. One problem is that gate oxides have been scaled as thin as they can go without leading to excessive leakage current. This is forcing a change from the planar MOSFET to the FinFET, which offers better electrostatic control at the same gate oxide thickness [9]. Another scaling challenge is caused by the failure of the subthreshold swing to scale. A maximum off-current is specified. Given that the subthreshold swing must be a little greater than 60 mV/decade and that the current increases linearly above threshold, it takes about $V_{DD} = 1$ V to achieve the required I_{DS} in the on-state. The result is that voltage scaling has stopped and power densities are increasing. Several innovative transistor structures are being explored to address these challenges [10].

10.7 Punch through

When 2D effects become strong, the transistor can "punch through", which means that the drain electric field has punched through to the source. Current then flows from the source to the drain with the gate having little effect. Figure 10.8 shows the transfer characteristics for three different situations. The first case (on the left in the figure) was for a device that suffers only a little from 2D electrostatics. The subthreshold swing is only a little bigger than 60 mV/decade, and the DIBL is modest. When 2D electrostatics are stronger (in the center of the figure), the subthreshold swing degrades noticeably and DIBL becomes quite large (e.g. $>$ 100 mV/V). When 2D electrostatics dominate (on the right in the figure) the transistor performance severely degrades. The current has a weak dependence on gate voltage and DIBL is hard to define because the subthreshold characteristics under low and high V_{DS} are not even close to parallel.

Figure 10.10 shows how 2D electrostatics affect the output characteristics of a transistor. For the long channel device (on the left in the figure), the drain current is constant in saturation, and the output resistance approaches infinity. For a short channel MOSFET, the output resistance is much lower. The reason is easy to appreciate. The current is proportional to $(V_{GS} - V_T)$ and V_T decreases with increasing drain voltage due to DIBL. It is not obvious that the 2D electrostatics in subthreshold (where DIBL is measured and where the mobile charge in the channel is negligible) should be the same above threshold where the mobile charge in the channel is large. It is found, however, that for well-designed transistors, the same DIBL parameter describes 2D electrostatics below and above threshold [11, 12]. Finally, the IV characteristics on the right of Fig. 10.10 are for a transistor that is punched through. Even in the "saturation" region, the drain voltage has a very large effect on the current.

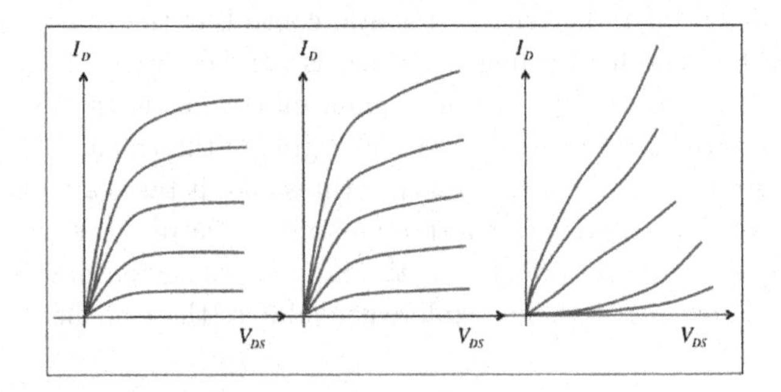

Fig. 10.10 Illustration of how 2D electrostatics affects the output characteristics of a MOSFET, Left: A long channel device with nearly infinite output resistance. Center: A short channel device with a much lower output resistance. Right: A device that suffers from punch though.

Punch through occurs when the electric field from the drain reaches all the way through to the source. To first order, this occurs when the drain depletion region reaches the source depletion region, as illustrated in the sketch on the left of Fig. 10.11. As illustrated on the right of the figure, the boundaries of the depletion region can have complicated profiles because of 2D doping profiles and 2D electrostatics. The depletion regions can meet at the surface (as on the left) or in the bulk (as on the right); the result is either *surface punchthrough* or *bulk punch through.*

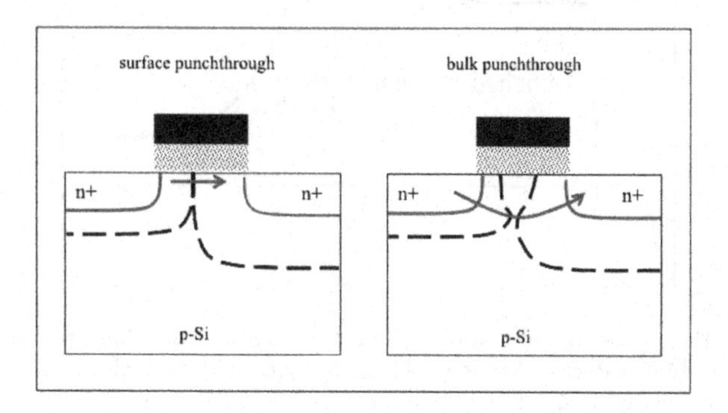

Fig. 10.11 Cross-sectional sketches illustrating the depletion region boundaries for surface punch-through (Left) and bulk punch-through (Right).

The criterion that $L > W_S + W_D$ to avoid punch-through is only a crude estimate of when punch through occurs. The energy band diagram in Fig. 10.12 gives a better explanation. Complete punch through occurs when the drain potential reaches through and doesn't just lower the barrier a bit, but completely removes it. Current can then flow from source to drain with no help from the gate voltage. From another perspective, we can define punch through as occurring when the drain control of the current is as strong as the gate control. According to eqn. (10.27) for the capacitor model, this occurs when $C_{G-VS} = C_{D-VS}$. The actual voltage at which punch through occurs can only be determined by a numerical solution to the 2D Poisson equation for the particular device structure of interest.

10.8 Discussion

In this lecture we have discussed how two-dimensional electrostatics degrades the performance of short channel transistors. How does an electrostatically well-behaved MOSFET operate? Figure 10.13 summarizes the operation of an electrostatically "well-tempered MOSFET" under high gate and drain bias.

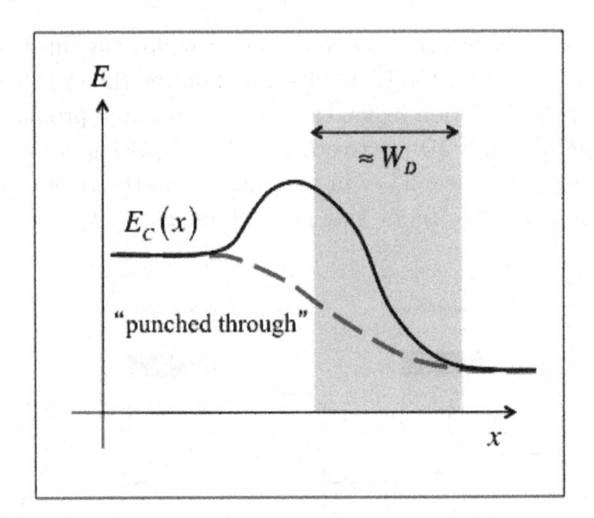

Fig. 10.12 Energy band illustration of punch through. Solid line: well-behaved device. Dashed line: a device that is punched through. The shaded region indicates the boundaries of the drain depletion region for a well-behaved device.

In a well-behaved MOSFET, there is a region near the beginning of the channel where the potential in the channel is strongly controlled by the gate voltage. In this region, dE_C/dx (the lateral electric field) is small. This region of strong gate control is necessary to shield the top of the barrier from the influence of the drain potential and keep DIBL low.

The potential at the top of the barrier controls the height of the barrier and, therefore, the drain current of the MOSFET. Ideally, this potential is only controlled by the gate voltage (i.e. as in eqn. (10.25)). In practice, the drain voltage always has some effect on the potential at the top of the barrier (as in eqn. (10.26)) — especially in short channel MOSFETs. The goal of the transistor designer is to ensure that the inversion layer charge at the top of the barrier is given by the classical, 1D result,

$$Q_n(x=0) = -C_G\left(V_{GS} - V_T\right) , \tag{10.41}$$

where $x = 0$ is the location of the top of the barrier. It is reasonable to expect that the 1D MOS electrostatics, which leads to eqn. (10.41) applies at the top of the barrier because $d^2\psi(x)/dx^2 = 0$ at this location, and the 2D Poisson equation reduces to a 1D Poisson equation. The assumption of 1D electrostatics is not exact because 2D electrostatics makes V_T a function

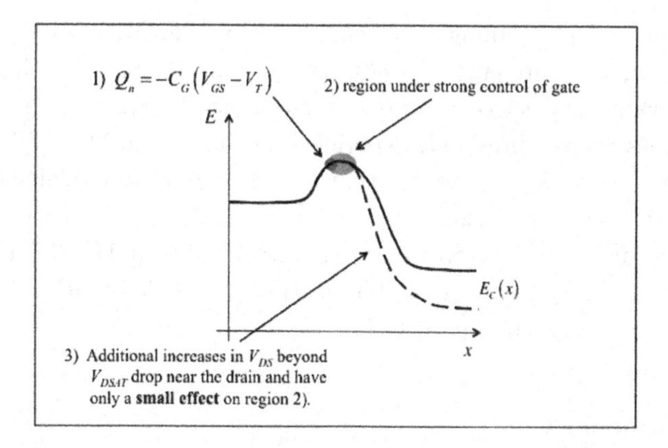

Fig. 10.13 Illustration of an electrostatically well-behaved MOSFET operating under high gate and drain voltages. As shown by the dashed line, in the saturation region, increases in drain voltage increase the potential (lower the condition band) in most of the channel — except near the beginning of the channel where the potential is mostly controlled by the gate voltage.

of drain voltage

$$V_T = V_{T0} - \delta V_{DS} \,, \tag{10.42}$$

where δ is the DIBL parameter.

The current in a MOSFET under high drain bias is due to electrons that surmount the barrier, diffuse across the short low-field region near the beginning the channel, and then enter the high field region at the drain end of the channel. The low-field region is a bottleneck that limits the drain current. This picture of how a MOSFET operates is essentially the way a bipolar transistor operates with the source playing the role of the emitter, the low-field region near the beginning the of the channel acting as the base, and the high field region near the drain operating as the collector. In fact, the analogy between the MOSFET and the bipolar transistor is very close and has long been appreciated [13].

Under low drain bias, the current is proportional to V_{DS} but for high drain bias, the current in a well-designed transistor shows much less dependence on V_{DS}. In a long channel device, the current actually saturates. This behavior occurs because the strong gate control shields the potential near the beginning of the channel from the influence of the drain potential. Increases in V_{DS} beyond the saturation voltage, V_{DSAT} mainly increase the potential (and electric field) near the drain end of the channel. In a well-designed transistor the drain voltage has only a small effect on the potential near the beginning of the channel — this is DIBL above threshold and results in the finite output conductance of the MOSFET. While there is no reason that 2D electrostatics below threshold should be the same as 2D electrostatics above threshold, experience in fitting the MIT VS model to well-behaved transistors suggests that it is [11] and numerical simulations support this conclusion [12].

The picture of the electrostatics of a well-designed MOSFET outlined above will be used in the next few lectures to understand the essential physics of the nanoscale MOSFET.

10.9 Summary

Two-dimensional electrostatics degrade the performance of transistors and produce: 1) a subthreshold swing that is greater than the fundamental limit of 60 mV/decade, 2) a shift of the transfer characteristic, $\log_{10} I_{DS}$ vs. V_{GS} to the left for increasing drain voltage (i.e. DIBL), 3) a threshold voltage that is a function of gate length and drain voltage, and 4) a low output resistance. When 2D electrostatics are strong, the gate can lose control of the drain current and the device is said to be punched through. Because these effects can be strong in short channel device, they are referred to as *short channel effects*. As transistors are scaled to smaller and smaller dimensions, the main challenge to the transistor designer is to control short channel effects. To properly treat them, numerical simulations are necessary. In this lecture, we discussed the essential physical ideas that can be used to interpret detailed simulations and experimental measurements.

10.10 References

The gradual channel approximation applies to long channel MOSFETs for which the effects of 2D electrostatics are minimal. In MOSFET modeling, effects arising from 2D electrostatics are then added to the long channel model. This approach is discussed in several textbooks.

[1] Robert F. Pierret, *Semiconductor Device Fundamentals, 2nd Ed.*, Addison-Wesley Publishing Co, 1996.

[2] Ben Streetman and Sanjay Banerjee, *Solid State Electronic Devices, 6th Ed.*, Prentice Hall, 2005.

[3] Y. Tsividis and C. McAndrew, *Operation and Modeling of the MOS Transistor*, 3rd Ed., Oxford Univ. Press, New York, 2011.

[4] Y. Taur and T. Ning, *Fundamentals of Modern VLSI Devices*, 2nd Ed., Oxford Univ. Press, New York, 2013.

The computation of the geometric screening length for various MOSFET geometries is discussed in the following papers. The third paper, discusses the capacitor model for 2D electrostatics.

[5] D. J. Frank, Y. Taur, and H.-S. P. Wong, "Generalized scale length for two-dimensional effects in MOSFETs," *IEEE Electron Device Lett.*, **19**, pp. 385-387, Oct. 1998.

[6] Jing Wang, Paul Solomon and Mark Lundstrom, "A General approach for the performance assessment of nanoscale silicon field effect transistors," *IEEE Transactions on Electron Dev.*, **51**, pp. 1361-1365, 2004.

[7] Risho Koh, Haruo, and Hiroshi Matsumoto "Capacitance network model of the short channel effect for a 0.1 μm fully depleted SOI MOSFET," *Jpn. J. Appl. Phys.* **35**, pp. 996-1000, 1996.

The classic paper on constant electric field scaling is by Robert Dennard and colleagues. The paper by Ieong et al. discusses the challenges of scaling MOSFETs to their ultimate limits.

[8] Robert H. Dennard, F.H. Gaensslen, H.-N. Yu, V. L. Rideout, E. Bassous, and A.R. LeBlanc, "Design of Ion-Implanted MOSFETS with Very Small Physical Dimensions," *IEEE J. Solid-State Circuits*, **51**, pp. 256-264, 1974.

For a discusion of the FinFET and other transistor structures suitable for scaling to very short channel lengths, see:

[9] Xuejue Huang, Wen-Chin Lee, Charles Kuo, Digh Hisamoto, Leland Chang, Jakub Kedzierski, Erik Anderson, Hideki Takeuchi, Yang-Kyu Choi, Kazuya Asano, Vivek Subramanian, Tsu-Jae King, Jeffrey Bokor and Chenming Hu, "Sub 50-nm FinFET: PMOS," Technical Digest, International Electron Devices Meeting, pp. 67-70, 1999.

[10] Meikei Ieong, Bruce Doris, Jakub Kedzierski, Ken Rim and Min Yang, "Silicon Device Scaling to the Sub-10-nm Regime," *Science* **306**, pp. 2057-2060, 2004.

The MIT Virtual Source Model, which provides a framework for these lectures, is described in:

[11] A. Khakifirooz, O.M. Nayfeh, and D.A. Antoniadis, "A Simple Semi-empirical Short-Channel MOSFET Current Voltage Model Continuous Across All Regions of Operation and Employing Only Physical Parameters," *IEEE Trans. Electron. Dev.*, **56**, pp. 1674-1680, 2009.

This paper presents a good review and critique of ways to model 2D electrostatics.

[12] Qian Xie, Jun Xu, and Yuan Taur, "Review and Critique of Analytic Models of MOSFET Short-Channel Effects in Subthreshold," *IEEE Transactions on Electron Dev.*, **9**, pp. 1569- 1579, 2012.

Johnson describes the close relation of bipolar and field-effect trnsistors.

[13] E.O. Johnson, "The IGFET: A Bipolar Transistor in Disguise," *RCA Review*, **34**, pp. 80-94, 1973.

Lecture 11

The VS Model Revisited

11.1 Introduction

In Lecture 5, we introduced a simple, *top-of-the-barrier* or *Virtual Source* model to serve as a framework for our discussions. The model of Lecture 5 was a simplified version of the MIT Virtual Source model for nanotransistors [1]. It was derived using very simple, traditional arguments (in contrast to the MIT VS model, which was specifically developed to describe the physics of nanoscale transistors). As we proceed in these lectures, we'll refine the model ending up with the MIT VS model and a clear understanding of its physical underpinnings.

The MOSFET drain current is given by

$$I_{DS} = W|Q_n(x=0, V_{GS}, V_{DS})| \left\langle \upsilon_x(x=0, V_{GS}, V_{DS}) \right\rangle , \tag{11.1}$$

where $x = 0$ is the location of the virtual source, taken to be the top of the barrier. Current is continuous, so we choose to evaluate it where it is easiest to do so. At the top of the barrier in a well-designed MOSFET, $Q_n(V_{GS}, V_{DS}) \approx Q_n(V_{GS})$ as given by 1D MOS electrostatics. Only small corrections to account for DIBL are needed. After the discussion of the last few lectures, we have good understanding of $Q_n(x = 0, V_{GS}, V_{DS})$. In this lecture, we'll extend the VS model developed in Lecture 5 by including a better description of MOS electrostatics. In subsequent lectures, we will develop improved models for $\langle \upsilon_x(x = 0, V_{GS}, V_{DS}) \rangle$ in nanoscale MOSFETs.

11.2 VS model review

We developed the VS model in Lecture 5 from separate expressions for the strong inversion linear and saturation region currents,

$$\begin{aligned} I_{DLIN} &= \frac{W}{L}\mu_n \left|Q_n(V_{GS})\right| V_{DS} \\ I_{DSAT} &= W\upsilon_{sat} \left|Q_n(V_{GS})\right| , \end{aligned} \tag{11.2}$$

which are shown as the dashed lines in Fig. 11.1 (same as Fig. 5.1). The actual characteristic (solid line in Fig. 11.1) is produced by smoothly connecting the linear and saturation region currents by defining the drain voltage dependent average velocity to be

$$\begin{aligned} \langle \upsilon_x(V_{DS}) \rangle &= F_{SAT}(V_{DS})\upsilon_{sat} \\ F_{SAT}(V_{DS}) &= \frac{V_{DS}/V_{DSAT}}{\left[1 + (V_{DS}/V_{DSAT})^{\beta}\right]^{1/\beta}} , \end{aligned} \tag{11.3}$$

where F_{SAT} is the drain current saturation function, and

$$V_{DSAT} = \upsilon_{sat}L/\mu_n \, . \tag{11.4}$$

By including DIBL, the charge increases with drain voltage producing the finite output conductance.

In Lecture 5, we described the charge at the top of the barrier as

$$\begin{aligned} &Q_n(V_{GS}) = 0 && V_{GS} \leq V_T \\ &Q_n(V_{GS}) = -C_{ox}\left(V_{GS} - V_T\right) && V_{GS} > V_T \\ &V_T = V_{T0} - \delta V_{DS} \, . \end{aligned} \tag{11.5}$$

We understand now that C_{ox} should be replaced by $C_G(inv)$, which is the series combination of C_{ox} and the semiconductor capacitance, C_s, in inversion and that $C_G < C_{ox}$. We also understand how to describe $Q_n(V_{GS}, V_{DS})$ below threshold, so we can extend our VS model to include both subthreshold and above threshold conduction.

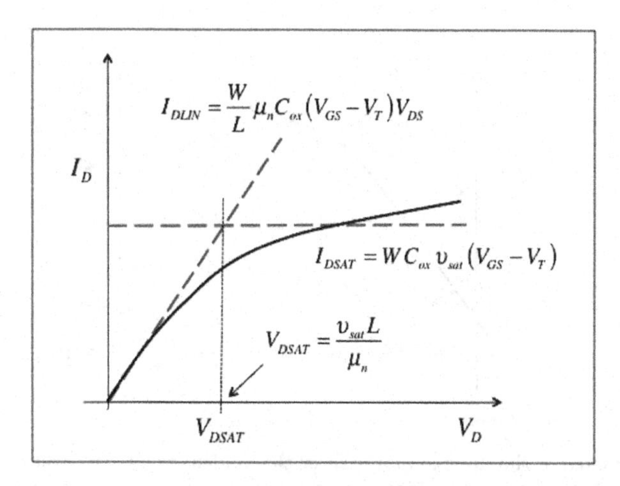

Fig. 11.1 Sketch of a common source output characteristic of an n-channel MOSFET at a fixed gate voltage (solid line). The dashed lines are the linear and saturation region currents as given by eqns. (11.2). (Same as Fig. 5.1.)

11.3 Subthreshold

For gate voltages below the threshold voltage, a MOSFET is said to be operating in the *subthreshold region.* Figure 11.2 is a sketch of $|Q_n(V_{GS})|$ vs. V_{GS} on both logarithmic and linear axes. Below threshold, we showed in Lecture 8 for a bulk MOSFET that the electron charge in C/cm^2 was given by eqn. (8.12) as

$$Q_n(V_{GS}) = -(m-1)C_{ox}\left(\frac{k_BT}{q}\right)e^{q(V_{GS}-V_T)/mk_BT} . \tag{11.6}$$

For an extremely thin SOI device, the corresponding result, eqn. (9.42), was similar but with $m = 1$. The key point is that below threshold, the electron charge varies as $\exp[q(V_{GS} - V_T)/mk_BT]$.

From eqns. (11.1) and (11.6), an equation for the subthreshold current of a bulk MOSFET is easy to obtain. The result is

$$I_{DS} = W(m-1)C_{ox}\left(\frac{k_BT}{q}\right)e^{q(V_{GS}-V_T)/mk_BT}\left\langle \upsilon_x(x=0)\right\rangle . \tag{11.7}$$

Recall that

$$m = 1 + \frac{C_\Sigma}{C_{ox}} , \tag{11.8}$$

where C_Σ is the total capacitance connected to the virtual source. In a bulk MOSFET, it is the sum of the gate capacitance and the capacitance to the bulk and to the source and drain.

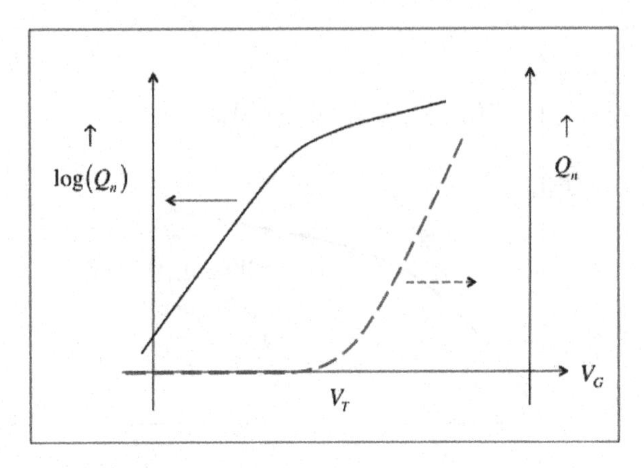

Fig. 11.2 Sketch of the inversion layer sheet charge density vs. gate voltage on a logarithmic axis (Left axis) and on linear axis (Right axis).

From the subthreshold drain current, we readily obtain the subthreshold swing as

$$SS = \left[\frac{\partial(\log_{10} I_D)}{\partial V_{GS}}\right]^{-1} = 2.3\, m k_B T/q \quad \text{V/decade}. \tag{11.9}$$

The units of subthreshold swing are volts per decade — the number of volts that the gate voltage must be increased to increase the drain current by a factor of 10. Subthreshold swings are usually quoted in millivolts per decade with less than 100 mV/decade being considered a good subthreshold swing.

Figure 11.3 shows why the subthreshold swing is such an important device metric. Applications usually require a low off-current, so that the circuit does not consume excessive standby power. For a specified off-current, the SS parameter determines how large a voltage must be applied to achieve a desired on-current. High on-currents allow fast operation of circuits because the capacitors in the circuit can be charged and discharged quickly. For a transistor with a lower SS, the required on-current can be achieved at a lower voltage. The circuit will operate at the same speed, but since power is proportional to V_{DD}^2, the circuit will dissipate much less power. With the billions of transistors now being placed on an integrated circuit, power has become a critical issue — both *active power* (while the circuit is operating), which is proportional to V_{DD}^2 and *standby power* (which is determined by the off-current)

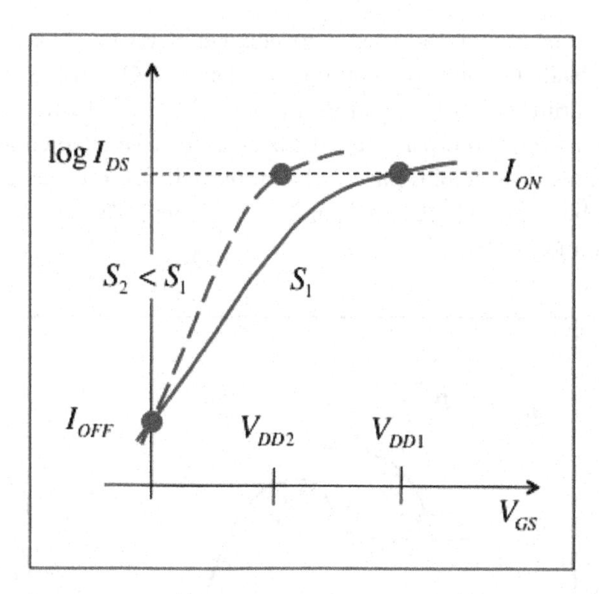

Fig. 11.3 Illustration of how the subthreshold swing determines the power supply voltage of a circuit. The on-current results when the maximum voltage, the power supply voltage, V_{DD}, is applied to the gate.

According to eqn. (11.8), the lowest the subthreshold swing can be is 60 mV/decade at room temperature. Fully depleted structures such as the Extremely Thin SOI MOSFET have $m = 1$ and are beneficial for achieving the lowest possible SS. Above threshold, the current varies approximately linearly with gate voltage, so the 60 mV/decade lower limit to SS places a lower limit to the power supply voltage than can be used. In practice, this lower limit is about 1 volt. Recall that the constant field (Dennard) scaling prescription requires that V_{DD} be scaled down each technology generation. Because there is a lower limit to SS, the power supply can no longer be scaled down, and the result is that the power dissipation of integrated circuits has become a critical issue [2].

Equation (11.8) clearly shows that SS has a lower limit (assuming that $m \geq 1$), but where does this physical limit come from? Figure 11.4 explains where. The drain current consists of electrons that are emitted from the source over the top of the barrier where they can then flow to the drain. The probability of this *thermionic emission* process is exponential with the barrier height, so the probability than an electron from the source can be emitted to the top of the barrier is

$$P_{S \to D} = e^{-E_B/k_B T}, \tag{11.10}$$

where E_B is height of the source to channel barrier. This exponential probability leads to the exponential dependence of Q_n on V_{GS} and fundamentally limits the subthreshold swing of a MOSFET to no less than 60 mV/decade at room temperature. To beat this thermionic emission limit, different physical principles have to be employed. For one example of how this might be achieved, see the papers by Appenzeller *et al.* [3] and by Salahuddin and Datta [4].

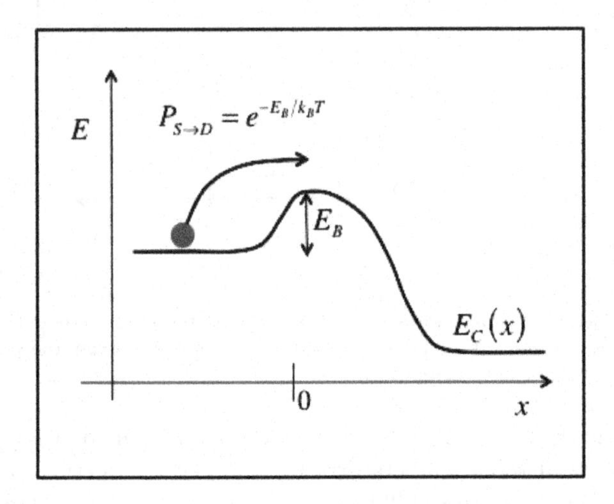

Fig. 11.4 Thermionic emission process of current flow in a MOSFET.

Finally, we should discuss the average channel velocity in subthreshold, $\langle \upsilon_x(x=0)\rangle$; it is not given by eqns. (11.3), which applies above threshold. In subthreshold, carriers diffuse across the channel; the velocity at the virtual source is

$$\langle \upsilon_x(x=0)\rangle = \frac{D_n}{L}\left(\frac{n_S(0)-n_S(L)}{n_S(0)}\right) . \tag{11.11}$$

A simple thermionic emission model gives $n_S(L)/n_S(0) = e^{-qV_{DS}/k_BT}$, so we conclude that

$$\langle \upsilon_x(x=0)\rangle = \frac{D_n}{L}\left(1-e^{-qV_{DS}/k_BT}\right) = \frac{k_BT}{q}\frac{\mu_n}{L}\left(1-e^{-qV_{DS}/k_BT}\right) . \tag{11.12}$$

Equation (11.12), not (11.3) is the correct expression for $\langle \upsilon(0) \rangle$ below threshold. Using this result in eqn. (11.7), we arrive at

$$\boxed{I_{DS} = \mu_n C_{ox} \frac{W}{L} (m-1) \left(\frac{k_B T}{q} \right)^2 e^{q(V_{GS}-V_T)/mk_BT} \left(1 - e^{-qV_{DS}/k_BT} \right)}, \tag{11.13}$$

which is the standard expression for the subthreshold current [5].

One final point should be mentioned. From the energy band diagrams in Fig. 11.4, it is not clear that electrons need to diffuse across the entire channel. It seems that they only need to diffuse across the low-field portion of the channel, and then the electric field can quickly sweep them across the rest of the channel. Accordingly, we expect that L in eqn. (11.13) should be replaced by ℓ, where $\ell < L$. While this is true, it is difficult in practice to clearly determine the pre-exponential factor, so eqn. (11.13) generally provides a satisfactory description of real devices [5].

11.4 Subthreshold to above threshold

Equation (11.6) gives $Q_n(V_{GS})$ below threshold, and in strong inversion, $Q_n(V_{GS}) = -C_G(inv)(V_{GS} - V_T)$, but the transition from subthreshold to strong inversion is a gradual one that we would like to treat. This is especially important for circuit simulation where the system of nonlinear equations is solved by Newton-Raphson iteration, which requires functions with smooth derivatives. A numerical treatment is possible by solving the Poisson-Boltzmann equation for $Q_n(\psi_S)$. This is the basis for so-called *surface potential models* [6].

It is also possible to describe $Q_n(V_{GS})$ empirically. One expression has been developed by Wright [7]:

$$Q_n(V_{GS}) = -m\, C_G(inv) \left(\frac{k_B T}{q} \right) \ln \left(1 + e^{q(V_{GS}-V_T)/mk_BT} \right) . \tag{11.14}$$

For $V_{GS} \ll V_T$, we can use the expansion $\ln(x) \approx 1 + x$ to write eqn. (11.14) as

$$Q_n(V_{GS}) = -m\, C_G(inv) \left(\frac{k_B T}{q} \right) e^{q(V_{GS}-V_T)/mk_BT} . \tag{11.15}$$

Comparing eqn. (11.15) to (11.6), we see that the two are close. In practice, this difference in pre-exponential factors is not critical, so the empirical expression is not too bad.

For $V_{GS} \gg V_T$, the exponential in the argument of the logarithm dominates, and eqn. (11.14) becomes

$$Q_n(V_{GS}) = -C_G(inv)\,(V_{GS} - V_T)\ , \tag{11.16}$$

which is the correct result. We conclude that an empirical expression like eqn. (11.14) can do a good job of describing $Q_n(V_{GS})$ from subthreshold to strong inversion. The MIT VS model uses a slightly extended version of eqn. (11.14) which does a better job of matching the transition from weak to strong inversion [1].

Finally, we should mention the close connection between the off-current and the on-current. We have seen that the off-current goes as $\exp[(V_{GS} - V_T)/mk_BT]$, and we know that the on-current goes as $(V_{GS} - V_T)$. The result is that

$$\ln(I_{OFF}) \propto I_{ON}\ . \tag{11.17}$$

The device designer might decrease V_T to increase the on-current, which improves circuit speed, but the result is an exponential increase in the off-current, which increases the standby power. Figure 11.5 is an example of how a technology can be characterized by a plot of $\log(I_{OFF})$ vs. I_{ON}. This is a fundamental trade-off that comes from the physics of the MOSFET.

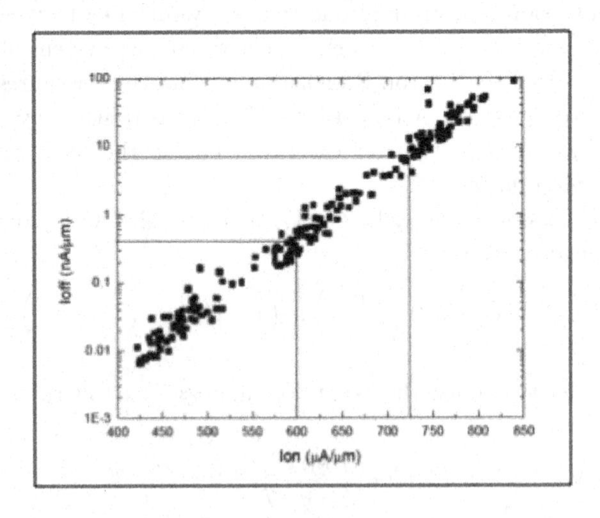

Fig. 11.5 Plot of $\log_{10}(I_{OFF})$ vs. I_{ON} for a 65 nm NMOS technology. (©IEEE 2005. Reprinted, with permission, from: A. Steegen, *et al.*, "65nm CMOS Technology for low power applications," *Intern. Electron Dev. Meeting*, Dec. 2005.)

11.5 Discussion

Equation (11.4) gives V_{DSAT} in strong inversion, but from eqn. (11.13), we see that V_{DSAT} is a few k_BT/q in the subthreshold region. The MIT VS model treats this difference in drain saturation voltages empirically. An empirical function is used to vary V_{DSAT} from its strong inversion value given by eqn. (11.4) to k_BT/q in the subthreshold region. Although this procedure is only heuristic, the typical error is less than 10% [1].

As was shown in Fig. 5.4, the VS model does an excellent job of fitting the IV characteristics of nanoscale transistors. This is surprising because the parameters, μ_n and υ_{sat} have clear physical meaning when the channel is many mean-free-paths long, but this is not the case in nanoscale transistors. Still, excellent fits result if we view these parameters as empirical parameters that can be adjusted to fit experimental data. We shall see, however, that there is more to it. As we develop our understanding of carrier transport at the nanoscale in subsequent lectures, we will see that a clear, physical meaning can be given to these parameters.

11.6 Summary

The last several lectures have discussed MOS electrostatics. One-dimensional MOS electrostatics bend the bands, lowers the energy barrier, and allows current to flow from the source to the drain. Two-dimensional MOS electrostatics degrade the performance of field-effect transistors by increasing the subthreshold swing and producing DIBL, which increases the output conductance and also reduces the threshold voltage of short channel devices. In general, numerical solutions are required to treat 2D MOS electrostatics, but the effects are readily understood in a qualitative sense.

Returning to eqn. (11.1), we now have a good understanding of $Q_n(x = 0, V_{GS}, V_{DS})$. Beginning in the next lecture, we'll develop a similar, physical understanding of $\langle \upsilon_x(x = 0, V_{GS}, V_{DS}) \rangle$.

11.7 References

The MIT Virtual Source Model, which provides a framework for these lectures, is described in:

[1] A. Khakifirooz, O.M. Nayfeh, and D.A. Antoniadis, "A Simple Semiempirical Short-Channel MOSFET Current Voltage Model Continuous

Across All Regions of Operation and Employing Only Physical Parameters," *IEEE Trans. Electron. Dev.*, **56**, pp. 1674-1680, 2009.

The design of MOS integrated circuits is power constrained. For a discussion of the issues involved see:

[2] D.J. Frank, R.H. Dennard, E. Nowak, P.M. Solomon, Y. Taur, and H.S.P. Wong, "Device scaling limits of Si MOSFETs and their application dependencies," *Proc. IEEE*, **89**, pp. 259-288, 2001.

Because the $SS \geq 60\ mV/decade$*, the power supply of a MOSFET cannot be much less than 1 volt. The result is a rather large power dissipation. To reduce the power supply well below 1 V, transistors that operate on physical principles that are different from MOSFETs must be developed. For two examples of how this might be accomplished, see the following two papers.*

[3] J. Appenzeller, Y.-M. Lin, J. Knoch, and Ph. Avouris, "Band-to-Band Tunneling in Carbon Nanotube Field-Effect Transistors," *Phys. Rev. Lett.*, **93**, pp. 196805-1-4, 2004.

[4] S. Salahuddin and S. Datta, "Use of Negative Capacitance to Provide Voltage Amplification for Low Power Nanoscale Devices," *Nano Lett.*, **8**, pp. 405-410, 2008.

For the conventional treatment of the subthreshold current, see Chapter 3 of:

[5] Y. Taur and T. Ning, *Fundamentals of Modern VLSI Devices*, 2nd Ed., Oxford Univ. Press, New York, 2013.

Surface potential models numerically describe $Q_n(V_{GS})$ *from subthreshold to above threshold. For an example of such a model, see:*

[6] G. Gildenblat, X. Li, W. Wu, H. Wang, A. Jha, R. van Langevelde, G. D. J. Smit, A. J. Scholten, and D. B. M. Klaassen, "PSP: An advanced surface-potential-based MOSFET model for circuit simulation," *IEEE Trans. Electron Devices*, **53**, pp. 1979-1993, 2006.

An empirical model that describes $Q_n(V_{GS})$ *from subthresholf to strong inversion has been presented by Wright.*

[7] G.T. Wright "Threshold modelling of MOSFETs for CAD of CMOS VLSI," *Electron. Lett.* **21**, pp. 221-222, 1985.

PART 3

The Ballistic MOSFET

Lecture 12

The Landauer Approach to Transport

12.1 Introduction

The drain current of a MOSFET is proportional to the product of charge and velocity. We have discussed the charge (MOS electrostatics); now it is time to discuss the average velocity. To compute the average velocity, we must understand carrier transport. The analysis of semiconductor devices traditionally begins with the drift-diffusion equation [1],

$$J_{nx} = n_S q \mu_n \mathcal{E}_x + q D_n \frac{dn_S}{dx} \quad \text{A/m}, \tag{12.1}$$

where J_{nx} is the 2D electron current in a thin sheet, n_S is the sheet electron density per m^2, μ_n is the electron mobility, $\mathcal{E}_x$, the electric field in the x-direction, and D_n the diffusion coefficient. Though suitable for long channel devices, eqn. (12.1) is not the best starting point for analyzing nanoscale devices.

In these lectures, we'll make use of the *Landauer approach*, in which the current is given by

$$\boxed{I = \frac{2q}{h}\int_{-\infty}^{+\infty} \mathcal{T}(E)M(E)(f_1(E) - f_2(E))dE \quad \text{Amperes}\,,} \tag{12.2}$$

where $\mathcal{T}(E)$ is the *transmission* at energy, E, $M(E)$, the number of *modes* (or channels) at energy, E, and $f_{1,2}(E)$ is the Fermi function of contact 1 or 2. Our goal in this lecture is to develop some familiarity with the Landauer approach before we apply it to transistors.

12.2 Qualitative description

Reference [2] discusses the derivation of eqn. (12.2) and its underlying physics. Reference [3] discusses applications. Our goal in this section is to convince ourselves that eqn. (12.2) makes sense.

Figure 12.1 is a cartoon illustration of a nanodevice. We assume that the two contacts are large and that inelastic electron-phonon scattering is strong so that electrons in the contacts are in thermodynamic equilibrium. In equilibrium, the probability that an electron state at energy, E, is occupied is given by the Fermi function,

$$f_{1,2}(E) = \frac{1}{1 + e^{(E - E_{F1,2})/k_B T}}\,, \tag{12.3}$$

where $E_{F1,2}$ is the *Fermi function* (also called the *electrochemical potential*) of contact 1 or 2.

If the voltages and temperatures of the two contacts are identical, then $f_1(E) = f_2(E)$, and according to (12.2), the current is zero. This makes sense because in this case, the probability that a state at energy, E, in the device is filled by an electron from contact 1 is the same as the probability that the state at that energy is filled by an electron from contact 2. Because the states in the device have the same probability of being filled by electrons from either contact, there is no flow of electrons from one contact to the other.

Next, consider the case for which $f_1(E) \neq f_2(E)$ and current flows. According to eqn. (12.3), this situation can occur in two ways. First, the temperatures of the two contacts could be different, which would give rise to *thermoelectric* effects [2, 3] that are not of interest to us in these lectures. The second possibility is that the voltages of the two contacts are different.

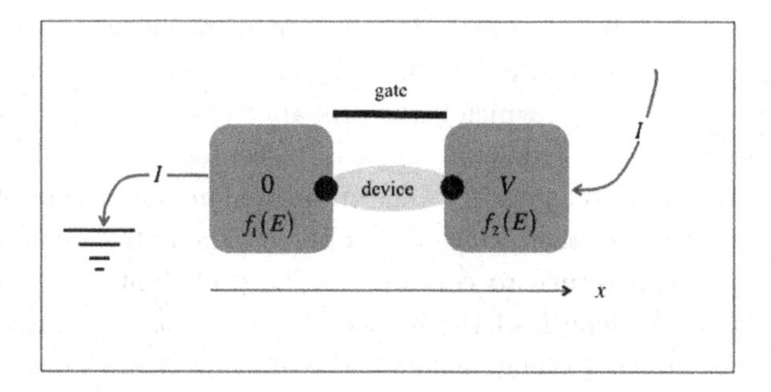

Fig. 12.1 Sketch of a generic nanodevice with two large contacts in thermodynamic equilibrium. If either the voltages or temperatures of the two contacts are different, then $f_1 \neq f_2$ over some energy range, and current can flow.

Assume that contact 1 is grounded and that a voltage, V, is applied to contact 2. Recall that applying a positive voltage to a contact lowers its Fermi level (electrochemical potential). In this case,

$$E_{F2} = E_{F1} - qV\,. \tag{12.4}$$

We will assume that even under bias, the probability that a state in the contacts is occupied is given by the equilibrium Fermi function, eqn. (12.3), but the two contacts have different Fermi levels. Strictly speaking, this cannot be true because when current flows the system is out of equilibrium, but the assumption is that the contacts are so large and heavily doped so that only a very small perturbation from equilibrium occurs.

When there is a voltage difference between the two contacts, then $f_1 \neq f_2$ over an energy range that is called the *Fermi window*. Figure 12.2 illustrates the concept of the Fermi window. On the left is the case for $T = 0$ K. In this case, $f_1(E) \neq f_2(E)$ over the energy range, qV below E_{F1}. On the right we show the case for $T > 0$ K. In this case, we also find $f_1(E) \neq f_2(E)$ over the range of energies that is mostly below E_{F1}. According to eqn. (12.2), only electrons in the Fermi window where $f_1(E) \neq f_2(E)$ contribute to the current.

Differences in the equilibrium Fermi functions of the two contacts cause current to flow, but eqn. (12.2) shows that the magnitude of the resulting current at energy, E, is proportional to the product, $\mathcal{T}(E)M(E)$. The quantity $M(E)$ is the number of channels (or modes) at energy, E. The number of channels is analogous to the number of lanes in a highway [2]. The more lanes (channels) the more traffic (current) can flow — provided that the channels lie inside the Fermi window. We expect $M(E)$ to depend on the density-of-states at energy, E and also on the velocity at energy, E, because for current to flow, the states must have a velocity. The quantity, $\mathcal{T}(E)$, is the transmission, which is the probability that electrons that enter a channel from contact 1 flow all the way to contact 2 without backscattering and returning to contact 1. The transmission is less than one in the presence of carrier *backscattering*. If an electron enters from contact 1 and backscatters, it can return to contact 1. The probability of backscattering depends on the length of the device, L, and on the average distance between backscattering events, which is the *mean-free-path for backscattering*, λ. When $L \ll \lambda$, $\mathcal{T} \to 1$, and when $L \gg \lambda$, $\mathcal{T} \to 0$. We are assuming in eqn. (12.2) that the probability that an electron transmits from contact 1 to contact 2 is equal to the probability that an electron at the same energy transmits from contact 2 to contact 1. It can be shown that this occurs when electron scattering is elastic so that electrons flow in parallel, non-interacting energy channels [2, 3]. In the Landauer Approach, we assume that scattering in the device is elastic, but strong inelastic scattering takes place in the two contacts.

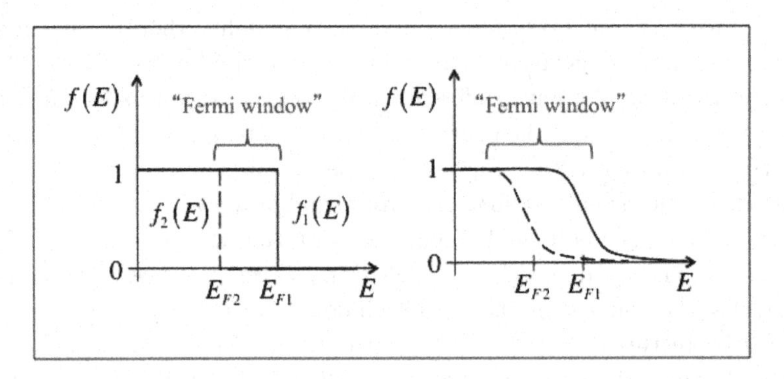

Fig. 12.2 Illustration of the Fermi window. Left: At $T = 0$ K and Right: at $T > 0$ K. A large bias on contact 2 is assumed; see Fig. 12.3 for the case of a small bias.

In summary, eqn. (12.2) is a simple description of carrier transport that works from the *ballistic limit* where there is no scattering and $\mathcal{T}(E) = 1$ to the *diffusive limit* where there is a lot of scattering and $\mathcal{T}(E) \ll 1$. The current at energy, E, is proportional to $\mathcal{T}(E)M(E)\big(f_1(E) - f_2(E)\big)$. To get the total current, we just add the contributions from each of the energy channels, which are assumed to be independent (i.e. there is no inelastic scattering to couple channels).

12.3 Large and small bias limits

The quantity, $(f_1(E) - f_2(E))$, plays an important role. In this section, we examine this quantity for large and small applied bias. When the voltage applied to contact 2 is large, then $f_1(E) \gg f_2(E)$ for all energies of interest, and eqn. (12.2) reduces to

$$I = \frac{2q}{h} \int \mathcal{T}(E) M(E) f_1(E) dE \quad \text{Amperes}. \tag{12.5}$$

We will use this equation to compute the saturation region current of a MOSFET.

When a small voltage is applied to contact 2, eqn. (12.2) also simplifies. To see how, consider Fig. 12.3, which shows the Fermi window for small bias. At $T = 0$ K, the Fermi window looks like a δ-function at $E = E_F$ (left side of Fig. 12.3). For $T > 0$ K, $(f_1(E) - f_2(E))$ is sharply peaked near the Fermi level (right side of Fig. 12.3).

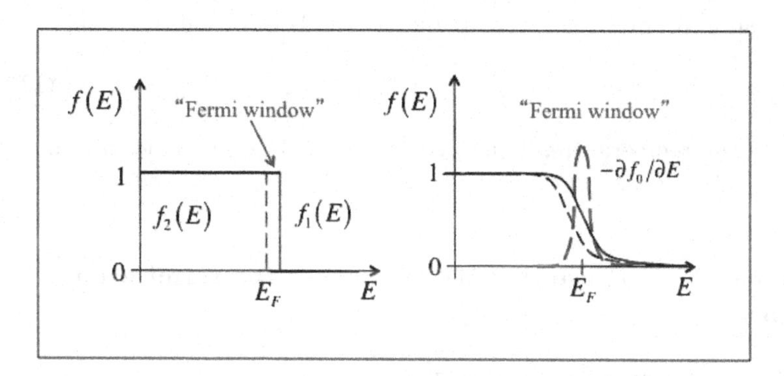

Fig. 12.3 Illustration of the Fermi window. Left: At $T = 0$ K. Right: At $T > 0$ K. In this example, a small bias on contact 2 is assumed. See Fig. 12.2 to compare with the case of a large bias.

For the small bias (*near-equilibrium*) case, we evaluate $f_2(E)$ by Taylor series expanding $f_1(E)$ as

$$f_2(E) \approx f_1(E) + \frac{\partial f_1}{\partial E_F} \delta E_F. \tag{12.6}$$

The only difference between f_1 and f_2 is a small difference of δE_F in their Fermi levels. Using eqn. (12.3), we find

$$f_1(E) - f_2(E) = -\left(\frac{\partial f_1}{\partial E_F}\right) \delta E_F = -\left(-\frac{\partial f_1}{\partial E}\right) \delta E_F. \tag{12.7}$$

(From the form of the Fermi function, eqn. (12.3), we see that $\partial f_1/\partial E_F = -\partial f_1/\partial E$.) By recalling that $\delta E_F = -qV$, we can write eqn. (12.7) as

$$f_1(E) - f_2(E) = q\left(-\frac{\partial f_0}{\partial E}\right)V\,, \tag{12.8}$$

where we have replaced f_1 with f_0, the equilibrium Fermi function because near equilibrium, $f_1(E) \approx f_2(E) \approx f_0(E)$. Using eqn. (12.8) in (12.2), we find the near-equilibrium current as

$$\boxed{\begin{aligned} &I = GV \quad \text{Amperes} \\ &G = \frac{2q^2}{h}\int \mathcal{T}(E)M(E)\left(-\frac{\partial f_0}{\partial E}\right)dE \quad \text{Siemens}\,. \end{aligned}} \tag{12.9}$$

We will use this equation to compute the linear region current of a MOSFET. Finally, note that the Fermi window,

$$W(E) \equiv \left(-\frac{\partial f_0}{\partial E}\right)\,, \tag{12.10}$$

plays an important role. The area under the Fermi window is one,

$$\int_{-\infty}^{+\infty} W(E)dE = 1\,, \tag{12.11}$$

and as the temperature, T, approaches zero, $W(E)$ becomes a δ-function at $E = E_F$.

Exercise 12.1: Prove that the area under the Fermi window is one.

Using eqn. (12.10) in (12.11), we find

$$\begin{aligned} \int_{-\infty}^{+\infty} W(E)dE &= \int_{-\infty}^{+\infty}\left(-\frac{\partial f_0}{\partial E}\right)dE \\ &= -\int_{-\infty}^{+\infty} df_0 = f_0(-\infty) - f_0(+\infty) = 1\,, \end{aligned}$$

where we have used eqn. (12.3) to evaluate $f_0(-\infty)$ and $f_0(+\infty)$.

For low temperatures, $W(E)$ is sharply peaked near the Fermi level. Since the area under the window function is one, we can treat the window function as a δ-function, $W(E) \approx \delta(E_F)$. For metals, the Fermi level lies in the middle of the conduction band, and the width of the window function (a few k_BT) is small compared to the energy range of interest — even at room temperature, so for metals, $W(E)$ can be treated as a δ-function.

12.4 Transmission

Consider the problem illustrated in Fig. 12.4 in which a steady-state flux of carriers, $F^+(x = 0)$, is injected into the left of a uniform region with no electric field, and a flux, $F^+(x = L)$ emerges from the right. No flux is injected from the right. We could resolve the injected and emerging fluxes into energy channels, but in this case, we'll just assume that an integration over energy channels has been done and that $F^+(x = 0)$ represents a thermal equilibrium distribution of fluxes at different energies.

We define the transmission to be the ratio of the emerging flux at $x = L$ to the incident flux at $x = 0$,

$$\mathcal{T} \equiv \frac{F^+(x = L)}{F^+(x = 0)}, \tag{12.12}$$

which is a number between zero and one. Some part of the injected flux transmits across the slab and some part backscatters and emerges from the left as $F^-(0)$. Assuming that there is no recombination or generation in the slab, then $F^+(0) = F^-(0) + F^+(L)$, from which we can show that $F^-(0) = (1 - \mathcal{T})F^+(0)$.

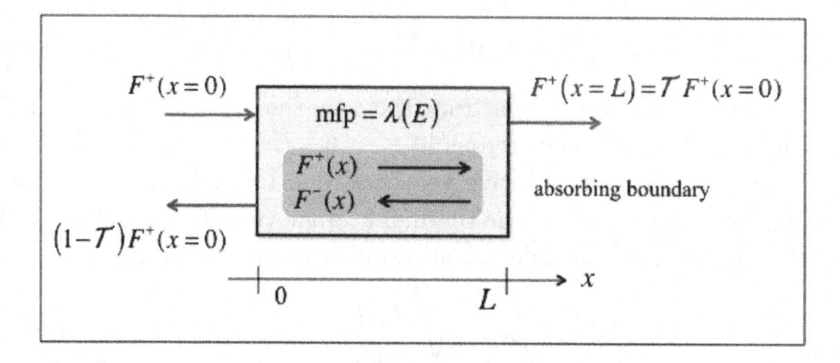

Fig. 12.4 Sketch of a model problem in which a flux of carriers is injected into a slab of length, L, at $x = 0$ and a fraction, $\mathcal{T}$, emerges from the left at $x = L$. It is assumed that there is no recombination or generation within the slab and that no flux is injected from the right.

The physical problem that Fig. 12.4 illustrates might be the diffusion of electrons across the base of a bipolar transistor. A flux of electrons is injected from the emitter into the beginning of the base at $x = 0$ and emerges at $x = L$ where it is collected by a reverse biased collector. If the collector voltage is large, then the collector acts as an *absorbing contact*; all electrons incident upon it are collected, and no electrons are injected back into the base.

Consider first the case of a thin base for which $L \ll \lambda$. Because the base is thin compared to the mean-free-path, almost all of the injected flux emerges at the right, and there is no backscattered flux, so $F^+(L) = F^+(0)$ and $F^-(0) = 0$. In this *ballistic limit*, the transmission is one,

$$\mathcal{T}_{ball} = 1\,. \tag{12.13}$$

Next, consider the *diffusive limit* for which $L \gg \lambda$. The region is many mean-free-paths long, so we expect the transmission to be small. This situation is the case for conventional, micrometer scale semiconductor devices.

To compute the transmission in the diffusive limit, note that the injected flux produces an electron concentration at $x = 0$ of $n(x = 0)$. If the slab is thick, then $n(x = L) \approx 0$. The net flux of carriers is given by *Fick's Law of diffusion* as

$$F = -D_n \frac{dn}{dx} = D_n \frac{n(x=0)}{L} = F^+(x = L)\,. \tag{12.14}$$

(The assumed linearity of $n(x)$ can be proved by solving the equations for $F^+(x)$ and $F^-(x)$.) In the diffusive limit that we are considering, the positive flux injected at $x = 0$ is

$$F^+(x = 0) = \frac{n(x=0)}{2} \upsilon_T\,, \tag{12.15}$$

where the factor of two comes from the fact that in the diffusive limit, approximately half of the electrons at $x = 0$ have positive velocities and approximately half have negative velocities due to backscattering within the slab. The velocity, υ_T, is the thermal average velocity of electrons with positive velocities, the so-called *unidirectional thermal velocity*,

$$\boxed{\upsilon_T = \sqrt{\frac{2k_BT}{\pi m^*}}\,.} \tag{12.16}$$

(Maxwell-Boltzmann statistics are assumed.) From eqns. (12.14) and (12.15), we find

$$\mathcal{T} = \frac{F^+(x = L)}{F^+(x = 0)} = \frac{F}{F^+(x = 0)} = \frac{D_n n(x=0)/L}{\upsilon_T n(x=0)/2} = \frac{2D_n}{\upsilon_T L}\,. \tag{12.17}$$

The diffusion coefficient is simply related to the unidirectional thermal velocity and the mean-free-path for backscattering [3],

$$\boxed{D_n = \frac{\upsilon_T \lambda}{2} \ \ \text{cm}^2/\text{s}\,,} \tag{12.18}$$

so (12.17) can be re-written as

$$\mathcal{T}_{\text{diff}} = \frac{\lambda}{L}. \tag{12.19}$$

As expected, the transmission in the diffusive limit is small because $L \gg \lambda$.

We have derived the transmission in the ballistic and diffusive limits, but modern devices often operate in the *quasi-ballistic* regime between these two limits. In general, the transmission is

$$\boxed{\begin{aligned} \mathcal{T} &= \frac{\lambda_0}{\lambda_0 + L} \\ \mathcal{T}(E) &= \frac{\lambda(E)}{\lambda(E) + L} \end{aligned}}. \tag{12.20}$$

The first equation assumes an energy-independent mean-free-path, λ_0, so the transmission is the same for all energy channels. The second equation refers to the transmission and mean-free-path in a specific energy channel. Equation (12.20) is clearly correct in the ballistic and diffusive limits, but it is also accurate between those two limits. It can be derived from a simple Boltzmann transport equation [2, 3].

Finally, we should point out that the mean-free-path, λ, is a specially defined mean-free-path for backscattering. Physically, it is the probability per unit length that a forward flux will be backscattered to a negative flux. More commonly, the mean-free-path is simply taken to be the average distance between scattering events,

$$\Lambda(E) \equiv \upsilon(E)\tau(E). \tag{12.21}$$

The mean-free-path for backscattering is defined in 2D as [2, 3]

$$\lambda(E) \equiv \frac{\pi}{2}\upsilon(E)\,\tau_m(E), \tag{12.22}$$

where τ_m is the momentum relaxation time. The momentum relaxation time is always greater than the scattering time, τ, so $\lambda > \Lambda$; the mean-free-path for backscattering is always longer than the mean-free-path for scattering.

Exercise 12.2: Derive the unidirectional thermal velocity.

The unidirectional thermal velocity plays an important role in transport. In equilibrium, the average velocity is zero, but the average velocity of only the electrons with velocities in the $+x$-direction is a positive quantity equal to the magnitude of the average velocity of electrons in the $-x$ direction. In this exercise, we will compute this velocity for 2D electrons in a parabolic band semiconductor.

Consider first, the average over angle. Figure 12.5 show a velocity vector in the $x-y$ plane at a specific energy and angle, θ, with the x-axis. For simple, parabolic energy bands, the magnitude of the velocity depends only on energy and is independent of angle. The x-directed velocity is $\upsilon(E)\cos\theta$. Assuming circular energy bands in 2D ($E = \hbar^2(k_x^2 + k_y^2)/2m^*$), the magnitude of the velocity, $\upsilon(E)$, is independent of angle. The average velocity in the $+x$ direction is

$$\langle \upsilon_x^+(E) \rangle = \frac{\int_{-\pi/2}^{+\pi/2} \upsilon(E)\cos\theta d\theta}{\pi} = \frac{2}{\pi}\upsilon(E)\,,$$

where the single brackets, $\langle\cdot\rangle$, denote an average over angle in the $x-y$ plane at a specific energy, E.

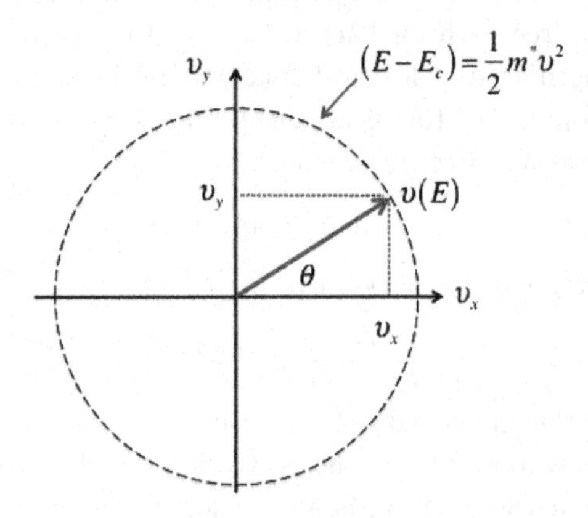

Fig. 12.5 Illustration of a velocity vector at energy, E, in the $x-y$ plane. For a parabolic energy band, the magnitude of the velocity (length of the vector) is determined by the energy and is independent of direction.

The quantity of interest is $\langle\langle \upsilon_x^+ \rangle\rangle$, where the double brackets denote an average over angle and energy. This quantity is determined from the integral,

$$\langle\langle v_x^+ \rangle\rangle = \frac{\int_{E_c}^{\infty} \langle v_x^+(E)\rangle D_{2D}(E) f_0(E) dE}{\int_{E_c}^{\infty} D_{2D}(E) f_0(E) dE}$$

$$= \frac{\int_{E_c}^{\infty} \frac{2}{\pi} \upsilon(E) D_{2D}(E) f_0(E) dE}{\int_{E_c}^{\infty} D_{2D}(E) f_0(E) dE} .$$

For parabolic bands

$$\upsilon(E) = \sqrt{\frac{2(E - E_c)}{m^*}} ,$$

$$D_{2D}(E) = g_v \frac{m^*}{\pi \hbar^2} ,$$

so we find

$$\langle\langle v_x^+ \rangle\rangle = \frac{\int_{E_c}^{\infty} \frac{2}{\pi} \sqrt{\frac{2(E-E_c)}{m*}} \left(g_v \frac{m^*}{\pi \hbar^2} \right) \frac{dE}{1+e^{(E-E_F)/k_B T}}}{\int_{E_c}^{\infty} \left(g_v \frac{m^*}{\pi \hbar^2} \right) \frac{dE}{1+e^{(E-E_F)/k_B T}}} .$$

After making the definitions,

$$\eta \equiv (E - E_c)/k_B T$$

$$\eta_F \equiv (E_F - E_c)/k_B T ,$$

we find

$$\langle\langle v_x^+ \rangle\rangle = \sqrt{\frac{2k_B T}{\pi m^*}} \times \frac{\frac{2}{\sqrt{\pi}} \int_0^{\infty} \frac{\eta^{1/2} d\eta}{1+e^{\eta - \eta_F}}}{\int_0^{\infty} \frac{d\eta}{1+e^{\eta - \eta_F}}} .$$

The numerator of the last factor can be recognized as a Fermi-Dirac integral of order 1/2 [5] and the denominator as a Fermi-Dirac integral of order 0 [5], so we have

$$\boxed{\langle\langle v_x^+ \rangle\rangle = \sqrt{\frac{2k_B T}{\pi m^*}} \frac{\mathcal{F}_{1/2}(\eta_F)}{\mathcal{F}_0(\eta_F)} ,} \tag{12.23}$$

which is the general result. Below threshold, we can assume Maxwell-Boltzmann statistics where the Fermi-Dirac integrals of any order approach $\exp(\eta_F)$ [5], so eqn. (12.23) becomes

$$\langle\langle v_x^+ \rangle\rangle = \upsilon_T = \sqrt{\frac{2k_B T}{\pi m^*}} ,$$

which is the desired result, eqn. (12.16).

12.5 Modes (channels)

The *distribution of modes*, $M(E)$, gives the number of channels at energy, E, through which current can flow. This quantity is derived and discussed in [2, 3]. In this section we discuss $M(E)$ for a 2D channel, as for the channel of a MOSFET.

We should expect $M(E)$ to be related to the density-of-states, because there must be a state in the channel for electrons to occupy, but the state must also have a velocity for current to flow. We conclude that

$$M(E) \propto \langle v_x^+(E)\rangle D(E)/4\,, \tag{12.24}$$

where $D(E)dE$ is the number of states between E and $E+dE$, and $\langle v_x^+(E)\rangle$ is the angle-averaged velocity in the direction of current flow (assumed to be the $+x$-direction in this case) for electrons at energy, E. The factor of four includes a factor of 2 for spin degeneracy; we need the density of states per spin, $D(E)/2$ because spin degeneracy is already included in eqn. (12.2) as the factor of 2 out front. Another factor of two occurs because only half of the states, $D(E)/2$, have a velocity in the direction of current flow. Dimensional analysis shows that the constant of proportionality must have the units of J-s, which are the units of Planck's constant, h. We conclude that

$$M(E) = \frac{h}{4}\langle v_x^+(E)\rangle D(E)\,. \tag{12.25}$$

For planar MOSFETs, carriers flow in a two-dimensional channel, so

$$\boxed{M_{2D}(E) = \frac{h}{4}\langle v_x^+(E)\rangle D_{2D}(E) \quad \mathrm{m}^{-1}\,,} \tag{12.26}$$

which gives the number of channels at energy, E, per unit width of the channel. Note that the number of states between E and $E+dE$ is given by $D(E)dE$, but the number of states per unit area is $D_{2D}(E)dE = D(E)dE/A$, where A is the area.

For parabolic energy bands, the two-dimensional density-of-states is [1]

$$D_{2D}(E) = g_v\left(\frac{m^*}{\pi\hbar^2}\right) \quad \mathrm{J}^{-1}\mathrm{m}^{-2}\,, \tag{12.27}$$

where g_v is the *valley degeneracy*. For a MOSFET, these 2D states lie in the conduction (valence) band at an energy above $E_C + \epsilon_1$ or below $E_V - \epsilon_1$, where ϵ_1 is the confinement energy for the lowest subband. If there are multiple subbands due to quantum confinement, each one will have a density-of-states like eqn. (12.27). According to eqn. (12.26), the distribution of channels in 2D is

$$\boxed{M_{2D}(E) = g_v \frac{\sqrt{2m^*(E - (E_C + \epsilon_1))}}{\pi \hbar} \quad \text{m}^{-1}} \tag{12.28}$$

where we have used $\langle v_x^+(E) \rangle = 2v(E)/\pi$. Figure 12.5 compares the 2D density-of-states with the 2D distribution of channels. Similarly, we can obtain $M(E)$ in 1D and 3D for parabolic energy bands, or in 2D for graphene by using the appropriate density-of-states and velocity [3].

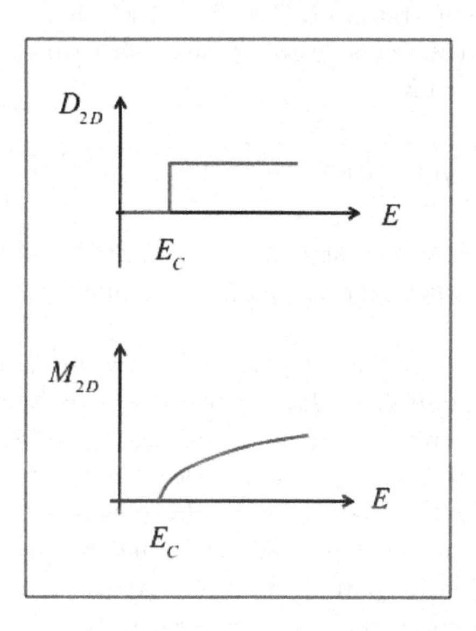

Fig. 12.6 Comparison of the 2D density-of-states and distribution of channels for parabolic energy bands. Top: $D_{2D}(E)$. Bottom: $M_{2D}(E)$.

To understand MOS transistors, we need to understand how the gate voltage controls the number of carriers in the channel and the resulting current that flows. To relate the carrier densities to the location of the Fermi level, we will make use of the density-of-states, as will be discussed in Sec. 12.7. To relate the current to the location of the Fermi level, we use the distribution of modes, as indicated in eqn. (12.2). So we need an understanding of both quantities, $D(E)$, and $M(E)$. This is analogous to the two different "effective masses" used in traditional semiconductor theory, the *density-of-states effective mass* and the *conductivity effective mass*.

12.6 Quantum of conductance

Now let's consider the conductance of a 2D channel at $T = 0$ K. We begin with the Landauer expression for the conductance, eqn. (12.9), and remember that the window function, $(-\partial f_0/\partial E)$, acts as a δ-function at $E = E_F$. Accordingly, eqn. (12.9) gives

$$G(T = 0\,\mathrm{K}) = \frac{2q^2}{h}\mathcal{T}(E_F)M(E_F)\,. \qquad (12.29)$$

If we assume ballistic transport, $\mathcal{T}(E_F) = 1$, which is not hard to achieve under low bias at low temperatures in short structures, then the ballistic conductance at $T = 0$ K is

$$G_B(T = 0\,\mathrm{K}) = \frac{2q^2}{h}M(E_F) = \frac{M(E_F)}{12.9\,\mathrm{k}\Omega}\,. \qquad (12.30)$$

In small structures, the number of modes (channels) is small and countable. We conclude that conductance is quantized in units of $2q^2/h$, which is one over 12.9 kilohms.

The fact that conductance is quantized is a well-established experimental fact. See, for example, Fig. 12.6, which shows experimental results. The resistor is a 2D electron gas formed at an interface of AlGaAs and GaAs. The width of the resistor is controlled electrostatically by reverse-biased Schottky junctions. The mobility of the electrons is very high (because the electrons reside in an undoped GaAs layer and because the temperature is low), so ballistic transport is expected. As the width was electrically varied, the measured conductance was seen to increase in discrete steps according to eqn. (12.30). Quantized conductance has been observed in many different systems. The experiment shown in Fig. 12.6 was done at low temperature to achieve near ballistic transport, but modern devices are so short that these effects are becoming important at room temperature in some systems.

Ballistic transport, $\mathcal{T}(E_F) = 1$, and quasi-ballistic transport, $\mathcal{T}(E_F) \lesssim 1$, are not uncommon in modern transistors, even at room temperature. Most useful devices are, however, large enough so that the discrete nature of the modes is not apparent. For transistors, we can usually treat $M(E)$ as a continuous quantity that is proportional to W and won't need to consider the discrete nature of $M(E)$. We will assume that $M(E) = WM_{2D}(E)$ as given by eqn. (12.28).

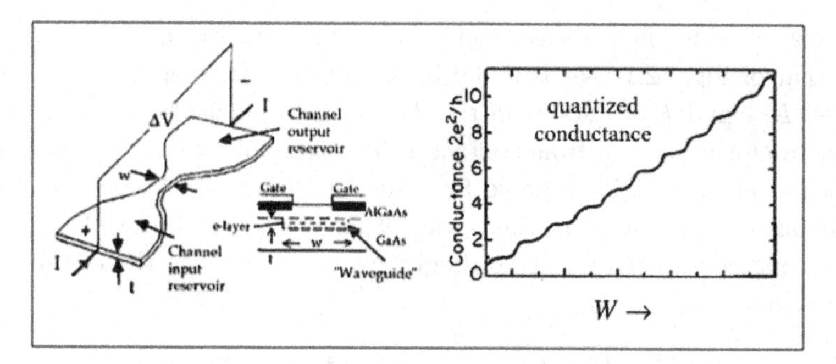

Fig. 12.7 Experiments of van Wees, et al. experimentally demonstrating that conductance is quantized. Left: sketch of the device structure. Right: measured conductance. (Data from: B. J. van Wees, et al., *Phys. Rev. Lett.* **60**, 848-850, 1988.) (Reproduced from: D. F. Holcomb, "Quantum electrical transport in samples of limited dimensions", *Am. J. Phys.*, **67**, pp. 278-297, 1999, with permission from the American Association of Physics Teachers.)

12.7 Carrier densities

According to conventional semiconductor theory, the density of electrons in the conduction band is an integral of the density-of-states at energy, E, times the probability that states at E are occupied [1]. The two-dimensional carrier density per m^2 (the sheet carrier density, is given by

$$n_S = \int_{E_C}^{\infty} D_{2D}(E) f_0(E) dE \,. \tag{12.31}$$

Using eqn. (12.27) for the 2D density-of-states and eqn. (12.3) for the equilibrium Fermi function, f_0, we find

$$\begin{aligned} n_S &= \int_{E_C}^{\infty} \left(g_v \frac{m^*}{\pi \hbar^2} \right) \frac{dE}{1 + e^{(E-E_F)/k_B T}} \\ &= \left(\frac{g_v m^* k_B T}{\pi \hbar^2} \right) \int_0^{\infty} \frac{d\eta}{1 + e^{\eta - \eta_F}} \,, \end{aligned} \tag{12.32}$$

where η and η_F were defined in Exercise 12.2.
The integral in eqn. (12.32) can be performed to find

$$\begin{aligned} n_S &= \left(\frac{g_v m^* k_B T}{\pi \hbar^2} \right) \ln \left(1 + e^{\eta_F}\right) \\ &= N_{2D} \ln \left(1 + e^{\eta_F}\right) = N_{2D} \mathcal{F}_0(\eta_F) \,, \end{aligned} \tag{12.33}$$

where N_{2D} is the two-dimensional effective density-of-states and $\mathcal{F}_0(\eta_F) = \ln\left(1 + e^{\eta_F}\right)$ is the Fermi-Dirac integral of order zero [4].

Now consider how we compute the carrier density in a small device like that of Fig. 12.1 that is under bias. In this case, there are two Fermi levels, E_{F1} and E_{F2}. States in the device are occupied by electrons that enter from contact 1 or from contact 2. The probability that a state in the device at energy, E, is occupied from the left contact is $f_1(E_{F1})$, and the probability that a state in the device at energy, E, is occupied from the right contact is $f_2(E_{F2})$. Accordingly, we can generalize the equilibrium expression, eqn. (12.33), to

$$\boxed{n_S = \int_{E_C}^{\infty} \left(\frac{D_{2D}(E)}{2} f_1(E) + \frac{D_{2D}(E)}{2} f_2(E) \right) dE\,.} \tag{12.34}$$

We have assumed that electrons stay in their energy channels, so that a state at E cannot be occupied by an electron in the device that scatters in from a different energy. We have also assumed that the two contacts are identical, so the states in the device divide into two equal components, one that is filled by electrons from contact 1 and the other from contact 2.

By working out eqn. (12.34), we find the 2D carrier density as

$$n_S = \frac{N_{2D}}{2} \mathcal{F}_0(\eta_F) + \frac{N_{2D}}{2} \mathcal{F}_0(\eta_F - qV/k_BT)\,, \tag{12.35}$$

where $\eta_F = (E_{F1} - E_c)/k_BT$. We see that the non-equilibrium carrier density is related to the density-of-states in a manner that is similar to the equilibrium relation; we just need to remember that there are two different Fermi levels and two different groups of states in the device, one in equilibrium with contact 1 and the other in equilibrium with contact 2.

As discussed earlier, the density-of-states is used to compute carrier densities while the distribution of modes is used to compute the current. Both $M(E)$ and $D(E)$ are needed to model a MOSFET.

12.8 Discussion

Our purpose in this lecture has been to get acquainted with the Landauer approach to transport, which we'll use to describe transport in nanoscale transistors. The approach works from the ballistic to diffusive limits, but let's apply the Landauer approach to a familiar problem, a 2D resistor under low bias, and see what happens. If the width, W, and the length, L, are large, then the traditional expression for the conductance is

$$G = \sigma_S \frac{W}{L} = n_S q \mu_n \frac{W}{L}\,, \tag{12.36}$$

where σ_S is the sheet conductivity in ohms. Equation (12.36) assumes that the resistor is many mean-free-paths long — i.e. that it operates in the diffusive limit. What does the Landauer approach give for the conductance?

To compute the conductance, we begin with eqn. (12.9), assume diffusive transport so that $\mathcal{T} = \lambda/L$, and assume parabolic energy bands, so that $M(E)$ is given by eqn. (12.28). We find

$$G = \frac{2q^2}{h} \int_{E_c}^{\infty} \left(\frac{\lambda(E)}{L} \right) \left(\frac{W g_v \sqrt{2m^*(E - E_c)}}{\pi\hbar} \right) \left(-\frac{\partial f_0}{\partial E} \right) dE \,. \qquad (12.37)$$

From the form of the Fermi function, eqn. (12.3), we observe that $\partial f_0/\partial E = -\partial f_0/\partial E_F$, which can be used to take the derivative out of the integral in eqn. (12.37) to write

$$G = \left[\frac{2q^2}{h} \left(\frac{g_v \sqrt{2m^*}}{\pi\hbar} \right) \frac{\partial}{\partial E_F} \int_{E_c}^{\infty} \frac{\lambda(E)\sqrt{E - E_c}}{1 + e^{(E-E_F)/k_BT}} dE \right] \frac{W}{L} \,. \qquad (12.38)$$

Now use the definitions for η and η_F from Exercise 12.2 and assume (just to keep the math simple) that $\lambda(E) = \lambda_0$ is independent of energy, so that eqn. (12.38) becomes

$$G = \left[\frac{2q^2}{h} \left(g_v \frac{\sqrt{2m^* k_B T}}{\pi\hbar} \right) \lambda_0 \frac{\partial}{\partial \eta_F} \int_0^{\infty} \frac{\eta^{1/2} d\eta}{1 + e^{\eta - \eta_F}} \right] \frac{W}{L} \,. \qquad (12.39)$$

The integral is $\sqrt{\pi}/2$ times the Fermi-Dirac integral of order $1/2$, $\mathcal{F}_{1/2}(\eta_F)$, so [5]

$$G = \left[\frac{2q^2}{h} \frac{\sqrt{\pi}}{2} \left(g_v \frac{\sqrt{2m^* k_B T}}{\pi\hbar} \right) \lambda_0 \frac{\partial \mathcal{F}_{1/2}(\eta_F)}{\partial \eta_F} \right] \frac{W}{L} \,. \qquad (12.40)$$

(A note of caution. Some authors define the Fermi-Dirac integral without the $2/\sqrt{\pi}$ factor, in which case it is usually written as a Roman F, $F_{1/2}(\eta_F)$ [4].)

Next, we make use of a property of Fermi-Dirac integrals, $\partial \mathcal{F}_j/\partial \eta_F = \mathcal{F}_{j-1}$ [4] to write

$$G = \left[\frac{2q^2}{h} \frac{\sqrt{\pi}}{2} \left(g_v \frac{\sqrt{2m^* k_B T}}{\pi\hbar} \right) \lambda_0 \mathcal{F}_{-1/2}(\eta_F) \right] \frac{W}{L} \,. \qquad (12.41)$$

To keep things simple, let's assume non-degenerate carrier statistics; under non-degenerate conditions, Fermi-Dirac integrals reduce to exponentials [4]. After re-arranging the factors in the brackets of eqn. (12.41), we find

$$G = \left[\frac{q^2}{k_B T} \left(\frac{g_v m^* k_B T}{\pi\hbar^2} \right) e^{\eta_F} \sqrt{\frac{2k_B T}{\pi m^*}} \lambda_0 \right] \frac{W}{L} \,. \qquad (12.42)$$

Now we can recognize some terms:

$$n_S = g_v\left(\frac{m^* k_B T}{\pi\hbar^2}\right)e^{\eta_F} = N_{2D}e^{\eta_F}$$
$$\upsilon_T = \sqrt{\frac{2k_B T}{\pi m^*}}$$
$$D_n = \frac{\upsilon_T \lambda_0}{2}$$
$$\frac{D_n}{\mu_n} = \frac{k_B T}{q},$$

where υ_T is the unidirectional thermal velocity (eqn. (12.16)) and D_n is the diffusion coefficient (eqn. (12.18)). Using these terms in eqn. (12.42), we finally obtain

$$G = n_S q \mu_n \frac{W}{L}, \tag{12.43}$$

where

$$\boxed{\mu_n = \frac{\upsilon_T \lambda_0}{2k_B T/q},} \tag{12.44}$$

Equation (12.43), the Landauer result, is identical to the conventional result in the diffusive limit. Equation (12.44) gives the mobility in terms of the mean-free-path for backscattering (assuming an energy-independent mean-free-path and non-degenerate carrier statistics).

This exercise shows that the Landauer result in the diffusive limit gives the expect conventional result, but the advantage of the Landauer approach is that it also works in the ballistic limit. What is the conductance in the ballistic limit? It is readily computed from eqn. (12.9) with $\mathcal{T} = 1$. Instead of eqn. (12.37), we find the ballistic conductance as

$$G_B = \frac{2q^2}{h}\int_{E_c}^{\infty}(1)\left(\frac{W g_v\sqrt{2m^*(E-E_c)}}{\pi\hbar}\right)\left(-\frac{\partial f_0}{\partial E}\right)dE\,. \tag{12.45}$$

This equation can be evaluated in the same way we treated the diffusive case; instead of eqn. (12.39), we find

$$G_B = \left[\frac{2q^2}{h}\left(g_v\frac{\sqrt{2m^* k_B T}}{\pi\hbar}\right)\frac{\partial}{\partial\eta_F}\int_0^{\infty}\frac{\eta^{1/2}d\eta}{1+e^{\eta-\eta_F}}\right]W\,. \tag{12.46}$$

After integrating and taking the non-degenerate limit, we find

$$G_B = n_S q\left(\frac{\upsilon_T}{2k_B T/q}\right)W\,. \tag{12.47}$$

As expected, the ballistic conductance is independent of the length, L. We can, however, write the ballistic conductance in the traditional form, eqn. (12.43), if we multiply and divide eqn. (12.47) by L and define a *ballistic mobility* [5],

$$\boxed{\mu_B \equiv \frac{v_T L}{2k_B T/q}\,.} \tag{12.48}$$

If the ballistic mobility is used in place of the actual mobility, μ_n in (12.43), we find the ballistic conductance. The ballistic mobility is given by the same expression as the traditional mobility of a bulk material, μ_n, except that the mean-free-path, λ_0, is replaced by the length of the resistor, L.

What is the physical significance of the ballistic mobility for a device in which there is no scattering? In a bulk semiconductor, the average distance between backscattering events is λ_0, and the mobility is a well-defined material parameter. In a ballistic resistor, there is no scattering in the device, but electrons in contacts 1 and 2 scatter frequently, so the distance between scattering events is the length of the device. It seems sensible to replace the actual mean-free-path by the length of the device, and that leads to the concept of a ballistic mobility. The ballistic mobility is just a way to write the conductance of a ballistic device in the traditional, diffusive form, eqn. (12.43), but it also has a clear, physical interpretation.

Modern devices often operate between the ballistic and diffusive limits. Again, it is easy to evaluate the conductance beginning with eqn. (12.9). In this case, eqn. (12.37) becomes

$$G = \frac{2q^2}{h}\int_{E_c}^{\infty}\left(\frac{\lambda(E)}{\lambda(E)+L}\right)\left(\frac{W g_v\sqrt{2m^*(E-E_c)}}{\pi\hbar}\right)\left(-\frac{\partial f_0}{\partial E}\right)dE\,. \tag{12.49}$$

Again, this expression is readily evaluated if we assume $\lambda(E) = \lambda_0$. We find that we can write the result in the traditional form, eqn. (12.43) if we replace the actual mobility by an *apparent mobility* that is given by

$$\boxed{\frac{1}{\mu_{app}} = \frac{1}{\mu_B} + \frac{1}{\mu_n}\,.} \tag{12.50}$$

The smaller of the two mobilities will limit the current in the device. As the length decreases, the ballistic mobility decreases according to eqn. (12.48), When $\lambda \gg L$, the ballistic mobility in eqn. (12.50) will dominate ($\mu_B \ll \mu_n$), and the apparent mobility will approach the ballistic mobility. If we were to use eqn. (12.43) to compute the conductance of a resistor that is short compared to the mean-free-path without including the ballistic mobility, we would find a conductance above the ballistic limit. The ballistic mobility must be included in the traditional expression in order to get physically sensible answers for short conductors.

This example shows that the Landauer approach gives the same answer as the traditional approach in the diffusive limit, but that it also works in the ballistic and quasi-ballistic regimes.

12.9 Summary

In this lecture, we have introduced the Landauer approach to carrier transport, which we will use to describe MOSFETs under low and high drain bias (i.e. near equilibrium and far from equilibrium). For long channel transistors, the results will reduce to conventional MOSFET theory, but we will also be able to describe short channel transistors that operate in the ballistic or quasi-ballistic limit.

We have only been able to introduce the Landauer approach and to point out that it is intuitive and sensible. The approach provides a simple, physical description of transport in so-called mesoscopic structures. Those interested in more discussion of the underlying physical assumptions should consult ref. [2], and those interested in a more complete discussion of applications, should consult [3]. The treatment in this lecture, will, however, be enough for us to understand the operation of nanoscale transistors.

12.10 References

For a description of the traditional approach to carrier transport, drift-diffusion equation, Drude equation for mobility, etc., see:

[1] Robert F. Pierret, *Advanced Semiconductor Fundamentals, 2nd Ed.*, Vol. VI, Modular Series on Solid-State Devices, Prentice Hall, Upper Saddle River, N.J., USA, 2003.

The Landauer approach to carrier transport at the nanoscale is discussed in Vols. 1 and 2 of this series.

[2] Supriyo Datta, *Lessons from Nanoelectronics*, 2nd Ed., PART A: Basic Concepts, World Scientific Publishing Company, Singapore, 2017.

[3] Mark Lundstrom and Changwook Jeong, *Near-Equilibrium Transport: Fundamentals and Applications*, World Scientific Publishing Company, Singapore, 2012.

For a quick summary of the essentials of Fermi-Dirac integrals, see:

[4] R. Kim and M.S. Lundstrom, "Notes on Fermi-Dirac Integrals," 3rd Ed., `https://www.nanohub.org/resources/5475`.

The concept of ballistic mobility is discussed by Shur.

[5] M. S. Shur, "Low ballistic mobility in submicron HEMTs," *IEEE Electron Device Lett.*, **23**, pp. 511-513, 2002.

Lecture 13

The Ballistic MOSFET

13.1 Introduction

In previous lectures we discussed the MOSFET as a barrier controlled device (Lecture 3), MOS electrostatics (Lectures 6-10), and transport (Lecture 12), now we are ready to put these concepts together in a model that describes the essential physics of nanoscale MOSFETs. We begin with ballistic MOSFETs. Real MOSFETs can be complicated [1] and detailed semiclassical simulations (that treat electrons as particles [2]) and quantum mechanical simulations (that treat electrons as waves [3]) are necessary to understand devices in detail. Our goal is different — it is to understand the principles of nanotransistors in a simple, physically sound way that is suitable for interpreting what we see in experiments and in detailed simulations and for device modeling. The basic principles apply to Si MOSFETs and to other MOSFETs such as III-V MOSFETs [4] and nanowire and carbon nanotube MOSFETs [5].

We will assume that the electron charge vs. gate voltage, $Q_n(V_{GS}, V_{DS})$, is known both below and above threshold. It may, for example, be described by a semiempirical expression, such as eqn. (11.14). In this lecture, we'll use the Landauer approach, eqn. (12.2), and assume ballistic transport, $\mathcal{T}(E) = 1$, so the drain current is given by

$$I_{DS} = \frac{2q}{h}\int M(E)\big(f_S(E) - f_D(E)\big)dE \quad \text{Amperes}, \tag{13.1}$$

where f_S is the Fermi function in the source and f_D the Fermi function in the drain.

When the drain voltage is large, then $f_S(E) \gg f_D(E)$ for all energies of interest, and the saturation current is given by

$$I_{DSAT} = \frac{2q}{h}\int M(E) f_S(E) dE \quad \text{Amperes}. \tag{13.2}$$

In the linear region, the drain to source voltage is small, $f_S \approx f_D$, so we can find the linear region current from eqn. (12.9) as

$$\begin{aligned} I_{DLIN} &= G_{ch} V_{DS} = V_{DS}/R_{ch} \quad \text{Amperes} \\ G_{ch} &= 1/R_{ch} = \frac{2q^2}{h}\int M(E)\left(-\frac{\partial f_0}{\partial E}\right) dE \quad \text{Siemens}, \end{aligned} \tag{13.3}$$

where G_{ch} (R_{ch}) is the channel conductance (resistance). By evaluating these equations, we will obtain the ballistic linear region current, the ballistic on-current, and the ballistic current from $V_{DS} = 0$ to $V_{DS} = V_{DD}$, but these equations were derived to describe a nanodevice like that in Fig. 12.1. How do we treat the MOSFET as a nanodevice?

13.2 The MOSFET as a nanodevice

In Lecture 12, we presented the Landauer approach as a way to describe a nanodevice illustrated schematically in Fig. 12.1. Figure 13.1 shows how we treat a MOSFET as a nanodevice. As discussed in Lecture 3, the MOSFET uses a gate voltage to modulate the height of an energy barrier. Figure 13.1 shows $E_c(x)$ vs. x with the source and drain Fermi levels, E_{FS} and E_{FD}, indicated. As discussed in Lecture 3, the magnitude of the drain current is determined by the height of the energy barrier and by the transmission across a short region of length, $\ell < L$, near the top of the barrier. If electrons injected from the source backscatter in this short region (the "bottleneck" region), they return to the source and do not contribute to the drain current. If they transmit across this short region, they are almost certain to exit from the drain. This occurs because the strong electric field in the drain end of the channel sweeps electrons across and out the drain — even if they backscatter in this region, they are likely to exit through the drain contact. The high-field region acts as a carrier collector, an absorbing contact. In this lecture, we assume that the transmission across the short region near the top of the barrier is one. The critical, bottleneck region is ballistic, but the entire device need not be ballistic.

When applying the Landauer approach to the MOSFET, we compute the density of electrons near the top of the barrier from the local density of states there, $LDOS = D_{2D}(E, x = 0)$, and the current from the number of channels at the top of the barrier, $M(E, x = 0)$ and the transmission, $\mathcal{T}(E)$, across the critical region of length, ℓ. Note that we do not attempt to spatially resolve the calculations, so we can't specify the detailed shape of $E_c(x)$ vs. x; to do that we need to solve the transport equations (e.g. drift-diffusion, Boltzmann, or quantum) self-consistently with the Poisson equation. Such simulations are needed to fully understand transistors, but insight into the essentials can be gained by focusing on the region near the top of the energy barrier.

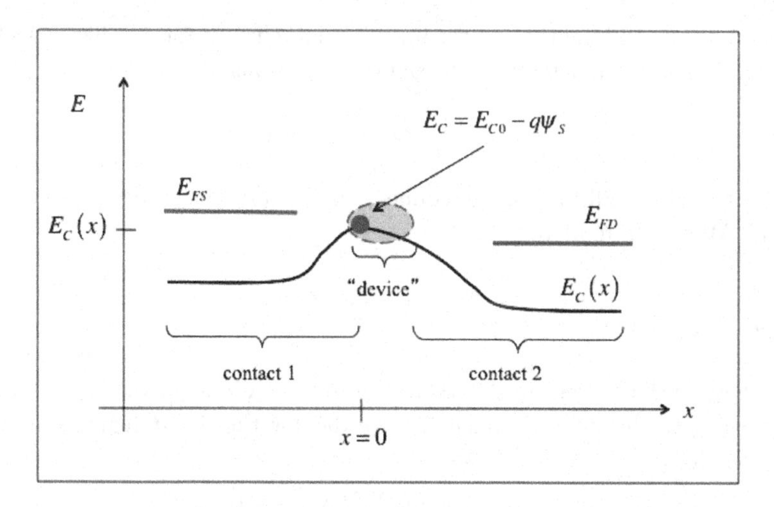

Fig. 13.1 Illustration of how a MOSFET is treated as a nanodevice of the type illustrated in Fig. 12.1. The short region of length, ℓ, is a bottleneck for current that begins at the top of the barrier. This short region is treated as a nanodevice.

13.3 Linear region

To evaluate the linear region current, we begin with eqn. (13.3), the ballistic channel conductance. For the distribution of channels, we use eqn. (12.28) to write

$$M(E) = WM_{2D}(E) = Wg_v \frac{\sqrt{2m^*(E - E_c(0))}}{\pi\hbar}\,, \tag{13.4}$$

where we use $E_c(0)$ to denote the bottom of the first subband. For the Fermi function, we use eqn. (12.3) with $E_F \approx E_{FS} \approx E_{FD}$. The integral can be evaluated as in Sec. 12.8; the result is just like eqn. (12.41) except that the diffusive transmission, λ_0/L, is replaced by unity, because we are assuming ballistic transport. The result is

$$I_{DLIN} = G_{ch}V_{DS} = \left[W\frac{2q^2}{h}\left(\frac{g_v\sqrt{2\pi m^* k_BT}}{2\pi\hbar}\right)\mathcal{F}_{-1/2}(\eta_F)\right]V_{DS}\,, \tag{13.5}$$

where

$$\eta_F = \left(E_{FS} - E_c(0)\right)/k_BT\,, \tag{13.6}$$

with $E_c(0)$ being the bottom of the conduction band at the top of the barrier.

Equation (13.5) is the correct linear region current for a ballistic MOSFET, but it looks much different from the traditional expression, eqn. (4.5),

$$I_{DLIN} = \frac{W}{L}|Q_n(V_{GS})|\mu_n V_{DS}\,. \tag{13.7}$$

In Lecture 15, we'll discuss the connection between the ballistic and traditional MOSFET models.

13.4 Saturation region

To evaluate the current in the saturation region, we begin with eqn. (13.2) and evaluate the integral much like we did for the linear region current. The result is

$$I_{DSAT} = W\frac{2q}{h}\left(\frac{g_v\sqrt{2m^* k_BT}}{\pi\hbar}\right)k_BT\frac{\sqrt{\pi}}{2}\mathcal{F}_{1/2}(\eta_F)\,. \tag{13.8}$$

Equation (13.8) is the correct saturated region current for a ballistic MOSFET, but it looks much different from the traditional velocity saturation expression, eqn. (4.7),

$$I_{DSAT} = W|Q_n(V_{GS}, V_{DS})|\upsilon_{sat} \,. \tag{13.9}$$

We'll discuss the connection between these two models in Lecture 15.

13.5 From linear to saturation

In the previous two sections, we derived the ballistic drain current in the linear (low V_{DS}) and saturation (high V_{DS}) regions. The virtual source model describes the drain current across the full range of V_{DS} by using an empirical drain saturation function to connect these two currents. We'll discuss this virtual source approach in Lecture 15. For the ballistic MOSFET, however, it is easy to compute drain current from low to high V_{DS}.

To evaluate the ballistic drain current for arbitrary drain voltage, we begin with eqn. (13.1) and evaluate the integral much like we did for the saturation region current. The result is

$$I_{DS} = W\frac{q}{h}\left(\frac{g_v\sqrt{2\pi m^* k_B T}}{\pi\hbar}\right) k_B T\left[\mathcal{F}_{1/2}(\eta_{FS}) - \mathcal{F}_{1/2}(\eta_{FD})\right] , \tag{13.10}$$

where

$$\begin{aligned} \eta_{FS} &= \left(E_{FS} - E_c(0)\right)/k_B T \\ \eta_{FD} &= \left(E_{FD} - E_c(0)\right)/k_B T = \left(E_{FS} - qV_{DS} - E_c(0)\right)/k_B T. \end{aligned} \tag{13.11}$$

13.6 Charge-based current expressions

Equation (13.10) is the correct current for a ballistic MOSFET at arbitrary V_{DS}, but it is not in terms of the inversion large charge, Q_n. To compute Q_n, we need to include the positive velocity electrons injected from the source that populate $+\upsilon_x$ states at the top of the barrier and negative velocity electrons injected from the drain that populate $-\upsilon_x$ states at the top of the barrier. For arbitrary V_{DS}, we find the inversion charge from

$$Q_n = -qn_S = -q\frac{N_{2D}}{2}\left[\mathcal{F}_0(\eta_{FS}) + \mathcal{F}_0(\eta_{FD})\right] , \tag{13.12}$$

which comes from eqn. (12.34).

We can now use eqn. (13.12) with (13.10) to express the drain current in terms of Q_n. After some algebra, we find

$$I_{DS} = W|Q_n(V_{GS}, V_{DS})|\, \upsilon_{inj}^{\text{ball}} \left[\frac{1 - \mathcal{F}_{1/2}(\eta_{FD})/\mathcal{F}_{1/2}(\eta_{FS})}{1 + \mathcal{F}_0(\eta_{FD})/\mathcal{F}_0(\eta_{FS})} \right]$$

$$Q_n(V_{GS}, V_{DS}) = -q\frac{N_{2D}}{2}\left[\mathcal{F}_0(\eta_{FS}) + \mathcal{F}_0(\eta_{FD})\right] \qquad (13.13)$$

$$\upsilon_{inj}^{\text{ball}} = \langle\langle \upsilon_x^+ \rangle\rangle = \sqrt{\frac{2k_BT}{\pi m*}}\frac{\mathcal{F}_{1/2}(\eta_{FS})}{\mathcal{F}_0(\eta_{FS})} = \upsilon_T \frac{\mathcal{F}_{1/2}(\eta_{FS})}{\mathcal{F}_0(\eta_{FS})}$$

$$\eta_{FD} = \eta_{FS} - qV_{DS}/k_BT \,.$$

Equations (13.13), which give the IV characteristic of a ballistic MOSFET, are the main result of this lecture. Equations of this kind were first derived by Natori [6] and later extended [7]. They give the drain current over the entire range of V_{DS}.

The ballistic IV characteristic would be computed as follows. First, we compute $Q_n(V_{GS}, V_{DS})$ from MOS electrostatics, perhaps using the semi-empirical expression, eqn. (11.14). Next we determine the location of the source Fermi level, η_{FS} by solving the second equation in (13.13) for η_{FS} given a value of $Q_n(V_{GS}, V_{DS})$. Then we determine $\upsilon_{inj}^{\text{ball}}$ from the third equation in (13.13). Finally, we determine the drain current at the bias point, (V_{GS}, V_{DS}) using the first equation in (13.13). Figure 13.2 shows the computed IV characteristics using parameters for an Extremely Thin SOI n-channel MOSFET taken from [8].

Exercise 13.1: Show that eqn. (13.13) gives the correct linear and saturation region currents.

Equations (13.13) gave the ballistic IV characteristic for arbitrary V_{DS} in terms of Q_n. For low V_{DS} and for high V_{DS}, eqn. (13.13) should reduce to eqns. (13.5) and (13.8) respectively. The point of this exercise is to demonstrate this.

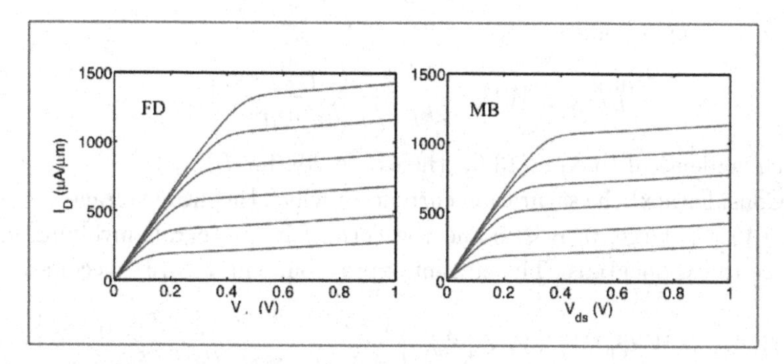

Fig. 13.2 Simulated IV characteristic of a ballistic MOSFET. Realistic parameters (including series resistance) for an ETSOI MOSFET were taken from [8]. An off-current of 100 nA/μm was assumed, which resulted in a threshold voltage of 0.44 V. Left: Fermi-Dirac statistics assumed. Right: Maxwell-Boltzmann statistics assumed. In both cases, a series resistance of $R_{SD} = R_S + R_D = 260\,\Omega - \mu$m was used, and the steps are from $V_{GS} = 0.5$ to $V_{GS} = 1.0$ V. (Figure provided by Xingshu Sun, Purdue University, August, 2014.)

Consider the linear region first. Since V_{DS} is small, $\eta_{FS} \approx \eta_{FD}$, the denominator of eqn. (13.13) becomes two, and we find

$$I_{DLIN}^{\text{ball}} = W|Q_n|\upsilon_{inj}^{\text{ball}} \left[\frac{1 - \mathcal{F}_{1/2}(\eta_{FD})/\mathcal{F}_{1/2}(\eta_{FS})}{2}\right].$$

Now multiply and divide by $\mathcal{F}_{1/2}(\eta_{FS})$ to find

$$I_{DLIN}^{\text{ball}} = W|Q_n|\upsilon_{inj}^{\text{ball}} \frac{1}{\mathcal{F}_{1/2}(\eta_{FS})} \left[\frac{\mathcal{F}_{1/2}(\eta_{FS}) - \mathcal{F}_{1/2}(\eta_{FD})}{2}\right].$$

Next, multiply and divide by $\eta_{FS} - \eta_{FD} = qV_{DS}/k_BT$ and find

$$I_{DLIN}^{\text{ball}} = W|Q_n| \frac{\upsilon_{inj}^{\text{ball}}}{2k_BT/q} \frac{1}{\mathcal{F}_{1/2}(\eta_{FS})} \left[\frac{\mathcal{F}_{1/2}(\eta_{FS}) - \mathcal{F}_{1/2}(\eta_{FD})}{\eta_{FS} - \eta_{FD}}\right] V_{DS}.$$

Because η_{FD} is just a little less than η_{FS}, we recognize the term in square brackets as a derivative of a Fermi-Dirac integral [9],

$$\left[\frac{\mathcal{F}_{1/2}(\eta_{FS}) - \mathcal{F}_{1/2}(\eta_{FD})}{\eta_{FS} - \eta_{FD}}\right] \approx \frac{\partial \mathcal{F}_{1/2}(\eta_{FS})}{\partial \eta_{FS}} = \mathcal{F}_{-1/2}(\eta_{FS}),$$

so the current becomes

$$I_{DLIN}^{\text{ball}} = W|Q_n|\frac{\upsilon_{inj}^{\text{ball}}}{2k_BT/q}\frac{\mathcal{F}_{-1/2}(\eta_{FS})}{\mathcal{F}_{1/2}(\eta_{FS})}V_{DS},$$

which is identical to eqn. (13.5), the expression for I_{DLIN}.

Consider next the saturation current for which the drain voltage is large. Since V_{DS} is large, $\eta_{FD} \ll 0$ and the Fermi-Dirac integrals involving η_{FD} reduce to exponentials. The current expression, eqn. (13.13), becomes,

$$I_{DSAT}^{\text{ball}} = W|Q_n(V_{GS}, V_{DS})|\upsilon_{inj}^{\text{ball}}\left[\frac{1 - e^{\eta_{FS} - qV_{DS}/k_BT}/\mathcal{F}_{1/2}(\eta_{FS})}{1 + e^{\eta_{FS} - qV_{DS}/k_BT}/\mathcal{F}_0(\eta_{FS})}\right].$$

For large V_{DS}, the term in the square brackets is seen to approach one, so the current for large drain voltage is

$$I_{DSAT}^{\text{ball}} = W|Q_n(V_{GS}, V_{DS})|\upsilon_{inj}^{\text{ball}},$$

which is identical to eqn. (13.8) for I_{DSAT}^{ball}.

Exercise 13.2: Derive the *IV* characteristics, analogous to eqns. (13.13), for a ballistic nanowire MOSFET.

In a nanowire MOSFET, the gate surrounds the channel leading to better electrostatics and therefore, lower DIBL and improved channel length scalability. Assume that the diameter of the nanowire is small, so that electrons behave as 1D carriers with only a single subband occupied. Derive the *IV* characteristics for a 1D MOSFET and compare the results to eqns. (13.13) for a 2D MOSFET.

Just as for a 2D MOSFET, we begin with eqn. (13.1), but instead of eqn. (13.4) for $M(E)$, we need the 1D distribution of channels. From eqn. (12.25), we find

$$M(E) = M_{1D}(E) = \frac{h}{4}\langle \upsilon_x^+(E)\rangle D_{1D}(E).$$

For a 1D semiconductor with parabolic energy bands, the 1D density-of-states is [5, 10]

$$D_{1D}(E) = g_v\frac{\sqrt{2m^*}}{\pi\hbar}\frac{1}{\sqrt{E - E_c}}. \tag{13.14}$$

There are no angles to average over in 1D, so

$$\langle \upsilon_x^+(E)\rangle = \upsilon(E).$$

Putting this all together, we find

$$\begin{aligned} M_{1D}(E) &= 0 \quad E < E_c \\ M_{1D}(E) &= g_v \quad E > E_c\,; \end{aligned} \tag{13.15}$$

the distribution of channels in 1D is constant for $E > E_c$ [10]. (Note that we are using E_c to denote the bottom of the lowest subband.)

Now we can integrate eqn. (13.1) using the 1D $M(E)$ to find

$$I_{DS} = \frac{q}{h} k_B T \left[\mathcal{F}_0(\eta_{FS}) - \mathcal{F}_0(\eta_{FD})\right],$$

which is the 1D analog of eqn. (13.10) for 2D electrons.

Next, we wish to express the drain current in terms of the electron charge. We begin with eqn. (12.34), but use the 1D density-of-states to write

$$n_L = \frac{N_{1D}}{2}\left[\mathcal{F}_{-1/2}(\eta_{FS}) + \mathcal{F}_{-1/2}(\eta_{FD})\right] \quad \text{m}^{-1}, \tag{13.16}$$

where the 1D effective density-of-states is

$$N_{1D} = \sqrt{\frac{2m^* k_B T}{\pi \hbar^2}} \quad \text{m}^{-1}. \tag{13.17}$$

In 2D, the carrier density is per m^2, but in 1D it is per m. Using this expression for n_L, we find the electron charge per unit length to be

$$Q_n = -q n_L = -q\frac{N_{1D}}{2}\left[\mathcal{F}_{-1/2}(\eta_{FS}) + \mathcal{F}_{-1/2}(\eta_{FD})\right] \quad \text{C/m},$$

which is the 1D analog of eqn. (13.12) for 2D. Now it just takes a little algebra to express the drain current in terms of Q_n. The result, analogous to eqn. (13.13) for 2D electrons, is

$$\begin{aligned}
I_{DS} &= |Q_n(V_{GS}, V_{DS})| \upsilon_{inj}^{\text{ball}} \left[\frac{1 - \mathcal{F}_0(\eta_{FD})/\mathcal{F}_0(\eta_{FS})}{1 + \mathcal{F}_{-1/2}(\eta_{FD})/\mathcal{F}_{-1/2}(\eta_{FS})}\right] \\
Q_n(V_{GS}, V_{DS}) &= -q\frac{N_{1D}}{2}\left[\mathcal{F}_{-1/2}(\eta_{FS}) + \mathcal{F}_{1/2}(\eta_{FD})\right] \\
\upsilon_{inj}^{\text{ball}} &= \langle\langle \upsilon_x^+ \rangle\rangle = \sqrt{\frac{2k_B T}{\pi m*}\frac{\mathcal{F}_0(\eta_{FS})}{\mathcal{F}_{-1/2}(\eta_{FS})}} = \upsilon_T \frac{\mathcal{F}_0(\eta_{FS})}{\mathcal{F}_{-1/2}(\eta_{FS})} \\
\eta_{FD} &= \eta_{FS} - qV_{DS}/k_B T\,.
\end{aligned} \tag{13.18}$$

Note that the unidirectional thermal velocity in the nondegenerate limit is the same in 1D as in 2D and 3D. For nondegenerate conditions, $\langle\langle v_x^+ \rangle\rangle = \upsilon_T$, but for degenerate conditions, $\langle\langle v_x^+ \rangle\rangle > \upsilon_T$.

Finally, we need to discuss how to compute $Q_n(V_{GS}, V_{DS})$. We could develop expressions analogous to eqn. (9.52) for the ETSOI MOSFET, or, if we are content with a simple, above threshold treatment, we could find the charge in C/m from

$$\begin{aligned} Q_n &= 0 \quad V_{GS} \leq V_T \\ Q_n &= -C_{ins}\left(V_{GS} - V_T\right) \quad V_{GS} > V_T \, , \end{aligned}$$

where

$$C_{ins} = \frac{2\pi\epsilon_{ins}}{\ln\left(\frac{2t_{ins}+t_{wire}}{t_{wire}}\right)} \quad \text{F/m} \, ,$$

with t_{wire} being the diameter of the nanowire.

This exercise shows that the derivation of the ballistic IV characteristics of a nanowire MOSFET proceeds much like that of the planar MOSFET and that the final expressions are similar.

13.7 Discussion

Equations (13.13) describes the IV characteristics of a ballistic MOSFET. Recall from eqn. (11.1) that we can always write the drain current of a MOSFET as the product of charge and velocity,

$$I_{DS} = W|Q_n\left(V_{GS}, V_{DS}\right)|\,\upsilon(0) \, . \tag{13.19}$$

By equating this equation to the drain current in (13.13), we get an expression for the average velocity of carriers at the top of the barrier (located at $x = 0$),

$$\begin{aligned} \upsilon(0) &= \upsilon_{inj}^{\text{ball}} \left[\frac{1 - \mathcal{F}_{1/2}(\eta_{FD})/\mathcal{F}_{1/2}(\eta_{FS})}{1 + \mathcal{F}_0(\eta_{FD})/\mathcal{F}_0(\eta_{FS})}\right] \\ \upsilon_{inj}^{\text{ball}} &= \sqrt{\frac{2k_BT}{\pi m^*}\frac{\mathcal{F}_{1/2}(\eta_{FS})}{\mathcal{F}_0(\eta_{FS})}} \, . \end{aligned} \tag{13.20}$$

In the next lecture, we will discuss this velocity and explain why the velocity saturates for high drain biases in a ballistic MOSFET.

The Fermi-Dirac integrals in the equations we've developed make these equations look complicated and can hide the underlying simplicity of the ballistic MOSFET. Consider the nondegenerate case, where the equations simplify. For a nondegenerate semiconductor,

$$E_F \ll E_c$$
$$\eta_F = (E_F - E_c)/k_BT \ll 0\,.$$

In the nondegenerate case, Fermi-Dirac integrals of any order, j, reduce to exponentials [9]

$$\mathcal{F}_j(\eta_F) \rightarrow e^{\eta_F}\,.$$

Accordingly, in the nondegenerate limit, eqn. (13.13) becomes

$$I_{DS} = W|Q_n(V_{GS}, V_{DS})|\,\upsilon_T\left(\frac{1 - e^{-qV_{DS}/k_BT}}{1 + e^{-qV_{DS}/k_BT}}\right)$$
$$\upsilon_T = \sqrt{\frac{2k_BT}{\pi m^*}}\,. \tag{13.21}$$

Equation (13.21) has a simple, physical interpretation in terms of thermionic emission over a barrier, as illustrated in Fig. 13.3. The net current, the drain current, is the difference between the current injected from the source, I_{LR}, and the current injected from the drain, I_{RL}. As discussed in Lecture 3, Sec. 6, a simple thermionic emission treatment gives eqn. (3.7), which is identical to eqn. (13.21). The derivation from the Landauer approach described in this lecture provides a prescription for computing υ_T and for extending the treatment to non-degenerate carrier statistics (e.g. eqn. (13.13) vs. (13.21)). The drain current saturates when I_{RL} becomes negligible compared to I_{LR}. This occurs when V_{DS} is greater than a few k_BT/q for nondegenerate conditions and for a somewhat higher voltage when Fermi-Dirac statistics are used.

13.8 Summary

In this lecture, we used the Landauer approach introduced in Lecture 12 to compute the IV characteristics of a ballistic MOSFET. We combined the Landauer expression for current, eqn. (13.1), with the constraint that MOS electrostatics must be satisfied. The result was a fairly simple model for the ballistic MOSFET as summarized in eqns. (13.13). For non-degenerate

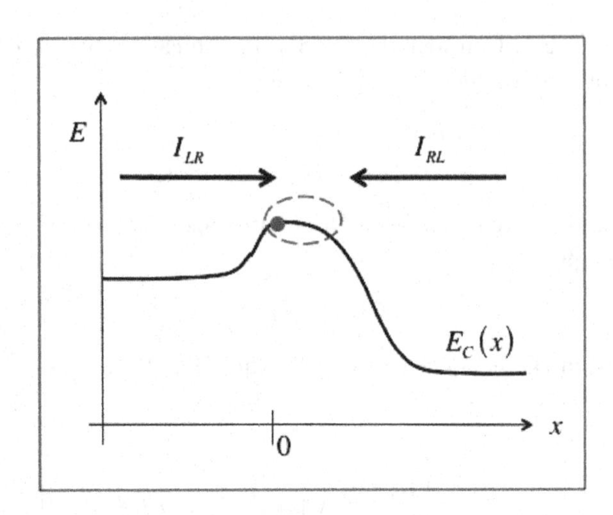

Fig. 13.3 Illustration of the two fluxes inside a ballistic MOSFET. I_{LR} is the current injected from the source, and I_{RL} is the current injected from the drain. The net drain current is the difference between the two, $I_{DS} = I_{LR} - I_{RL}$. For a well-designed MOSFET, however, MOS electrostatics also demands that the charge at the top of the barrier, $Q_n(0)$, is independent of the ratio of these two fluxes.

carrier statistics, the model simplifies to eqn. (13.21), which is identical to the thermionic emission model discussed in Sec. 3.6.

For a MOSFET operating in the subthreshold region, nondegenerate carrier statistics can be employed, so eqn. (13.21) can be used. Above threshold however, the conduction band at the top of the barrier is close to, or even below the Fermi level, so eqn. (13.13) should be used. Nevertheless, it is common in MOS device theory to assume nondegenerate conditions (i.e. to use Maxwell-Boltzmann statistics for carriers) because it simplifies the calculations and makes the theory more transparent. Also, in practice, there are usually some device parameters that we don't know precisely, so the use of nondegenerate carrier statistics with some empirical parameter fitting is common.

As discussed in earlier lectures, the drain current is the product of charge and velocity. In Lectures 6-10, we discussed the charge, $Q_n(V_{GS}, V_{DS})$, extensively because it is so important. Equations (13.20) describes the average velocity at the top of the barrier for arbitrary gate and drain voltages. Because understanding the average velocity is as important as understanding the charge, we devote the next lecture to a discussion of this topic.

13.9 References

The transport physics of nanoscale MOSFETs can be complex. See, for example, the following paper.

[1] M.V. Fischetti, T.P. O'Regan, N. Sudarshan, C. Sachs, S. Jin, J. Kim, and Y. Zhang, "Theoretical study of some physical aspects of electronic transport in n-MOSFETs at the 10-nm Gate-Length," *IEEE Trans. Electron Dev.*, **54**, pp. 2116-2136, 2007.

The two following references are examples of physically detailed MOSFET device simulation — the first semiclassical and the second quantum mechanical.

[2] D. Frank, S. Laux, and M. Fischetti, "Monte Carlo simulation of a 30 nm dual-gate MOSFET: How short can Si go?," Intern. Electron Dev. Mtg., pp. 553-556, Dec., 1992.

[3] Z. Ren, R. Venugopal, S. Goasguen, S. Datta, and M.S. Lundstrom "nanoMOS 2.5: A Two-Dimensional Simulator for Quantum Transport in Double-Gate MOSFETs," *IEEE Trans. Electron. Dev.*, **50**, pp. 1914-1925, 2003.

The most common MOSFETs for digital applications are made of silicon, but recently, III-V MOSFETs have attracted considerable interest. For a review, see:

[4] Jesus A. del Alamo, "Nanometre-scale electronics with III-V compound semiconductors," *Nature*, **479**, pp. 317-323, 2011.

For another treatment of ballistic MOSFETs — including nanowire and carbon nanotube MOSFETs, see:

[5] Mark Lundstrom and Jung Guo, *Nanoscale Transistors: Physics, Modeling, and Simulation*, Springer, New York, USA, 2006.

The theory of the ballistic MOSFET was first presented by Natori and later extended by Rahman, et al.

[6] K. Natori, "Ballistic metal-oxide-semiconductor field effect transistor," *J. Appl. Phys*, **76**, pp. 4879-4890, 1994.

[7] A. Rahman, J. Guo, S. Datta, and M. Lundstrom, "Theory of ballistic nanotransistors," *IEEE Trans. Electron Dev.*, **50**, pp. 1853-1864, 2003.

Silicon MOSFETs typically operate below the ballistic limit. The ballistic IV characteristics shown in Fig. 13.2 were computed with parameters (e.g. oxide thickness, series resistance, power supply, etc.) taken from the following paper.

[8] A. Majumdar and D.A. Antoniadis, "Analysis of Carrier Transport in Short-Channel MOSFETs," *IEEE Trans. Electron. Dev.*, **61**, pp. 351-358, 2014.

For the essentials of Fermi-Dirac integrals, see:

[9] R. Kim and M.S. Lundstrom, "Notes on Fermi-Dirac Integrals," 3rd Ed., `https://www.nanohub.org/resources/5475`.

The Landauer approach to carrier transport at the nanoscale is discussed in Vol. 2 of this series. Expressions for $D_{1D}(E)$ and $M_{1D}(E)$ can be found here.

[10] Mark Lundstrom and Changwook Jeong, *Near-Equilibrium Transport: Fundamentals and Applications*, World Scientific Publishing Company, Singapore, 2012.

Lecture 14

The Ballistic Injection Velocity

14.1 Introduction

The drain current of a MOSFET is the product of charge and velocity,

$$I_{DS} = W|Q_n\,(x=0, V_{GS}, V_{DS})\,|\,\upsilon(x=0, V_{GS}, V_{DS})\,. \tag{14.1}$$

In Lecture 13 we showed that by equating eqn. (14.1) to the ballistic drain current, eqn. (13.13), we obtain an expression for the average velocity of carriers at the top of the barrier (located at $x = 0$) as

$$\begin{aligned} \upsilon(x=0, V_{GS}, V_{DS}) &= \langle\langle \upsilon_x^+ \rangle\rangle \left[\frac{1-\mathcal{F}_{1/2}(\eta_{FD})/\mathcal{F}_{1/2}(\eta_{FS})}{1+\mathcal{F}_0(\eta_{FD})/\mathcal{F}_0(\eta_{FS})}\right] \\ \langle\langle \upsilon_x^+ \rangle\rangle = \upsilon_{inj}^{\text{ball}} &= \sqrt{\frac{2k_BT}{\pi m^*}\frac{\mathcal{F}_{1/2}(\eta_{FS})}{\mathcal{F}_0(\eta_{FS})}}\,. \end{aligned} \tag{14.2}$$

Equations (14.2) assume 2D electrons in the channel of a planar MOSFET. For 1D electrons in the channel of a nanowire MOSFET, different orders of the Fermi-Dirac integrals result.

The average velocity at the top of the barrier (the injection velocity) depends on both the gate and drain voltages. In this lecture, we'll discuss these dependencies. An important goal is to understand velocity saturation at high drain voltages in ballistic MOSFETs and then to understand how to compute the magnitude of the the saturated velocity. As shown in Fig. 13.2, the computed IV characteristics of ballistic MOSFETs display the signature of velocity saturation (the saturation current varies approximately linearly with $(V_{GS} - V_T)$), but it's clear that the cause of this saturation in a ballistic MOSFET cannot involve the scattering limited velocity, υ_{sat}, as discussed in Sec. 4.4. We will see that the velocity does, indeed, saturate in a ballistic MOSFET — for reasons that are easy to understand but that are much different than velocity saturation in bulk semiconductors under high electric field.

14.2 Velocity vs. V_{DS}

Equation (14.2) describes how the average velocity at the top of the barrier varies with voltage. For Maxwell-Boltzmann statistics, the equation simplifies to

$$\begin{aligned} \upsilon(0) &= \upsilon_T \left[\frac{1 - e^{qV_{DS}/k_BT}}{1 + e^{qV_{DS}/k_BT}} \right] \\ \upsilon_T &= \langle\langle \upsilon_x^+ \rangle\rangle = \upsilon_{inj}^{\text{ball}} = \sqrt{\frac{2k_BT}{\pi m^*}} \,. \end{aligned} \tag{14.3}$$

Figure 14.1 is a sketch of $\upsilon(0)$ vs. V_{DS} along with energy band diagrams for low V_{DS} and for high V_{DS}. For low V_{DS}, $\upsilon(0) \propto V_{DS}$, and for high V_{DS}, $\upsilon(0)$ saturates at υ_T.

The velocity vs. drain voltage sketched in Fig. 14.1 is much like the result from the traditional analysis; the velocity is proportional to the drain voltage for low voltage, and it saturates at high voltages. Note, however, that this velocity is at the top of the source to channel barrier, at $x = 0$. The velocity saturates at the source end of the channel, at the top of the barrier where the electric field is zero and not at the drain end of the channel where the electric field is high.

To understand the proportionality of the velocity to V_{DS} for small voltages, we expand the exponentials in eqn. (14.3) for small argument to find

$$\upsilon(0) = \frac{\upsilon_T}{2k_BT/q} V_{DS} \,. \tag{14.4}$$

Now multiply and divide by the channel length, L, to find

$$\upsilon(0) = \left(\frac{\upsilon_T L}{2k_BT/q} \right) \frac{V_{DS}}{L} \,. \tag{14.5}$$

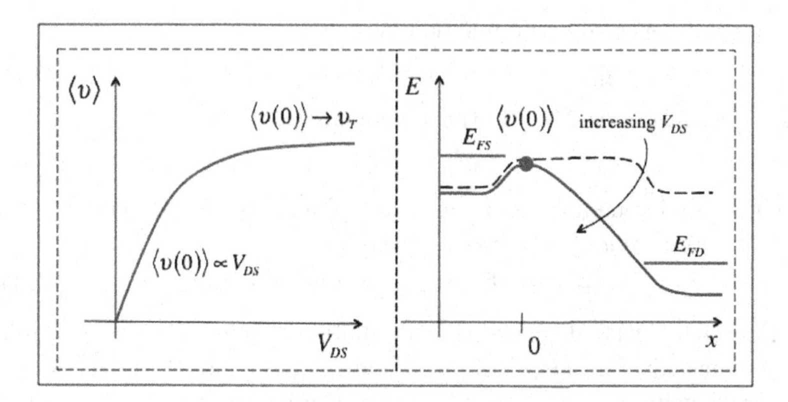

Fig. 14.1 Left: Sketch of the average velocity vs. V_{DS} according to eqn. (14.3). Right: Corresponding energy band diagrams under low and high V_{DS}. Maxwell-Boltzmann statistics are assumed.

The first term on the RHS can be recognized as the ballistic mobility from eqn. (12.48), and the second term is the electric field in the channel, $\mathcal{E}_x = V_{DS}/L$. The result is that the velocity for low V_{DS} can be written as

$$\upsilon(0) = \mu_B \mathcal{E}_x \,. \tag{14.6}$$

The low V_{DS} velocity in the ballistic MOSFET can, therefore, be written as in the traditional analysis where $\upsilon(0) = \mu_n \mathcal{E}_x$ with μ_n being the scattering limited velocity, but we need to replace μ_n with μ_B .

14.3 Velocity saturation in a ballistic MOSFET

According to eqn. (14.3), the average x-directed velocity at the top of the barrier saturates for high drain voltages. To see exactly how this occurs, we should examine the microscopic distribution of velocities in the $x-y$ plane of the channel. First, let's recall how they are distributed in a nondegenerate, bulk semiconductor in equilibrium. For a nondegenerate semiconductor, the Fermi function simplifies to

$$f_0(E) = \frac{1}{1 + e^{(E-E_F)/k_BT}} \rightarrow e^{(E_F-E)/k_BT} \,. \tag{14.7}$$

For electrons in a parabolic conduction band,

$$E = E_c + m^* \upsilon^2/2 \,, \tag{14.8}$$

so the nondegenerate Fermi function becomes

$$f_0(\upsilon) = e^{(E_F - E_c)/k_B T} \times e^{-m^* \upsilon^2 / 2k_B T} . \tag{14.9}$$

Since electrons move freely in the $x - y$ plane,

$$\upsilon^2 = \upsilon_x^2 + \upsilon_y^2 , \tag{14.10}$$

and the nondegenerate Fermi function reduces to the well-known *Maxwellian distribution* of velocities as given by

$$f_0(\upsilon_x, \upsilon_y) = e^{(E_F - E_c)/k_B T} \times e^{-m^* (\upsilon_x^2 + \upsilon_y^2)/2k_B T} . \tag{14.11}$$

Equation (14.11) describes the distribution of velocities in a nondegenerate semiconductor in equilibrium. Figure 14.2 is a plot of the Maxwellian velocity distribution. As expected, every positive velocity is balanced by a negative velocity, so the average velocity is zero in equilibrium.

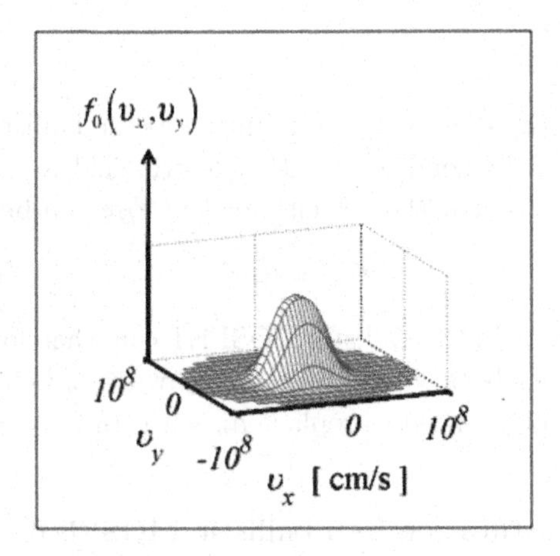

Fig. 14.2 Plot of the Maxwellian distribution of electron velocities in a nondegenerate semiconductor in equilibrium. (Reprinted from *Solid-State Electron.*, **46**, pp. 1899-1906, J.-H. Rhew, Zhibin Ren, and Mark Lundstrom, "A Numerical Study of Ballistic Transport in a Nanoscale MOSFET," ©2002, with permission from Elsevier.)

A ballistic MOSFET under high drain bias is far from equilibrium, so we expect the distribution of carrier velocities to be much different from the equilibrium distribution shown in Fig. 14.2. Figure 14.3 shows the results of a numerical solution of the Boltzmann Transport Equation for a 10 nm channel length ballistic MOSFET. The gate voltage is high, so the source to channel energy barrier is low. As indicated on the right of Fig. 14.3, we seek to understand the distribution of carrier velocities at the top of the barrier as the drain voltage increases from $V_{DS} = 0$ V to $V_{DS} = V_{DD}$.

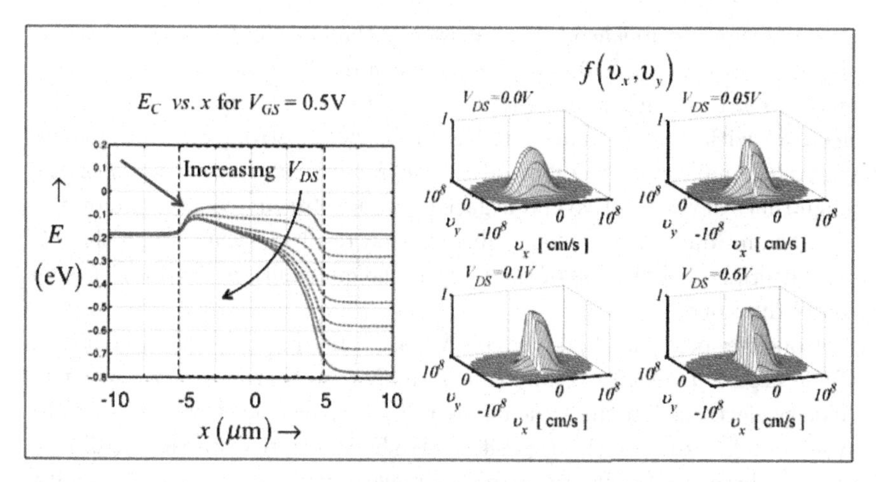

Fig. 14.3 Results of numerical simulations of a ballistic MOSFET. Left: $E_c(x)$ vs. x at a high gate voltage for various drain voltages. Right: The velocity distributions at the top of the barrier at four different drain voltages. (Reprinted from *Solid-State Electron.*, **46**, pp. 1899-1906, J.-H. Rhew, Zhibin Ren, and Mark Lundstrom, "A Numerical Study of Ballistic Transport in a Nanoscale MOSFET," ©2002, with permission from Elsevier.)

In a ballistic MOSFET, the distribution of velocities at the top of the barrier consists of two components, a positive velocity component injected from the source and a negative velocity component injected from the drain. These two components are given by

$$\begin{aligned} f^+(v_x > 0, v_y) &= e^{(E_{FS}-E_c(0))/k_BT} \times e^{-m^*(v_x^2+v_y^2)/2k_BT} \\ f^-(v_x < 0, v_y) &= e^{(E_{FD}-E_c(0))/k_BT} \times e^{-m^*(v_x^2+v_y^2)/2k_BT}, \end{aligned} \tag{14.12}$$

where E_{FS} is the Fermi level in the source, and $E_{FD} = E_{FS} - qV_{DS}$ is the Fermi level in the drain. As V_{DS} increases, the magnitude of $f^-(v_x, v_y)$ decreases.

On the right side of Fig. 14.3 is a plot of the velocity distributions at four different drain voltages. Consider first the $V_{DS} = 0$ case where the velocity distribution has an equilibrium shape and $v(0) = 0$. Since $V_{DS} = 0$, no current flows and the MOSFET is in equilibrium, so the observation of an equilibrium distribution of velocities is not a surprise. It is interesting, however, to ask how equilibrium is established, since it is the exchange of energy between electrons and phonons (via electron-phonon scattering) that brings the electron and phonon systems into equilibrium at a single temperature, T. There is no scattering in a ballistic MOSFET, so how is equilibrium established? The answer is that at the top of the barrier, all electrons with $\upsilon_x > 0$ came from the source, in which strong electron-phonon scattering maintains equilibrium. Also at the top of the barrier, all electrons with $\upsilon_x < 0$ came from the drain, where strong electron-phonon scattering maintains equilibrium. Since $E_{FS} = E_{FD}$ at $V_{DS} = 0$, the magnitudes of the positive and negative components are equal, so an overall equilibrium Maxwellian velocity distribution results. Even though there is no scattering near the top of the barrier, the distribution of velocities is an equilibrium one.

Consider next the $V_{DS} = 0.05$ V case. In this case, the magnitude of the negative velocity component is smaller, so there are fewer negative velocity electrons but the same number of positive velocity electrons, so the average x-directed velocity is positive. We have seen that for this small V_{DS} regime, the average velocity increases linearly with V_{DS}. The $V_{DS} = 0.1$ V velocity distribution shows an even smaller negative velocity component, so the average x-directed velocity is even larger. Finally, the $V_{DS} = 0.6$ V velocity distribution shows no negative velocity electrons because the drain Fermi level has been lowered so much that the probability of a negative velocity state at the top of the barrier being occupied is negligibly small. The average x-directed velocity is as large as it can be; further increases in the drain voltage will not increase the velocity — the velocity has saturated.

Figure 14.3 explains the velocity vs. drain voltage characteristic we derived in eqn. (14.3), but there is a subtle point that should be discussed. A careful look at the hemi-Maxwellian distribution for $V_{DS} = 0.6$ V shows that it is larger than the positive half of the equilibrium distribution for $V_{DS} = 0$ V. Apparently, the positive half of the distribution increased even though the source Fermi level, E_{FS}, did not change. A careful look at the left figure in Fig. 14.3 shows that $E_c(0)$ is pushed down for increasing V_{DS}. This is a result of MOS electrostatics in a well-designed MOSFET.

In a well-designed MOSFET, the charge at the top of the barrier, $Q_n(0)$, depends only (or strongly) on the gate voltage and does not change substantially with increasing drain voltage (i.e. the DIBL is low). As the population of negative velocity electrons decreases with increasing V_{DS}, more positive velocity electrons must be injected to balance the charge on the gate. Since the source Fermi level does not change, eqn. (14.12) shows that $E_c(0)$ must decrease in order to increase the charge injected from the source and satisfy MOS electrostatics.

Finally, we note that the overall shapes of the velocity distributions for $V_{DS} > 0$ are much different from the equilibrium shape, but each half has an equilibrium shape. Scattering is what returns a system to equilibrium, and there is no scattering in the channel of a ballistic MOSFET. The ballistic device is very far from equilibrium, but each half of the velocity distribution is in equilibrium with one of the two contacts.

14.4 Ballistic injection velocity

The ballistic injection velocity, $\upsilon_{\text{inj}}^{\text{ball}} = \langle\langle \upsilon_x^+ \rangle\rangle$, is an important device parameter — it plays the role of υ_{sat} in the traditional velocity saturation model. As indicated in Fig. 14.3, $\langle\langle \upsilon_x^+ \rangle\rangle$ is the average x-directed velocity of the hemi-Maxwellian (or Fermi-Dirac) velocity distribution that occurs at the top of the barrier under high drain bias. It is the angle-averaged x-directed velocity at a specific energy, E, which is then averaged over energy and is given by eqn. (14.2). Equation (14.2) was derived indirectly, by deriving the current and then writing it as the product of charge and velocity. It was derived directly in Exercise 12.2. Figure 14.4 is a plot of the ballistic injection velocity vs. inversion layer density, n_S for electrons in Si at $T = 300$ K. (As discussed in the next section, we assume that only the lowest subband in the conduction band is occupied, so that the appropriate effective mass is $m^* = 0.19\,m_0$, and the valley degeneracy is $g_v = 2$). Under high drain bias, the inversion layer density at the top of the barrier is

$$n_S = \frac{N_{2D}}{2}\mathcal{F}_0(\eta_F) = \frac{g_v m^* k_B T}{2\pi\hbar^2}\mathcal{F}_0(\eta_F)\,. \tag{14.13}$$

For a given n_S, eqn. (14.13) is solved to find η_F, which is then used in eqn. (14.2) to compute $\upsilon_{\text{inj}}^{\text{ball}}$. At 300 K, the 2D density-of-states for (100) Si has a numerical value of

$$N_{2D} = 2.05 \times 10^{12}\,\text{cm}^{-2}\,. \tag{14.14}$$

For $n_S < N_{2D}$, the semiconductor can by considered to be non-degenerate and for $n_S > N_{2D}$, Fermi-Dirac statistics becomes important.

As shown in Fig. 14.4 for $n_S \ll 10^{12}\,\text{cm}^{-2}$, the semiconductor is non-degenerate; the Fermi-Dirac integrals in eqn. (14.2) reduce to exponentials, so

$$\langle\langle \upsilon_x^+ \rangle\rangle = \upsilon_{inj}^{\text{ball}} \to \upsilon_T = \sqrt{\frac{2k_BT}{\pi m*}} = 1.2 \times 10^7\,\text{cm/s}\,. \tag{14.15}$$

For $n_S > 10^{12}\,\text{cm}^{-2}$, the semiconductor becomes degenerate and $\upsilon_{\text{inj}}^{\text{ball}}$ increases. This occurs because as states near the bottom of the band are occupied, the Fermi level rises in energy and higher velocity states are occupied. The increase in injection velocity explains why the computed IV characteristics for a ballistic MOSFET shown in Fig. 13.2 show higher currents for Fermi-Dirac statistics — Fermi-Dirac statistics lead to higher velocities. The dependence of the injection velocity on gate voltage is weak, so both the Maxwell-Boltzmann and Fermi-Dirac cases in Fig. 13.2 show saturation currents that increase about linearly with V_{GS}. Thus, in both cases, we would conclude from the IV characteristic that we are dealing with a velocity saturated MOSFET. As we will see in Lecture 17, scattering

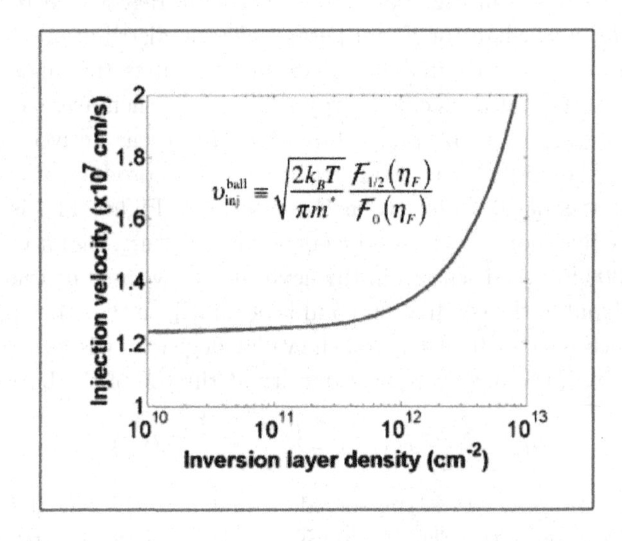

Fig. 14.4 Ballisitic injection velocity vs. sheet carrier density for 2D electrons in (100) Si. A single subband with an effective mass of $0.19m_0$ and room temperature are assumed.

in real MOSFETs reduces the injection velocity, but the ballistic injection velocity discussed here provides an upper limit to the injection velocity in a MOSFET.

Exercise 14.1: Ballistic injection velocity in the fully degenerate limit.

It is particularly easy to compute the ballistic injection velocity at $T = 0$ K where $f_0(E) = 1$ for $E < E_F$ and $f_0 = 0$ for $E > E_F$. We proceed as in Exercise 12.2 and write

$$v_{inj}^{\text{ball}} = \langle\langle v_x^+ \rangle\rangle = \frac{2}{\pi} \frac{\int_{E_c}^{\infty} \sqrt{2(E-E_c)/m^*} f_0(E) dE}{\int_{E_c}^{\infty} f_0(E) dE} = \frac{2}{\pi} \frac{\text{NUM}}{\text{DEN}} .$$

Beginning with the numerator, we have

$$\text{NUM} = \sqrt{\frac{2}{m^*}} \int_{E_c}^{\infty} (E-E_c)^{1/2} f_0(E) dE = \sqrt{\frac{2}{m^*}} \int_{E_c}^{E_F} (E-E_c)^{1/2} (1) dE ,$$

which is easily evaluated to find

$$\text{NUM} = \sqrt{\frac{2}{m^*}} \left(\frac{2}{3} (E_F - E_c)^{3/2} \right) .$$

Next, we turn to the denominator

$$\text{DEN} = \int_{E_c}^{\infty} f_0(E) dE = \int_{E_c}^{E_F} (1) dE = (E_F - E_c) .$$

Using these results we find the ballistic injection velocity as

$$v_{inj}^{\text{ball}} = \langle\langle v_x^+ \rangle\rangle = \frac{2}{\pi} \frac{\sqrt{\frac{2}{m^*}} \frac{2}{3} (E-E_c)^{3/2}}{(E_F - E_c)} = \frac{4}{3\pi} \sqrt{\frac{2}{m^*}} (E_F - E_c)^{1/2} .$$

It is useful to express this result in terms of the *Fermi velocity*, the velocity of electrons at the Fermi level, which is found from

$$\frac{1}{2} m^* v_F^2 = (E_F - E_c) .$$

The Fermi velocity is found to be

$$v_F = \sqrt{\frac{2(E_F - E_c)}{m^*}} . \tag{14.16}$$

Finally, we find the ballistic injection velocity in terms of the Fermi velocity to be

$$\boxed{v_{inj}^{\text{ball}} = \frac{4}{3\pi} v_F .} \tag{14.17}$$

As expected, the ballistic injection velocity is less that the Fermi velocity because it is the average velocity of all electrons below the Fermi level.

Exercise 14.2: Ballistic injection velocity for a realistic MOSFET.

Consider an n-channel Si MOSFET at $T = 300$ K biased in the on-state with an inversion layer density of $n_S = 10^{13}\,\text{cm}^{-2}$. Assume a (100) oriented wafer with only the bottom subband occupied. What is the ballistic injection velocity?

If the semiconductor is non-degenerate (which is not likely with n_S so high), the result would be eqn. (14.15),

$$\upsilon_{inj}^{\text{ball}} = \upsilon_T = \sqrt{\frac{2k_BT}{\pi m*}} = 1.2 \times 10^7\,\text{cm/s}\,.$$

Assuming that only the lowest subband is occupied, the correct expression is eqn. (14.2). To evaluate this expression, we need η_F, which is obtained by solving eqn. (14.13), from which we find

$$\eta_F = \log\left[e^{n_S/(N_{2D}/2)} - 1\right) = 9.76\,,$$

Using this result in eqn. (14.2), we find

$$\upsilon_{inj}^{\text{ball}} = \sqrt{\frac{2k_BT}{\pi m^*}}\frac{\mathcal{F}_{1/2}(\eta_F)}{\mathcal{F}_0(\eta_F)} = 1.2 \times 10^7\,\text{cm/s} \times \frac{\mathcal{F}_{1/2}(9.76)}{\mathcal{F}_0(9.76)}$$

$$= 1.2 \times 10^7\,\text{cm/s} \times \frac{23.2}{9.8} = 2.4 \times 10^7\,\text{cm/s}\,.$$

Note that this result is twice the result obtain with non-degenerate statistics.

As this calculation and Fig. 14.4 show, degenerate carrier statistics increase the value of the injection velocity considerably. For a typical Si MOSFET, however, the actual ballistic injection velocities are lower because multiple subbands (some with higher effective masses in the $x - y$ plane) are likely to be occupied and because quantum confinement increases the effect mass due to conduction band nonparabolicity. When quantitative predictions are needed, careful attention to bandstructure is required.

14.5 Discussion

When calculating the ballistic injection velocity for electrons in (100) Si, we assumed an effective mass of $0.19m_0$ and a valley degeneracy of $g_v = 2$.

The conduction band of Si has six equivalent valleys, and their constant energy surfaces are ellipsoids described by effective masses of $m_l^* = 0.91m_0$ and $m_t^* = 0.19m_0$. As discussed in Sec. 9.2, however, quantum confinement lifts the degeneracy of these six valleys. The lowest two subbands are degenerate with $g_v = 2$ and a circular constant energy in the $x-y$ plane with $m^* = 0.19m_0$. In the simple examples considered in this lecture (for example, when calculating the ballistic injection velocity), we have assumed that only the bottom, unprimed subband (for which the mass in the confinement direction is $m^* = m_l^*$ and the mass in the $x-y$ plane is $m^* = m_t^*$) is occupied. If higher subbands are occupied, the different subband energies and the different effective masses in the x and y directions much be accounted for. The total sheet carrier density is the sum of the contribution from each occupied subband, and the ballistic injection velocity is the average velocity in the occupied subbands.

14.6 Summary

We have discussed in this lecture the velocity vs. drain voltage characteristic of a ballistic MOSFET as well as the velocity vs. gate voltage (inversion charge) characteristic. It may, at first, be surprising that the velocity saturates with increasing drain voltage in the absence of carrier scattering, but as discussed in this lecture, the physics is easy to understand. While the velocity saturates in a ballistic MOSFET, it does not saturate near the drain end of the channel where the electric field and scattering are the highest — it saturates near the source end of the channel — at the top of be source to channel barrier where the electric field is zero.

We also discussed the saturated velocity itself, which is known as the ballistic injection velocity. This velocity provides an upper limit to the injection velocity in a MOSFET. For $n_S \ll N_{2D}/2$, the ballistic injection velocity is constant, but for $n_S \gtrsim N_{2D}/2$, it increases with increasing n_S. We discussed some simple, first order calculations of the ballistic injection velocity, but in practice, the calculation can be more complex. Quantum confinement can increase the effective mass. Multiple subbands with different effective masses can be populated, and strain, which also changes the effective mass may be present intentionally or as a byproduct of the fabrication process. Nevertheless, the basic considerations discussed in this lecture provide a clear starting point for more involved calculations.

14.7 References

For an introduction to semiconductor fundamentals such as density-of-states and quantum confinement, see:

[1] Robert F. Pierret, *Advanced Semiconductor Fundamentals, 2nd Ed.*, Vol. VI, Modular Series on Solid-State Devices, Prentice Hall, Upper Saddle River, N.J., USA, 2003.

The following online course discusses bandstructure fundamentals and topics such as the density-of-states.

[2] Mark Lundstrom, "ECE 656: Electronic Transport in Semiconductors," Purdue University, Fall 2015, //https://www.nanohub.org/groups/ece656_f15.

For a quick summary of the essentials of Fermi-Dirac integrals, which are needed to compute the ballistic injection velocity, see:

[3] R. Kim and M.S. Lundstrom, "Notes on Fermi-Dirac Integrals," 3rd Ed., https://www.nanohub.org/resources/5475.

An online tool to compute Fermi-Dirac integrals is available at:

[4] Xingshu Sun, Mark Lundstrom, and R. Kim, "FD integral calculator," https://nanohub.org/tools/fdical.

Lecture 15

Connecting the Ballistic and VS Models

15.1 Introduction

Equations (13.13) gave the IV characteristics for ballistic MOSFETs. Equations (5.9) gave the IV characteristics according to the virtual source model. The connection between these two models is the subject of this lecture.

For any model of the IV characteristics, the drain current is the product of charge and velocity,

$$I_{DS} = W|Q_n\,(x=0, V_{GS}, V_{DS})|\,\upsilon(x=0, V_{GS}, V_{DS})\,. \tag{15.1}$$

We begin by computing $Q_n(V_{GS}, V_{DS})$ from MOS electrostatics, perhaps using the semi-empirical expression, eqn. (11.14),

$$\begin{aligned} Q_n(V_{GS}, V_{DS}) &= -m\,C_G(\text{inv})\left(\frac{k_BT}{q}\right)\ln\left(1+e^{q(V_{GS}-V_T)/mk_BT}\right) \\ V_T &= V_{T0} - \delta V_{DS}\,. \end{aligned} \tag{15.2}$$

Next, the average velocity at the top of the barrier must be determined. As discussed in the next two sections, it is done differently in the ballistic and in the VS models.

15.2 Review of the ballistic model

We summarize the ballistic model as follows. The current is given by eqn. (15.1). The charge at a given bias, (V_{GS}, V_{DS}) is determined by MOS electrostatics, perhaps by using eqn. (15.2). To determine the velocity, we first need to determine the location of the Fermi level (unless Maxwell-Boltzmann carrier statistics are used). The Fermi level is determined from the known inversion charge,

$$Q_n(V_{GS}, V_{DS}) = -q\frac{N_{2D}}{2}\left[\mathcal{F}_0(\eta_{FS}) + \mathcal{F}_0(\eta_{FD})\right] , \qquad (15.3)$$

where

$$\eta_{FS} = (E_{FS} - E_c(0))/k_BT \quad \eta_{FD} = \eta_{FS} - qV_{DS}/k_BT \, . \qquad (15.4)$$

Next, we determine the ballistic injection velocity

$$v_{inj}^{\text{ball}} = \langle\langle v_x^+ \rangle\rangle = \sqrt{\frac{2k_BT}{\pi m*}\frac{\mathcal{F}_{1/2}(\eta_{FS})}{\mathcal{F}_0(\eta_{FS})}} , \qquad (15.5)$$

and then the average velocity at the given drain and gate voltages from eqn. (13.20),

$$\upsilon(x = 0, V_{GS}, V_{DS}) = \upsilon_{inj}^{\text{ball}}\left[\frac{1 - \mathcal{F}_{1/2}(\eta_{FD})/\mathcal{F}_{1/2}(\eta_{FS})}{1 + \mathcal{F}_0(\eta_{FD})/\mathcal{F}_0(\eta_{FS})}\right] . \qquad (15.6)$$

Finally, we compute the drain current at the bias point, (V_{GS}, V_{DS}) from eqn. (15.1). Series resistance (always important in practical devices) would be included as discussed in Sec. 5.4. This procedure is how the IV characteristics shown in Fig. 13.2 were computed.

15.3 Review of the VS model

The Virtual Source model begins with eqns. (15.1) and (15.2), but then computes the average velocity differently. As discussed in Sec. 5.2,

$$\upsilon(x = 0, V_{GS}, V_{DS}) = F_{SAT}(V_{DS})\upsilon_{sat} , \qquad (15.7)$$

where the drain voltage dependence of the average velocity is given by the empirical drain saturation function,

$$F_{SAT}(V_{DS}) = \frac{V_{DS}/V_{DSAT}}{\left[1 + (V_{DS}/V_{DSAT})^{\beta}\right]^{1/\beta}} , \qquad (15.8)$$

with

$$V_{DSAT} = \frac{\upsilon_{sat} L}{\mu_n} . \tag{15.9}$$

The VS drain current at the bias point, (V_{GS}, V_{DS}) is determined from eqn. (15.1) using the charge from eqn. (15.2) and the average velocity from eqns. (15.7)-(15.9). Series resistance would be included as discussed in Sec. 5.4.

The VS model is a semi-empirical model used to fit measured IV characteristics. Since the parameters in the model are physical, we learn something about the device by fitting measured data to the model. There are only a few device-specific input parameters to this model: $C_G(\text{inv})$, V_T, m, μ_n, υ_{sat}, and L. The parameter, β, in eqn. (15.8) is also a fitting parameter, but it does not vary much within a class of devices. (Another empirical parameter, α, in the charge expression will be discussed in Sec. 19.2.) To fit the measured characteristics of small MOSFETs, the parameters for long channel MOSFETs, μ_n and υ_{sat}, have to be adjusted:

$$\mu_n \rightarrow \mu_{app} \quad \upsilon_{sat} \rightarrow \upsilon_{inj} . \tag{15.10}$$

As we'll discuss in this lecture, the apparent mobility, μ_{app}, and the injection velocity, υ_{inj}, are not just fitting parameters — they have physical significance.

15.4 Connection

Figure 15.1 is a plot of the IV characteristics of a ballistic MOSFET as computed from eqns. (13.21), which assume Maxwell-Boltzmann carrier statistics. Appropriate device parameters were taken from [1], including the series resistance. Since the VS model is empirical, we fit it to the computed ballistic IV. The fitted parameters give $\mu_{app} = 654\,\text{cm}^2/\text{V}-\text{s}$ and $\upsilon_{inj} = 1.24 \times 10^7\,\text{cm/s}$. The parameter, β, in eqn. (15.8) was set to 2.9 (it typically varies between 1.6-2.0 for realistic Si MOSFETs operating below the ballistic limit). The physical meaning of β is not clear — it is simply a fitting parameter in the empirical F_{SAT} function used to describe the transition from linear to saturation region. The parameters, μ_{app} and $\upsilon_{inj}^{\text{ball}}$ do, however, have a clear, physical meaning.

To establish the physical meaning of μ_{app} and $\upsilon_{inj}^{\text{ball}}$, we need to relate the VS model to the Landauer model. We'll first compare linear region currents, then saturation region currents, and then briefly discuss the overall shape of the IV characteristic in Sec. 15.6.

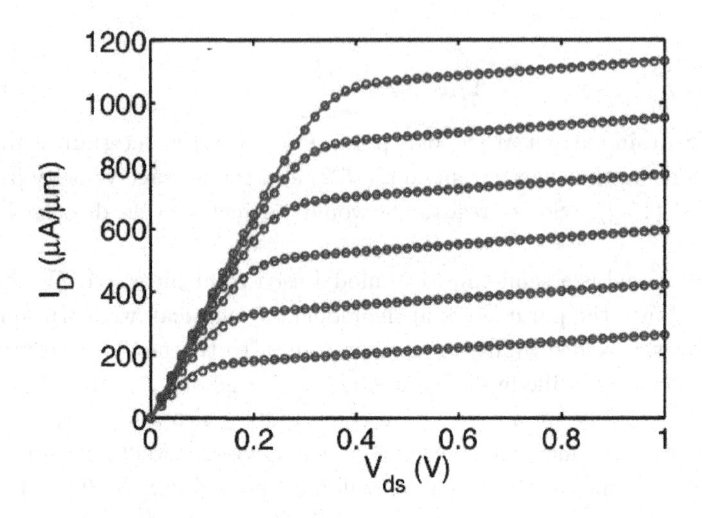

Fig. 15.1 Simulated *IV* characteristics of a ballistic MOSFET (lines). Realistic parameters for an ETSOI MOSFET were taken from [1] — including a series resistance of $R_{SD} = R_S + R_D = 260\,\Omega - \mu$m. An off-current of 100 nA/μm was assumed, which resulted in a threshold voltage of 0.44 V. Also shown is a VS model fit to the ballistic IV characteristic (symbols). The steps are from $V_{GS} = 0.5$ to $V_{GS} = 1.0$ V. (Figure provided by Xingshu Sun, Purdue University, August, 2014. Used with permission.)

Linear region: ballistic vs. VS

In Lecture 13, eqn. (13.5), we found the linear region ballistic current to be

$$I_{DLIN}^{\text{ball}} = \left[W \frac{2q^2}{h} \left(\frac{g_v \sqrt{2\pi m^* k_B T}}{2\pi\hbar} \right) \mathcal{F}_{-1/2}(\eta_F) \right] V_{DS} \,, \tag{15.11}$$

where

$$\eta_F = \left(E_{FS} - E_c(0) \right) / k_B T \,, \tag{15.12}$$

with $E_c(0)$ being the bottom of the conduction band at the top of the barrier.

For small drain bias, $F_{SAT} \rightarrow V_{DS}/V_{DSAT}$ and $\upsilon(x = 0, V_{GS}, V_{DS}) \rightarrow \mu_n V_{DS}/L$. From eqn. (15.1), the linear region drain current in the VS model becomes

$$I_{DLIN} = \frac{W}{L} |Q_n(V_{GS})| \, \mu_n V_{DS} \,, \tag{15.13}$$

which is also the result from traditional MOSFET theory. To fit the VS equation to the ballistic IV, we must adjust μ_n to the appropriate apparent mobility, μ_{app} so that eqns. (15.11) and (15.13) give the same answer. What is the physical significance of this fitted mobility?

Although eqns. (15.11) from the ballistic model and (15.13) from the VS look quite different, they are actually very similar when viewed in the right way. For example, we expect the linear region current to vary with the inversion layer charge, $Q_n(V_{GS})$, which is determined by MOS electrostatics. This is apparent in the traditional expression, eqn. (15.13), but not so apparent in the Landauer expression for the ballistic current, eqn. (15.11). Note that the magnitude of Q_n determines the location of the Fermi level (η_F), and that η_F appears in eqn. (15.11), so the dependence on Q_n is in eqn. (15.11), but only implicitly — we'd like to make this dependency explicit.

In the linear region, the relation between inversion charge and Fermi level is

$$Q_n = -qn_S = -qN_{2D}\mathcal{F}_0(\eta_F) = -q\left(\frac{g_v m^* k_B T}{\pi\hbar^2}\right)\mathcal{F}_0(\eta_F)\,. \qquad (15.14)$$

(This equation is the same as eqn. (13.12) with $\eta_{FS} \approx \eta_{FD} = \eta_F$.) Now let's write the ballistic I_{DLIN} as

$$\begin{aligned} I_{DLIN}^{\text{ball}} &= Q_n\left[\frac{G_{ch}}{Q_n}\right]V_{DS} \\ &= |Q_n|\left[\frac{W\frac{2q^2}{h}\left(\frac{g_v\sqrt{2\pi m^* k_B T}}{2\pi\hbar}\right)\mathcal{F}_{-1/2}(\eta_F)}{q\left(\frac{g_v m^* k_B T}{\pi\hbar^2}\right)\mathcal{F}_0(\eta_F)}\right]V_{DS}\,. \qquad (15.15) \end{aligned}$$

With a little algebra, this expression can be re-expressed as

$$I_{DLIN}^{\text{ball}} = W|Q_n(V_{GS})|\left[\frac{\upsilon_{inj}^{\text{ball}}}{2(k_B T/q)}\frac{\mathcal{F}_{-1/2}(\eta_F)}{\mathcal{F}_{+1/2}(\eta_F)}\right]V_{DS}\,, \qquad (15.16)$$

where $\upsilon_{\text{inj}}^{\text{ball}} = \langle\langle \upsilon_x^+ \rangle\rangle$ is the unidirectional thermal velocity as given by eqn. (12.23) (which is also (15.5)). Equation (15.16) is identical to eqn. (15.11); it just makes the connection to Q_n explicit.

Equation (15.16) still looks different from the traditional expression for the linear region current, eqn. (15.13). To make it look similar, we multiply and divide eqn. (15.16) by L and find

$$I_{DLIN}^{\text{ball}} = \frac{W}{L}|Q_n(V_{GS})| \left[\left(\frac{v_{inj}^{\text{ball}} L}{2(k_B T/q)} \right) \frac{\mathcal{F}_{-1/2}(\eta_F)}{\mathcal{F}_{+1/2}(\eta_F)} \right] V_{DS} \,. \tag{15.17}$$

Note that the dimensions of the quantity in square brackets are $\text{m}^2/\text{V}-\text{s}$, the dimensions of mobility. Accordingly, we define the ballistic mobility as [2]

$$\boxed{\mu_B \equiv \left(\frac{v_{inj}^{\text{ball}} L}{2(k_B T/q)} \right) \frac{\mathcal{F}_{-1/2}(\eta_F)}{\mathcal{F}_{+1/2}(\eta_F)} \,,} \tag{15.18}$$

which is the generalization of eqn. (12.48) for Fermi-Dirac statistics. Finally, we write the linear region current in the ballistic limit as

$$I_{DLIN}^{\text{ball}} = \frac{W}{L}|Q_n(V_{GS})| \, \mu_B V_{DS} \,, \tag{15.19}$$

which is exactly the traditional expression for the linear region current except that the mobility has been replaced by the ballistic mobility [2].

To summarize, we have shown that the ballistic linear region current, eqn. (15.11), can be written in the traditional (VS) form, eqn. (15.13), if we replace the scattering limited mobility, μ_{n} by the ballistic mobility, μ_B, as in eqn. (15.19). The physical significance of the ballistic mobility was discussed in Sec. 12.8.

Saturation region: ballistic vs. VS

In Lecture 13, we found the saturated ballistic current to be (eqn. (13.8))

$$I_{DSAT}^{\text{ball}} = W \frac{2q}{h} \left(\frac{g_v \sqrt{2m^* k_B T}}{\pi \hbar} \right) k_B T \frac{\sqrt{\pi}}{2} \mathcal{F}_{1/2}(\eta_F) \,. \tag{15.20}$$

Equation (15.20) is the correct saturated region current for a ballistic MOSFET, but it looks much different from the traditional velocity saturation expression, eqn. (4.7),

$$I_{DSAT} = W|Q_n(V_{GS}, V_{DS})| v_{sat} \,. \tag{15.21}$$

To fit eqn. (15.21) to the ballistic IV, we regard v_{sat} as a fitting parameter called the injection velocity, v_{inj}. What is the physical significant of this fitted velocity?

Again, we expect the on-current to be proportional to Q_n, so we can re-write the on-current from eqn. (15.20) as

$$I_{DSAT}^{\text{ball}} = Q_n \left[\frac{W \frac{2q}{h} \left(\frac{g_v \sqrt{2m^* k_B T}}{\pi \hbar} \right) k_B T \frac{\sqrt{\pi}}{2} \mathcal{F}_{1/2}(\eta_F)}{Q_n} \right] . \tag{15.22}$$

The next step is to relate Q_n to η_F, as we did with eqn. (15.14), but we need to be careful. Under high drain bias, only half of the states at the top of the barrier are occupied (see Fig. 14.3). This occurs because the positive velocity states continue to be occupied by positive velocity electrons that are injected from the source, but for high drain bias, the negative velocity states are empty because negative velocity electrons come from the drain where the Fermi level is low, so the probability of electrons from the drain having an energy greater than the energy at the top of the barrier is very small. Accordingly, we must divide the effective density-of-states in eqn. (15.14) by two because only one-half of the states are occupied,

$$Q_n = -qn_S = -q \frac{N_{2D}}{2} \mathcal{F}_0(\eta_F) = -q \left(\frac{g_v m^* k_B T}{2\pi \hbar^2} \right) \mathcal{F}_0(\eta_F) . \tag{15.23}$$

From eqns. (15.22) and (15.23), we find, after a little algebra,

$$I_{DSAT}^{\text{ball}} = W|Q_n| \left\langle \left\langle v_x^+ \right\rangle \right\rangle = W|Q_n| v_{inj}^{\text{ball}} , \tag{15.24}$$

where $v_{inj}^{\text{ball}} = \langle\langle v_x^+ \rangle\rangle$ is the ballistic injection velocity, which is the uni-directional thermal velocity as given by eqn. (15.5). Equation (15.24) is identical to eqn. (15.20); it just makes the connection to Q_n explicit.

To summarize, we have shown that the ballistic saturated region current, eqn. (15.20), can be written in the traditional form, eqn. (15.21) if we replace the scattering limited saturation velocity, v_{sat} by the injection velocity, v_{inj}, as in eqn. (15.24). The value of the injection velocity is the thermal average velocity at which electrons are injected from the source, $v_{inj}^{\text{ball}} = \langle\langle v_x^+ \rangle\rangle$. The physics of velocity saturation in ballistic MOSFETs was discussed in Lecture 14.

Exercise 15.1: Show that the VS fitting parameters used in Fig. 15.1 are the expected values.

When fitting the VS model to the computed ballistic *IV* characteristics in Fig. 15.1, μ_{app} and v_{inj} were simply adjusted to produce the best fit. How do the fitted parameters compare to the expected parameters?

As discussed in this section, the mobility in the VS model should be the ballistic mobility as given by eqn. (15.18), and the velocity should be the ballistic injection velocity as given by eqn. (15.5). Assuming numbers appropriate for (100) Si, ($m^* = 0.19m_0$), and Maxwell-Boltzmann statistics, we find

$$\upsilon_{inj}^{\text{ball}} = \upsilon_T = \sqrt{\frac{2k_BT}{\pi m*}} = 1.2 \times 10^7 \text{ cm/s},$$

$$\mu_B \equiv \left(\frac{\upsilon_{inj}^{\text{ball}} L}{2(k_BT/q)} \right) = 692 \text{ cm}^2/\text{V} - \text{s},$$

These results are quite close to the fitted parameters of $\upsilon_{inj} = 1.24 \times 10^7$ cm/s and $\mu_{app} = 654 \text{ cm}^2/\text{V} - \text{s}$.

15.5 Comparison with experimental results

To examine how well the ballistic theory of the MOSFET describes real transistors, we compare measured results to ballistic calculations for two different cases. The first is an $L = 30$ nm Si MOSFET — an ETSOI MOSFET from [1]. The second is an $L = 30$ nm III-V FET known as a high-electron mobility FET (or HEMT) [3]. The results for the Si MOSFET are shown in Fig. 15.2 and for the III-V HEMT in Fig. 15.3. Each figure shows the computed ballistic current (assuming Maxwell-Boltzmann statistics and including series resistance) along with the measured results and the VS fit to the measured results.

The VS fits to the measured results provide us with three key device parameters: i) the gate-voltage independent series resistance, ii) the apparent mobility, and iii) the injection velocity. The results are summarized below.

Si ETSOI MOSFET:

$$R_{SD} = R_S + R_D = 260\,\Omega - \mu\text{m}$$
$$\mu_{app} = 220\,\text{cm}^2/\text{V} - \text{s}$$
$$\upsilon_{inj} = 0.82 \times 10^7 \text{ cm/s}$$

$$\upsilon_T = 1.14 \times 10^7 \text{ cm/s}$$
$$\mu_n = 350\,\text{cm}^2/\text{V} - \text{s}$$
$$\mu_B = 658\,\text{cm}^2/\text{V} - \text{s}.$$

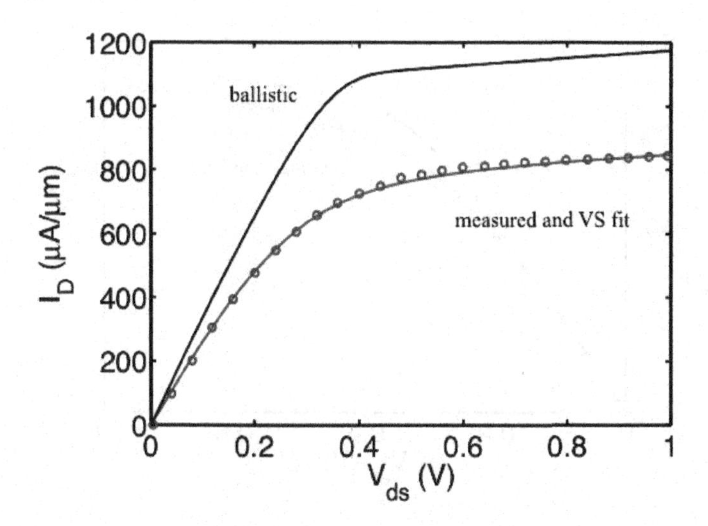

Fig. 15.2 Simulated *IV* characteristics of a ballistic ETSOI Si MOSFET (top line). Realistic parameters (including series resistance) for an ETSOI MOSFET were taken from [1]. The gate voltage is $V_{GS} = 0.5$ V. (Although this is an n-channel device, the threshold voltage is less than zero, so a substantial current flows for $V_{GS} = 0$ V.) Also shown in this figure are the measured characteristics for a 30 nm channel length device [1] (bottom line) and the fitted result for the VS model (symbols). (Figure and VS fits provided by Xingshu Sun, Purdue University, August, 2014. Used with permission.)

III-V HEMT:

$$R_{SD} = R_S + R_D = 434\,\Omega - \mu\text{m}$$
$$\mu_{app} = 1800\,\text{cm}^2/\text{V} - \text{s}$$
$$v_{inj} = 3.85 \times 10^7\,\text{cm/s}$$

$$v_T = 4.24 \times 10^7\ \text{cm/s}$$
$$\mu_n = 12,500\,\text{cm}^2/\text{V} - \text{s}$$
$$\mu_B = 2446\,\text{cm}^2/\text{V} - \text{s}$$

Also listed for comparison, are the the commuted unidirectional thermal velocities assuming Maxwell-Boltzmann carrier statistics (assuming $m^* = 0.22m_0$ for Si and $m^* = 0.016m_0$ for the III-V HEMT), the independently measured effective mobilities (the scattering limited mobility in a long channel FET), and the computed ballistic mobilities (from eqn. (15.18) assuming Maxwell-Boltzmann statistics). The apparent mobility is a fitting parameter in the VS model. We see that for both the Si and III-V FET,

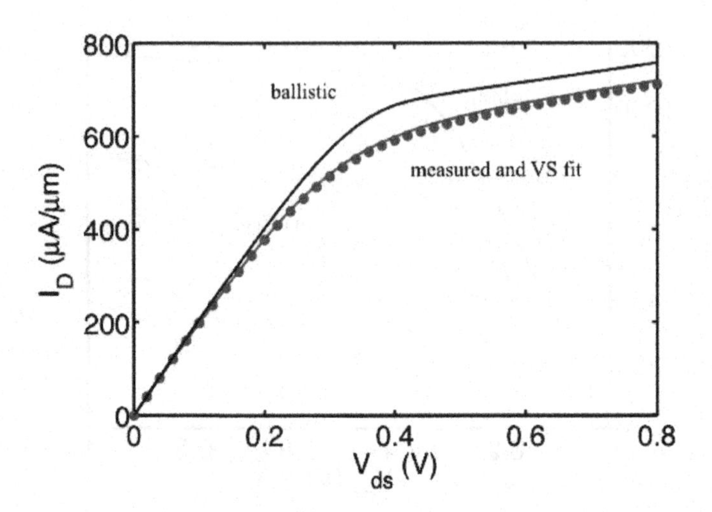

Fig. 15.3 Simulated IV characteristics of a ballistic III-V HEMT. Realistic parameters (including series resistance) were taken from [3]. The gate voltage is $V_{GS} = 0.5$ V. (Although this is an n-channel device, the threshold voltage is less than zero, so a substantial current flows for $V_{GS} = 0$ V.) Also shown in this figure are the measured characteristics for a 30 nm channel length device [3] (bottom line) and the fitted result for the VS model (symbols). (Figure and VS fits provided by Xingshu Sun, Purdue University, August, 2014. Used with permission.)

it is smaller than the smaller of the scattering-limited and ballistic mobilities. (Smaller than the ballistic mobilties in these examples.) In Sec. 18.4, we'll see that even in the presence of scattering, the apparent mobility is a well-defined physical quantity — not just a fitting parameter.

The ratio of the measured on-current to the computed ballistic on current is a measure of how close to the ballistic limit the transistor operates. From the data plotted in Figs. 15.2 and 15.3, we find

Si ETSOI MOSFET:

$$B = \frac{I_{ON}(\text{meas})}{I_{ON}(\text{ball})} = 0.73\,.$$

III-V HEMT:

$$B = \frac{I_{ON}(\text{meas})}{I_{ON}(\text{ball})} = 0.96\,.$$

The ballistic on-current ratios suggest that Si MOSFETs operate fairly close to the ballistic limit and that III-V FETs operate essentially at the ballistic limit. Note also that the apparent mobility deduced from the VS model is relatively close to the scattering limited mobility (μ_n) for Si but $\mu_{app} \ll \mu_n$ for the III-V FET. This is also an indication that the Si MOSFET operates below the ballistic limit but that the III-V FET is close to the ballistic limit. Note also that the injection velocities deduced from the VS fits are below the ballistic injection velocity (υ_T) in both cases.

Finally, we should comment on our use of Maxwell-Boltzmann carrier statistics for the analysis discussed above. Above threshold, it is more appropriate to use Fermi-Dirac statistics, but other complications such as band nonparabolicity and the occupation of multiple subbands should also be considered. Careful analyses should consider these effects, but Maxwell-Boltzmann statistics are often used to analyze experimental data and generally produce sensible results.

15.6 Discussion

We have seen in this lecture that one can clearly relate the linear region and saturation region currents of the VS model to the corresponding results for the ballistic model. We now understand why the scattering limited mobility that describes long channel transistors needs to be replaced by a ballistic mobility that comprehends ballistic transport. As also shown in this lecture, the saturation velocity in the traditional model corresponds to the ballistic injection velocity in the ballistic model. Figures 15.1–15.3 also show that the ballistic theory predicts larger currents than are observed in practice and that the shape of the ballistic I_D vs. V_{DS} characteristic is distinctly different from the measured characteristics (the transition from linear to saturation regions occurs over a smaller range of drain voltages). It turns out that the shape of the transition between the linear and saturation region depends on the drain voltage dependence of scattering. To understand this, we need to understand carrier scattering in field-effect transistors. Understanding scattering will also help us understand why the injection velocity is below the ballistic injection velocity and how to interpret the apparent mobility in the presence of scattering. Scattering is the focus of the next few lectures.

15.7 Summary

In this lecture, we have shown that the ballistic model of Lecture 13 can be clearly related to the VS model. By simply replacing the scattering limited mobility, μ_n, in the VS model with the correct ballistic mobility, the correct ballistic linear region current is obtained. By simply replacing the high-field, scattering limited bulk saturation velocity, v_{sat}, with the ballistic injection velocity, $v_{\text{inj}}^{\text{ball}}$, the correct ballistic on-current is obtained. But we also learned that the ballistic model predicts larger currents than are observed in real devices. This is due to carrier scattering, so developing an understanding of carrier scattering in nanoscale FETs is the subject of the next few lectures.

15.8 References

The ballistic IV characteristic shown in Fig. 15.1 was computed with parameters (e.g. oxide thickness, power supply, but with zero series resistance assumed) taken from the following paper. The ballistic IV characteristics shown in Fig. 15.2 were computed with series resistances taken from the corresponding VS model fits.

[1] A. Majumdar and D.A. Antoniadis, "Analysis of Carrier Transport in Short-Channel MOSFETs," *IEEE Trans. Electron. Dev.*, **61**, pp. 351-358, 2014.

The concept of ballistic mobility is discussed by Shur.

[2] M. S. Shur, "Low ballistic mobility in submicron HEMTs," *IEEE Electron Device Lett.*, **23**, pp. 511-513, 2002.

The ballistic IV characteristics of the III-V HEMT shown in Fig. 15.3 were computed with parameters taken from the following paper.

[3] D. H. Kim, J. A. del Alamo, D. A. Antoniadis, and B. Brar, "Extraction of virtual-source injection velocity in sub-100 nm III-V HFETs," in Int. Electron Dev. Mtg., (IEDM), Technical Digest, pp. 861-864, 2009.

PART 4

Transmission Theory of the MOSFET

Lecture 16

Carrier Scattering and Transmission

16.1 Introduction

To compute the IV characteristic of a ballistic MOSFET, we began with eqn. (13.1), which assumed that the transmission, $\mathcal{T}(E)$, was one. Carrier scattering from charged impurities, lattice vibrations, etc., reduces the transmission. To compute the IV characteristics in the presence of scattering, eqn. (13.1) becomes

$$I_{DS} = \frac{2q}{h} \int \mathcal{T}(E)\; M(E)\big(f_S(E) - f_D(E)\big)dE \quad \text{Amperes}\,, \tag{16.1}$$

Figure 16.1 shows schematically how a carrier trajectory in a ballistic MOSFET compares to one in which scattering occurs. As shown on the left for a ballistic MOSFET, electrons are injected from the source (where they scatter frequently) into the channel (where they don't scatter at all) and then exit by entering the drain (where they scatter frequently). The potential drop in the channel accelerates electrons, so they gain kinetic energy. The kinetic energy is deposited in the drain.

On the right of Fig. 16.1, a carrier trajectory in the presence of scattering is shown. Note that some scattering events are *elastic*, the carrier changes direction but the energy does not change. Some scattering events are *inelastic* — both the direction of motion and the energy of the electron change. For example, electrons can gain energy by absorbing a lattice vibration (a *phonon*), and they can lose energy by exciting a lattice vibration (generating a phonon). For the particular trajectory shown, the electron injected from the source exits through the drain, but scattering is a stochastic process, and for some carrier trajectories, electrons injected from the source backscatter and return to the source. The transmission from the source to drain, which is the ratio of the flux of electrons injected from the source to the flux that exits at the drain, is clearly reduced by scattering.

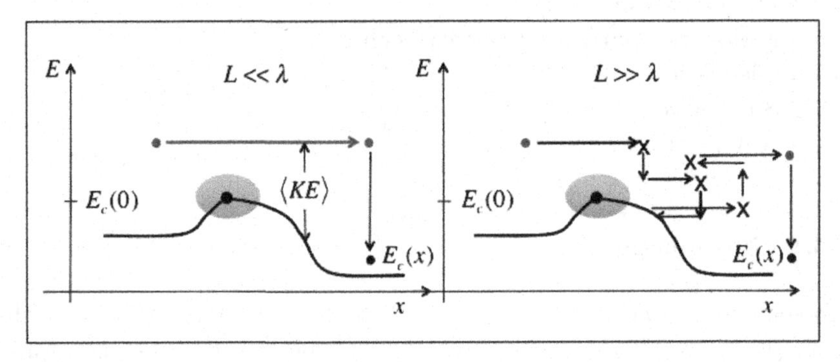

Fig. 16.1 Illustration of a ballistic (left) and quasi-ballistic (right) MOSFET. In each case, we show a carrier trajectory for an electron injected from the source at a specific energy, E. Scattering is a stochastic process, so the trajectory on the right is just one of a large ensemble of possible trajectories.

Nanoscale MOSFETs are neither fully ballistic ($\mathcal{T}(E) = 1$) nor fully diffusive ($\mathcal{T}(E) \ll 1$); they operate in a *quasi-ballistic* regime where $\mathcal{T}(E) \lesssim 1$. Our goal in this lecture is to understand some fundamentals of carrier scattering — then we will be prepared to evaluate eqn. (16.1) for the quasi-ballistic MOSFET. This lecture is a brief discussion of some fundamentals of scattering. For more on the physics of carrier scattering in semiconductors, see Chapter 2 in [1], and for a more extensive discussion of transmission, see Lecture 6 in [2].

16.2 Characteristic times and lengths

A good way to gain an understanding of scattering is through some characteristic times, such as the average time between collisions, τ, (the scattering rate, $1/\tau$, is the probability per unit time of a scattering event). One can also define characteristic lengths, like the *mean-free-path*, Λ, the average distance between scattering events ($1/\Lambda$ is the probability per unit length of scattering). In general, these characteristic times and lengths depend on the carrier's energy. We'll often be interested in average scattering times or mean-free-paths, where the average is taken over the physically relevant distribution of carrier energies.

Figure 16.2 illustrates three important characteristic times. Consider a beam of electrons with crystal momentum, $\vec{p}(E) = p(E)\hat{x}$ injected into a semiconductor at time $t = 0$. Assume that the electrons' energy, E, is much greater than the equilibrium energy, $3k_BT/2$. After a time, $\tau(E)$, every electron will, on average, have scattered once. The quantity, $\tau(E)$, is the average scattering time ($1/\tau(E)$, is the average scattering rate). Note that we are assuming that all of the states to which the electrons scatter are empty and that there is no in-scattering of electrons from other states. We might call $\tau(E)$ the out-scattering time for electrons with energy, E.

As shown in Fig. 16.2, it is also possible to define other characteristic times. For example, the dominant scattering mechanism might be elastic and anisotropic, so that scattering events don't change the energy and deflect an electron only a little. In that case, after a time, $\tau(E)$, the electrons still carry a significant x-directed momentum, and their energy (the average length of the vectors) is nearly the same as the injected energy. At a later time, the *momentum relaxation time*, $\tau_m(E)$, the initial momentum will have been relaxed and no net x-directed momentum remains, but the average energy can still be close to the injected energy if the dominant scattering mechanisms are elastic. Finally, at a still longer time, the *energy relaxation time*, $\tau_E(E)$, the injected electrons will have shed their excess energy and are then in equilibrium with no net momentum and with an energy that is equal to the lattice energy. Typically,

$$\tau_E(E) \gg \tau_m(E), \tau(E)\,, \tag{16.2}$$

because it generally takes several inelastic scattering events to shed the injected excess energy. When the scattering is isotropic (equal probability for an electron to be scattered in any direction), then $\tau(E) = \tau_m(E)$ [1].

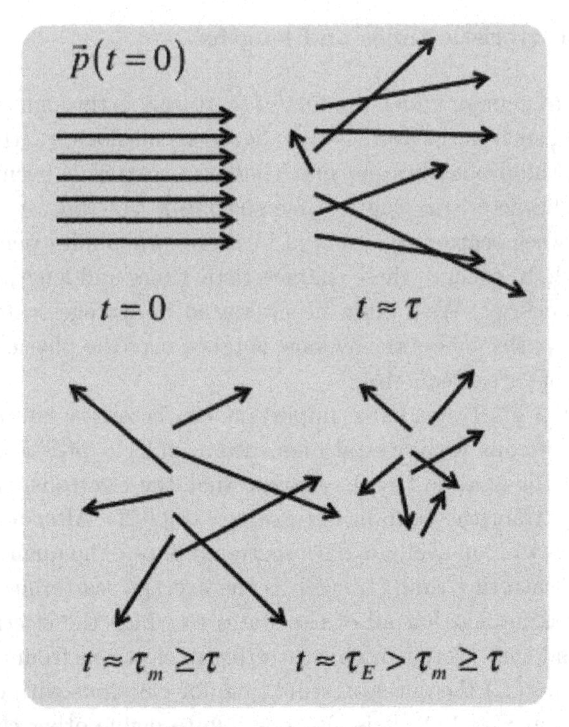

Fig. 16.2 Sketch illustrating the characteristic times for carrier scattering. An ensemble of carriers with momentum directed along one axis is injected at $t = 0$. Carriers have, on average, experienced one collision at $t = \tau(E)$. The momentum of the initial ensemble has been relaxed to zero at $t = \tau_m(E)$, and the energy has relaxed to its equilibrium value at $t = \tau_E(E)$. The length of the vectors is related to their energy. (After Lundstrom, [1]).

We can also define characteristic lengths for scattering, such as the mean-free-path, the mean-free-path for momentum relaxation, and the mean-free-path for energy relaxation. The mean-free-path,

$$\Lambda(E) = \upsilon(E)\tau(E)\,, \tag{16.3}$$

is simply the average distance between scattering events.

16.3 Scattering rates vs. energy

Characteristic scattering times are readily evaluated from the microscopic transition rate from state, $\vec{p}$ to state $\vec{p}\,'$. The transition rate, $S(\vec{p} \rightarrow \vec{p}\,')$, is the probability per unit time that an electron in state $\vec{p}$ will scatter to the state $\vec{p}\,'$. If the transition rate is known, the characteristic times are readily evaluated. For example, the out-scattering rate is the probability that an electron in state, $\vec{p}$ will scatter to any other state (assuming that the state is empty),

$$\frac{1}{\tau} = \sum_{\vec{p}\,'} S(\vec{p} \rightarrow \vec{p}\,') \,. \tag{16.4}$$

For the momentum relaxation rate, we need to weight by the fractional change in x-directed momentum for each scattering event

$$\frac{1}{\tau_m} = \sum_{\vec{p}\,'} S(\vec{p} \rightarrow \vec{p}\,') \frac{\Delta p_x}{p_x} \,. \tag{16.5}$$

Similarly, to find the energy relaxation rate, we would weight by the fractional change in carrier energy for each scattering event.

We see that if the microscopic transition rate, $S(\vec{p} \rightarrow \vec{p}\,')$, is known then the characteristic times and lengths relevant for transport calculations can be evaluated. For a discussion of how this is done for some common scattering mechanisms, see [1]; we will simply state a few key results here.

According to eqn. (16.4), the out-scattering rate is related to the number of final states at energy, $E(\vec{p}\,')$, available for the scattered electron. Specific scattering mechanisms may select out specific final states (see the discussion of charged impurity scattering below), but in the simplest case, the scattering rate should be proportional to the density of final states. For isotropic, elastic scattering of electrons in the conduction band, we find

$$\frac{1}{\tau(E)} = \frac{1}{\tau_m(E)} \propto D(E - E_c) \,, \tag{16.6}$$

where $D(E - E_c)$ is the density of states. For isotropic, inelastic scattering in which an electron absorbs or emits an energy, $\hbar\omega$ (e.g. from a phonon), we find

$$\frac{1}{\tau(E)} = \frac{1}{\tau_m(E - E_c)} \propto D(E \pm \hbar\omega - E_c) \,. \tag{16.7}$$

For simple, parabolic energy bands, analytical expressions for the scattering times can be developed [1], but for more complex band structures, a numerical sum over the final states is needed.

For semiconductor work, the scattering times are often written in *power law* form as

$$\tau_{m(E)} = \tau_{mo} \left(\frac{E - E_c}{k_B T} \right)^s \,, \tag{16.8}$$

where s is a characteristic exponent that describes the particular scattering mechanism. For example, acoustic phonon scattering can be considered to be nearly elastic and isotropic at room temperature. The scattering rate should be proportional to the density-of-states, which for 3D electrons with parabolic energy bands is proportional to $(E - E_c)^{1/2}$, so the scattering time should be proportional to $(E - E_c)^{-1/2}$. The characteristic exponent for acoustic phonon scattering is $s = -1/2$. For 2D electrons, the density-of-states is independent of energy, so the characteristic exponent is $s = 0$, and for 1D electrons, the density-of-states is proportional to $(E - E_c)^{-1/2}$, so the characteristic exponent for power law scattering is $s = +1/2$. It is not always possible to write the scattering time in power law form, but when it is possible, it simplifies calculations.

When the scattering involves an electrostatic interaction, as for charged impurity scattering or phonon scattering in polar materials, the dependence of scattering time on energy is different. As illustrated in Fig. 16.3, randomly located charges introduce fluctuations into the bottom of the conduction band, $E_c(\vec{r})$, which can scatter carriers. High energy carriers, however, do not feel this fluctuating potential as much as low energy carriers, so for charged impurity (and polar phonon) scattering, we expect that $1/\tau(E)$ will decrease (the scattering time, $\tau(E)$, will increase) as the carrier energy increases. The scattering time can be written in power law form with a characteristic exponent of $s = +3/2$ for 3D electrons [1]. For nonpolar phonon scattering, the scattering time decreases with energy, but for charged impurity and polar phonon scattering, it increases with energy.

One final point about charged impurity scattering should be mentioned — it is anisotropic. Most electrons are far away from the charged impurity, so their trajectories are deflected only a little. The result is that the momentum relaxation time for charged impurity scattering is significantly longer than the scattering time, $\tau_m(E) \gg \tau(E)$.

The carrier mean-free-path can also be written in power law form. From eqn. (16.3) and recalling that for parabolic energy bands $\upsilon(E) \propto (E - E_c)^{1/2}$, we find

$$\Lambda(E) = \upsilon(E)\tau(E) \propto (E - E_c)^{1/2} \left(\frac{E - E_c}{k_b T}\right)^s = \Lambda_o \left(\frac{E - E_c}{k_B T}\right)^r , \quad (16.9)$$

where $r = s + 1/2$ is the characteristic exponent for the mean-free-path. For acoustic phonon scattering in 3D, $s = -1/2$, so $r = 0$ — the mean-free-path is independent of energy. For acoustic phonon scattering in 2D, $s = 0$, so $r = 1/2$ — the mean-free-path increases with energy.

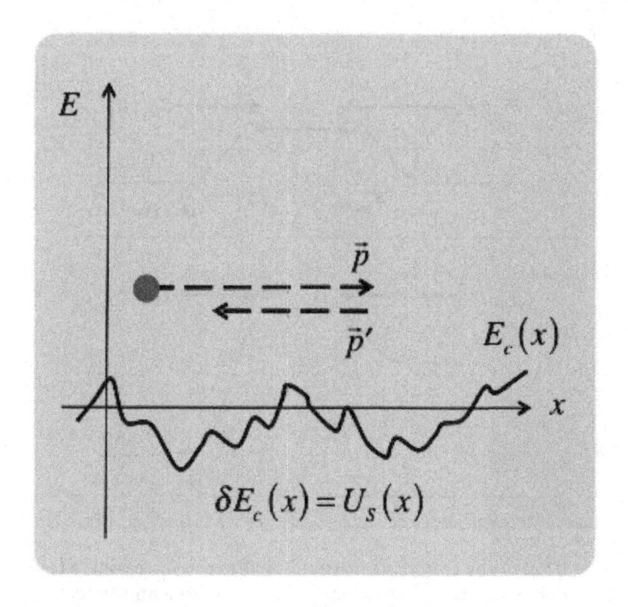

Fig. 16.3 Illustration of charged impurity scattering. High energy carriers feel the perturbed potential less than low energy carriers and are, therefore, scattered less. From Lundstrom and Jeong [2].

16.4 Transmission

Figure 16.4 illustrates the difference between the transmission from source to drain, $\mathcal{T}_{SD}(E)$, and the transmission from drain to source, $\mathcal{T}_{DS}(E)$. The quantity, $\mathcal{T}_{SD}(E)$, is the ratio of the steady-state flux of electrons that exits at the drain to the flux injected at the source; $\mathcal{T}_{DS}(E)$ is similarly defined for injection from the drain. For zero (or small) drain bias, we expect the two transmissions to be equal, $\mathcal{T}_{SD}(E) \approx \mathcal{T}_{DS}(E) = \mathcal{T}(E)$. This case is illustrated on the top of Fig. 16.4. The case of large drain bias is shown on the bottom of Fig. 16.4. In this case, it is not at all clear that $\mathcal{T}_{SD}(E)$ should be equal to $\mathcal{T}_{DS}(E)$, but it can be shown that for elastic scattering the two are equal. For inelastic scattering, however, the two transmissions can be quite different with $\mathcal{T}_{DS}(E) \ll \mathcal{T}_{SD}(E)$.

For modeling current in MOSFETs, the fact that $\mathcal{T}_{DS}(E) \ll \mathcal{T}_{SD}(E)$ for large drain bias does not matter much, because for large drain bias, the magnitude of the flux injected from the drain is very small anyway. So we will assume that we only need to compute one transmission function, $\mathcal{T}(E)$, and that it describes transmission in either direction.

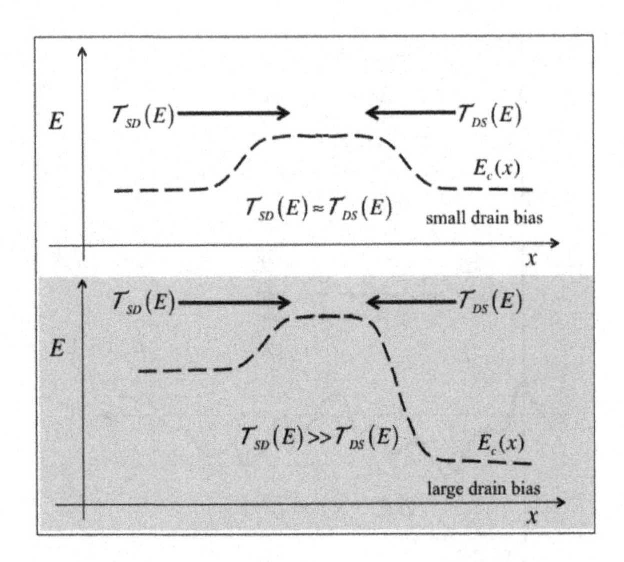

Fig. 16.4 Illustration of the two transmission functions — from source to drain and from drain to source. For $\mathcal{T}_{SD}$, we inject a flux from the source and determine the fraction that exits from the drain. For $\mathcal{T}_{DS}$, we inject a flux from the drain and determine the fraction that exits from the source. Top: Low drain bias. Bottom: High drain bias.

In Sec. 12.4, we argued that the transmission is related to the mean-free-path for backscattering by

$$\mathcal{T}(E) = \frac{\lambda(E)}{\lambda(E) + L} . \tag{16.10}$$

(Notice that we are using a lower case λ for the mean-free-path now instead of the upper case Λ as in eqn. (16.3). As discussed in Lecture 12, λ, the mean-freepath for backscattering, is the mean-free-path to use in the formula for transmission.) It is relatively easy to derive the transmission [2], but it is also easy to see that it makes sense.

Equation (16.10) describes the transmission from the ballistic to diffusive limits. When the slab is short compared to a mean-free-path, then

$$\mathcal{T}(E) = \frac{\lambda(E)}{\lambda(E) + L} \to 1 \quad (L \ll \lambda(E)) , \tag{16.11}$$

and transport across the slab is ballistic. When the slab is long compared to a mean-free-path, then

$$\mathcal{T}(E) = \frac{\lambda}{\lambda + L} \to \frac{\lambda}{L} \ll 1 \quad (L \gg \lambda) . \tag{16.12}$$

Equation (16.10) described the transmission of carriers across a region with no electric field. What happens if there is a strong electric field in the slab, as illustrated in Fig. 16.5? This sketch shows a short region with a large potential drop. An equilibrium flux of electrons is injected from the left. The injected carriers quickly gain kinetic energy, and their scattering rate increases. Simulating electron transport across short, high-field regions like this where effects such a *velocity overshot* occur is one of the most challenging problems in semiclassical carrier transport theory [1]. Computing the average velocity versus position is a difficult problem, but detailed simulations show that in terms of transmission, the end result is simple [3]. It is found that if the injected carriers penetrate just a short distance into the high field region without scattering, then even if they do subsequently scatter, they are bound to emerge from the right [3]. Even when there is a significant amount of scattering, the transmission is nearly one because the high electric field sweeps carriers across and out the right contact. The region acts as a nearly perfect carrier collector — the absorbing contact shown in Fig. 12.4.

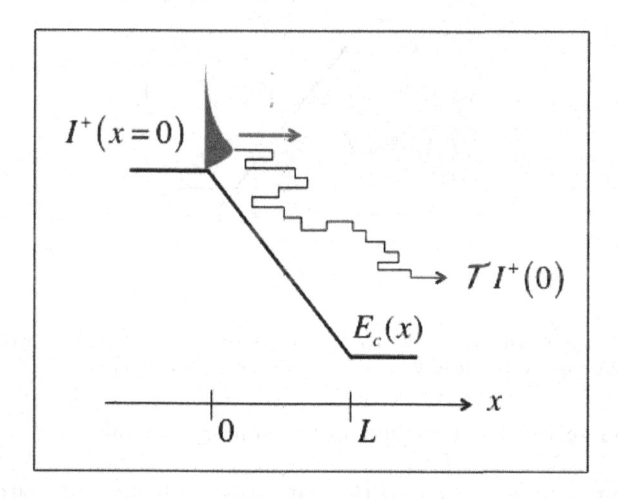

Fig. 16.5 Illustration of an electron trajectory in a short region with a high electric field. Electrons are injected from an equilibrium distribution at the left, and most exit at the right even if they scatter several times within the region. See [3] for a discussion of these results.

In a well-designed MOSFET under high drain bias, the electric field is low near the top of the source to channel barrier and high near the drain. To understand what happens for such electric field profiles, consider the model structure sketched in Fig. 16.6. Here, we model the channel profile as a short, constant potential region of length, L_1, and mean-free-path, λ_1, followed by a high-field region of length, L_2. The transmission across the first region is $\mathcal{T}_1 = \lambda_1/(\lambda_1 + L_1)$, and the transmission across the second region is $\mathcal{T}_2 \approx 1$. The composite transmission of the entire structure is $\mathcal{T} \approx \lambda_1/(\lambda_1 + L_1)$. The important point is that transmission across a structure with an initial low electric field followed by a high electric field is controlled by the length of the low field region. In practice, when the electric field varies smoothly with position, it may be difficult to precisely specify the length of the low field region [4, 5], but this simple picture provides a clear explanation for what detailed simulations confirm.

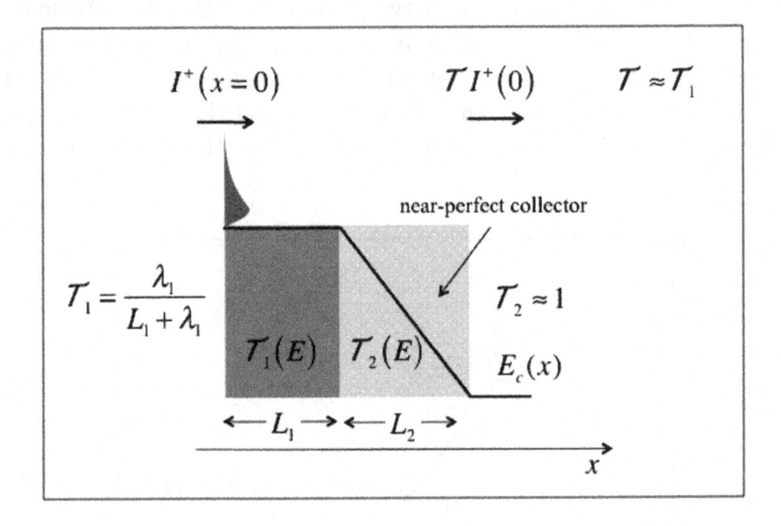

Fig. 16.6 A model channel profile that illustrates electron transmission across a region with an initial low electric field followed by a region with high electric field.

We summarize this discussion of transmission as follow.

(1) Transmission is related to the mean-free-path for backscattering according to $\mathcal{T} = \lambda/(\lambda + L)$.
(2) Ballistic transport occurs when $\mathcal{T} \to 1$, which happens when $L \ll \lambda$.
(3) Diffusive transport occurs when $\mathcal{T} \to \lambda/L \ll 1$, which happens when $L \gg \lambda$.
(4) Regions with a high electric field are good carrier collectors, $\mathcal{T} \approx 1$.
(5) For a structure in which the electric field varies from low to high (as in the channel of a MOSFET under high drain bias), the transmission is controlled by the low field region.

16.5 Mean-free-path for backscattering

In this lecture, we introduced two mean-free-paths. The mean-free-path, Λ, as defined in eqn. (16.3) is the average distance between scattering events. This is what most people mean when they refer to "mean-free-path." The quantity, $1/\Lambda$, is the probability per unit length of scattering. For our purposes, however, λ, the *mean-free-path for backscattering*, is a more relevant mean-free-path. The quantity, $1/\lambda$, is the probability per unit length that a forward (positive) flux will backscatter to a reverse (negative) flux. The transmission, eqn. (16.10) is expressed in terms of the mean-free-path for backscattering, λ. What is the relation between λ and Λ?

Figure 16.7 illustrates scattering in 1D (perhaps a nanowire MOSFET). Assume that scattering is isotropic and that the average time between scattering events is τ. If a forward-directed flux scatters after a time, τ, it has equal probability of scattering in a forward or reverse directions. Only backscattering, which happens on average after a time, 2τ matters for the current. Accordingly, the mean-free-path for backscattering in 1D is

$$\lambda(E) = 2v(E)\tau_m = 2\Lambda\,, \tag{16.13}$$

where we have used the momentum relaxation time because we assumed isotropic scattering for which $\tau_m = \tau$.

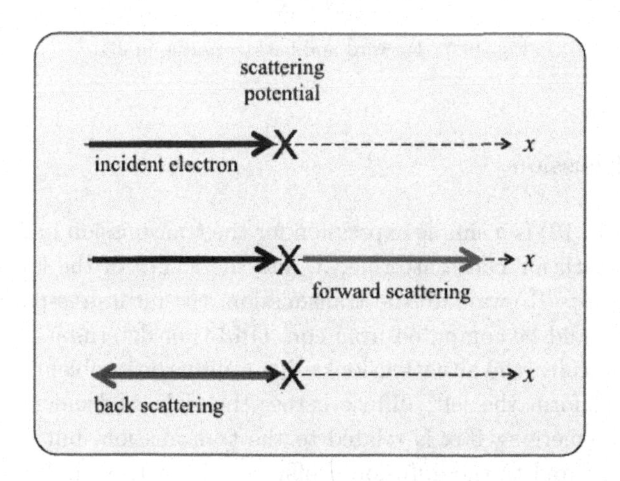

Fig. 16.7 Forward and backscattering in 1D. (From Lundstrom and Jeong [2])

In 2D and 3D, the definition of the backscattering mean-free-path involves an average over angles as illustrated in Fig. 16.8 for 2D. (See Lectures

6 and 7 in [2] for a short discussion and reference [6] for a more extensive discussion.) In 2D, the result is

$$\lambda(E) = \frac{\pi}{2} \upsilon(E) \tau_m = \frac{\pi}{2} \Lambda . \tag{16.14}$$

(In 3D, the numerical factor out front is 4/3 [2, 6].) In order to calculate transmissions properly, it is important to be aware of the distinction between λ and Λ.

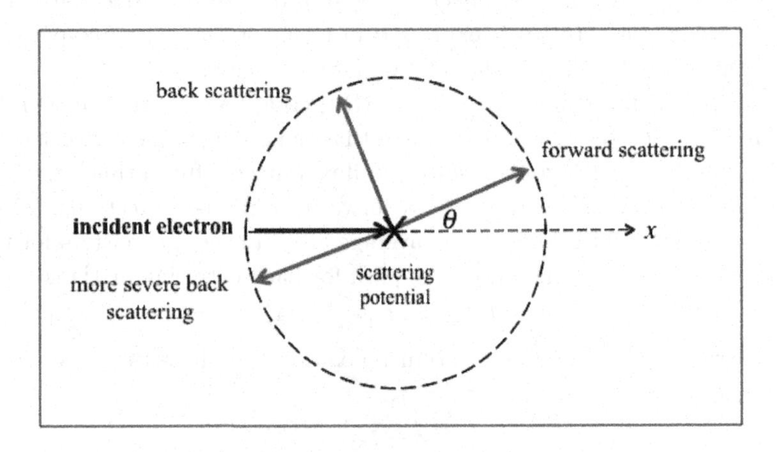

Fig. 16.8 Forward and backscattering in 2D.

16.6 Discussion

Equation (16.10) is a simple expression for the transmission in terms of the mean-free-path for backscattering, λ, and the length of the low field part of a structure. To evaluate the transmission, the mean-free-path must be known. It could be computed from eqn. (16.14) or determined experimentally. Classically, the situation looks like a diffusion problem — particles are injected form the left, diffuse across the slab, and emerge from the right. The emerging flux is related to the transmission, but classically it should be related to the diffusion coefficient. In fact, a careful analysis of this problem [2] shows that there is a simple relation between the diffusion coefficient and the average mean-free-path for backscattering.

$$D_n = \frac{\upsilon_T \langle \lambda \rangle}{2} , \tag{16.15}$$

where $\langle\lambda\rangle$ is the energy-averaged mean-free-path. This remarkably simple relation provides a way to determine the mean-free-path experimentally. (Note that this expression assumes non-degenerate carrier statistics. For the more general case, see [2].)

It is easier to measure mobilities than diffusion coefficients, so it is often easy to find data for the measured mobility. Fortunately, there is a relation between the diffusion coefficient and mobility, the Einstein relation:

$$\frac{D_n}{\mu_n} = \frac{k_B T}{q} . \tag{16.16}$$

This relation only apples near equilibrium, but electrons in the low field region, which determines the transmission of a structure, are typically near equilibrium.

So there is a simple way to estimate the mean-free-path for backscattering from the measured mobility. First, use the Einstein relation to determine the diffusion coefficient from eqn. (16.16) and then determine the mean-free-path for backscattering from eqn. (16.15). To include carrier degeneracy, which is expected to be important above threshold, see the discussion in [2].

Exercise 16.1: Mean-free-path and transmission in a 22 nm MOSFET.

Consider an $L = 22$ nm n-channel Si MOSFET at $T = 300$ K biased in the linear region. Assume a (100) oriented wafer with only the bottom subband occupied. The mobility is $\mu_n = 250\,\text{cm}^2/\text{V}-\text{s}$. What is the mean-free-path for backscattering? What is the transmission?

First, determine the diffusion coefficient from the given mobility and eqn. (16.16).

$$D_n = \mu_n \frac{k_B T}{q} = 6.5 \text{ cm}^2/\text{s} .$$

Next, assume $m^* = 0.19 m_0$ so that $\upsilon_T = 1.23 \times 10^7$ cm/s and determine the mean-free-path for backscattering from eqn. (16.15),

$$\langle\lambda\rangle = \frac{2 D_n}{\upsilon_T} = \frac{2 \times 6.5}{1.2 \times 10^7} = 10.5 \text{ nm} .$$

Finally, we determine the transmission from eqn. (16.10)

$$\mathcal{T} \approx \frac{\langle\lambda\rangle}{\langle\lambda\rangle + L} = \frac{10.5}{10.5 + 22} = 0.32 .$$

(The relation is only approximately true because the expression above is not energy averaged transmission from (16.10).) This result suggests that the MOSFET will operate at about one-third of its ballistic limit in the linear region. Under high drain bias, the carriers are more energetic, and we should expect more scattering. Surprisingly, we'll find that the MOSFET operates closer to the ballistic limit under high drain bias than under low drain bias.

16.7 Summary

This lecture began with a short primer on carrier scattering in semiconductors, and then we presented a simple relation for the transmission in terms of the length of the region and mean-free-path for backscattering. For a derivation of this result and for more discussion about the mean-free-path for backscattering, readers should consult [2]. The key results of this lecture can be summarized as follows. For 2D carriers:

$$\boxed{\begin{aligned} \mathcal{T}(E) &= \frac{\lambda(E)}{\lambda(E)+L} \\ \lambda(E) &= \frac{\pi}{2}\upsilon(E)\tau_m(E) \\ \langle\lambda\rangle &= \frac{2D_n}{\upsilon_T}\,. \end{aligned}} \tag{16.17}$$

In the first equation above, L is the length of the initial low-field part of the structure. The factor $\pi/2$ in the second equation accounts for the angle averaging for the mean-free-path for backscattering in 2D. The last equation is a simple way to estimate the average mean-free-path for backscattering, $\langle\lambda\rangle$, from the measured diffusion coefficient (when non-degenerate conditions can be assumed). With these simple concepts, we are ready to consider in the next lecture how backscattering affects the performance of a MOSFET.

16.8 References

A more complete discussion of carrier scattering in semiconductors can be found in Chapter 2 of:

[1] Mark Lundstrom, *Fundamentals of Carrier Transport, 2nd Ed.*, Cambridge Univ. Press, Cambridge, U.K., 2000.

More discussion of transmission can be found in Lecture 6 of:

[2] Mark Lundstrom and Changwook Jeong, *Near-Equilibrium Transport: Fundamentals and Applications*, World Scientific Publishing Company, Singapore, 2012.

Peter Price used Monte Carlo simulation to study electron transport in short semiconductors with a high electric field.

[3] Peter J. Price, "Monte Carlo calculation of electron transport in solids," *Semiconductors and Semimetals*, **14**, pp. 249-334, 1979.

The following papers discuss some of the issues involved in computing transmission in realistic MOSFETs.

[4] P. Palestri, D. Esseni S. Eminente, C. Fiegna, E. Sangiorgi, and L. Selmi,, "Understanding Quasi-Ballistic Transport in Nano-MOSFETs: Part I — Scattering in the Channel and in the Drain," *IEEE Trans. Electron. Dev.*, **52**, pp. 2727-2735, 2005.

[5] R. Clerc, P. Palestri, L. Selmi, and G. Ghibaudo, "Impact of carrier heating on backscattering in inversion layers," *J. Appl. Phys.* **110**, 104502, 2011.

The definition of mean-free-path for backscattering is discussed by Jeong.

[6] Changwook Jeong, Raseong Kim, Mathieu Luisier, Supriyo Datta, and Mark Lundstrom, "On Landauer vs. Boltzmann and Full Band vs. Effective Mass Evaluation of Thermoelectric Transport Coefficients," *J. Appl. Phys.*, **107**, 023707, 2010.

Lecture 17

Transmission Theory of the MOSFET

17.1 Introduction

In Lectures 13-15, we discussed the ballistic MOSFET, and in Lecture 16, we discussed carrier scattering and transmission. Now we are ready to develop a model for nanoscale MOSFETs that includes scattering. Scattering makes modeling transport difficult, and scattering in a MOSFET can be complex [1, 2]. Nevertheless, we shall see that the basic principles are easy to understand and to use for analyzing experimental data or for understanding detailed simulations.

In this lecture, we'll use the Landauer approach, eqn. (12.2), but instead of assuming ballistic transport ($\mathcal{T}(E) = 1$), as in Lecture 13, we'll retain the transmission coefficient, so that the drain current is given by

$$I_{DS} = \frac{2q}{h} \int \mathcal{T}(E) M(E) \big(f_S(E) - f_D(E)\big) dE \quad \text{Amperes}, \qquad (17.1)$$

where f_S is the Fermi function in the source and f_D the Fermi function in the drain. When the drain voltage is large, then $f_S(E) \gg f_D(E)$ for all energies of interest, and the saturation current is given by

$$I_{DSAT} = \frac{2q}{h} \int \mathcal{T}(E) M(E) f_S(E) dE \quad \text{Amperes}. \tag{17.2}$$

In the linear region, the drain to source voltage is small, $f_S \approx f_D$, so we can find the linear region current from eqn. (12.9) as

$$\begin{aligned} I_{DLIN} &= G_{ch} V_{DS} \quad \text{Amperes} \\ G_{ch} &= \frac{2q^2}{h} \int \mathcal{T}(E) M(E) \left(-\frac{\partial f_0}{\partial E} \right) dE \quad \text{Siemens}, \end{aligned} \tag{17.3}$$

where G_{ch} is the channel conductance. By evaluating these equations, we will obtain the linear region current, the on-current, and the current from $V_{DS} = 0$ to $V_{DS} = V_{DD}$. To simplify the calculations, we'll assume that the mean-free-path (and, therefore, the transmission) is independent of energy,

$$\mathcal{T}(E) = \frac{\lambda(E)}{\lambda(E) + L} \rightarrow \mathcal{T} = \frac{\lambda_0}{\lambda_0 + L}. \tag{17.4}$$

As a result, the final expressions we obtain for $\mathcal{T}$ should be regarded as an appropriately averaged transmission.

17.2 Review of the ballistic MOSFET

The ballistic *IV* characteristic of a MOSFET was derived in Lecture 13, and the final result was summarized in eqns. (13.13). The drain current can be written as

$$I_{DS} = W|Q_n(V_{GS}, V_{DS})| F_{SAT} \upsilon_{inj}^{\text{ball}}. \tag{17.5}$$

The drain voltage saturation function is given by

$$\begin{aligned} F_{SAT} &= \left[\frac{1 - \mathcal{F}_{1/2}(\eta_{FD})/\mathcal{F}_{1/2}(\eta_{FS})}{1 + \mathcal{F}_0(\eta_{FD})/\mathcal{F}_0(\eta_{FS})} \right] \quad \text{(Fermi – Dirac, FD)} \\ F_{SAT} &= \left[\frac{1 - e^{-qV_{DS}/k_BT}}{1 + e^{-qV_{DS}/k_BT}} \right] \quad \text{(Maxwell – Boltzmann, MB)} \\ \eta_{FS} &= (E_{FS} - E_c(0))/k_BT \quad \eta_{FD} = \eta_{FS} - qV_{DS}/k_BT. \end{aligned} \tag{17.6}$$

The ballistic injection velocity is

$$
\begin{aligned}
v_{inj}^{\text{ball}} &= v_T \frac{\mathcal{F}_{1/2}(\eta_{FS})}{\mathcal{F}_0(\eta_{FS})} && \text{(FD)} \\
v_{inj}^{\text{ball}} &= v_T && \text{(MB)} \\
v_T &= \sqrt{\frac{2k_BT}{\pi m^*}} \, .
\end{aligned}
\tag{17.7}
$$

The linear region ballistic currents for Fermi-Dirac and for Maxwell-Boltzmann statistics are

$$
\begin{aligned}
I_{DLIN} &= W|Q_n(V_{GS}, V_{DS})| \left(\frac{v_{inj}^{\text{ball}}}{2k_BT/q} \right) \frac{\mathcal{F}_{-1/2}(\eta_{FS})}{\mathcal{F}_{1/2}(\eta_{FS})} V_{DS} \quad \text{(FD)} \\
I_{DLIN} &= W|Q_n(V_{GS}, V_{DS})| \left(\frac{v_T}{2k_BT/q} \right) V_{DS} \quad \text{(MB)} \, ,
\end{aligned}
\tag{17.8}
$$

where the expression for the ballistic injection velocity from eqn. (17.7) must be used. Finally, the saturation current is

$$
\begin{aligned}
I_{DSAT} &= W|Q_n(V_{GS}, V_{DS})| \, v_{inj}^{\text{ball}} \quad \text{(FD)} \\
I_{DSAT} &= W|Q_n(V_{GS}, V_{DS})| \, v_T \quad \text{(MB)} \, .
\end{aligned}
\tag{17.9}
$$

Finally, the charge at the top of the barrier is given by

$$
Q_n(V_{GS}, V_{DS}) = -q \frac{N_{2D}}{2} \left[\mathcal{F}_0(\eta_{FS}) + \mathcal{F}_0(\eta_{FD}) \right] \, . \tag{17.10}
$$

(For Maxwell-Boltzmann carrier statistics, the Fermi-Dirac integrals reduce to exponentials.)

One might guess that to include scattering, we only need to multiply the above equations by the average transmission, $\mathcal{T}$. We'll find that this is true for the linear current but not for the saturation current and not for Q_n.

17.3 Linear region

To evaluate the linear region current in the presence of scattering, we begin with eqn. (12.9), the channel conductance. For the distribution of channels, we use eqn. (13.4). For the Fermi function, we use eqn. (12.3) with $E_F \approx E_{FS} \approx E_{FD}$. The integral can be evaluated as in Sec. 12.8; the result is just like eqn. (12.41) except that the transmission in the diffusive limit,

λ_0/L, is replaced by $\mathcal{T}$ to describe transport from the ballistic to diffusive regimes. The result is

$$I_{DLIN} = \mathcal{T}\left[W\frac{2q^2}{h}\left(\frac{g_v\sqrt{2\pi m^* k_B T}}{2\pi\hbar}\right)\mathcal{F}_{-1/2}(\eta_F)\right]V_{DS}\,, \tag{17.11}$$

which is just the ballistic linear region current, eqn. (13.5), multiplied by the transmission.

Equation (17.11) is the correct linear region current for a MOSFET operating from the ballistic to diffusive limits, but it looks much different from the traditional expression, eqn. (4.5),

$$I_{DLIN} = \frac{W}{L}\,\mu_n C_{ox}\left(V_{GS} - V_T\right)V_{DS}\,. \tag{17.12}$$

In the next lecture, we'll discuss the connection between the Landauer and traditional MOSFET models.

17.4 Saturation region

To evaluate the current in the saturation region, we begin with eqn. (17.2) and evaluate the integral much like we did for the linear region current. The result is

$$I_{DSAT} = \mathcal{T}W\frac{2q}{h}\left(\frac{g_v\sqrt{2m^* k_B T}}{\pi\hbar}\right)k_B T\frac{\sqrt{\pi}}{2}\mathcal{F}_{1/2}(\eta_F)\,, \tag{17.13}$$

which is just the ballistic saturation current multiplied by the transmission. Equation (17.13) is the correct saturated region current, but it looks much different from the traditional velocity saturation expression, eqn. (6.7),

$$I_{DSAT} = WC_{ox}\left(V_{GS} - V_T\right)\upsilon_{sat}\,. \tag{17.14}$$

We'll discuss the connection between these two models in the next lecture.

17.5 From linear to saturation

In the previous two sections, we derived the ballistic drain current in the linear (low V_{DS}) and saturation (high V_{DS}) regions. The virtual source model describes the drain current across the full range of V_{DS} by connecting these two currents using an empirical drain saturation function. We'll discuss this virtual source approach in Lecture 18. In the Landauer approach, however, we can derive an expression for the drain current from low to high V_{DS}. We do so in this section, but it can be complicated to use the full range expression in practice because to do so properly requires consideration of 2D electrostatics and the drain voltage dependent transmission — as will be discussed in Sec. 17.7.

To evaluate the drain current for arbitrary drain voltage, we begin with eqn. (17.1) and evaluate the integral much like we did for the saturation region current. The result is

$$\begin{aligned} I_{DS} &= \mathcal{T} W \frac{q}{h} \left(\frac{g_v \sqrt{2\pi m^* k_B T}}{\pi \hbar} \right) k_B T \left[\mathcal{F}_{1/2}(\eta_{FS}) - \mathcal{F}_{1/2}(\eta_{FD}) \right] \\ \eta_{FS} &= \left(E_{FS} - E_c(0) \right) / k_B T \quad \eta_{FD} = \eta_{FS} - q V_{DS} / k_B T \, . \end{aligned} \tag{17.15}$$

Equation (17.15) is just the ballistic result, eqn. (13.10) multiplied by the transmission. We leave it as an exercise to show that eqn. (17.15) reduces to eqn. (17.11) for small V_{DS} and to (17.13) for large V_{DS}. We see that the drain current in the presence of scattering is just $\mathcal{T}$ times the ballistic current. When we write the current in terms of charge, however, we will see that the result is not that simple.

Finally, note that we have assumed 2D electrons in this lecture. Deriving the corresponding results for 1D electrons in a nanowire MOSFET is a good exercise.

17.6 Charge-based current expressions

Equation (17.15) is the correct current for a Landauer MOSFET at arbitrary V_{DS}, but it is not written in terms of the inversion large charge, Q_n. When writing expressions for the drain current of a MOSFET, it is generally preferable to express them in terms of Q_n, because Q_n is largely determined by MOS electrostatics. To compute Q_n, we need to include the positive velocity electrons injected from the source that populate $+v_x$ states at the top of the barrier and the negative velocity electrons injected from the drain that populate $-v_x$ states at the top of the barrier. In a ballistic MOSFET, the result is eqn. (13.13). In the presence of backscattering, this changes because we must account for all of the ways the states at the top of the barrier can be populated. As illustrated in Fig. 17.1, we still have a ballistic flux injected from the source to the top of the barrier, but there is also a backscattered flux that returns to the source. The magnitude of the ballistic flux injected from the drain is reduced by the transmission to the top of the barrier. The result is that eqn. (17.10) must be changed to

$$Q_n = -q \frac{N_{2D}}{2} \left[\mathcal{F}_0(\eta_{FS}) + (1 - \mathcal{T}) \mathcal{F}_0(\eta_{FS}) + \mathcal{T} \mathcal{F}_0(\eta_{FD}) \right] . \tag{17.16}$$

The first term is the ballistic contribution injected from the source. Its magnitude depends on source Fermi level. The second term is the contribution of the backscattered flux. Since it came from the source, it also depends on the source Fermi level. The third term due to the ballistic flux injected from the drain reduced by the transmission; its magnitude depends on the drain Fermi level.

We can now use eqn. (17.16) with (17.15) to express the drain current in terms of Q_n. First, we multiple and divide eqn. (17.15) by $|Q_n|$,

$$I_{DS} = \mathcal{T} W \frac{|Q_n|}{|Q_n|} \left(\frac{q}{h} \frac{g_v \sqrt{2\pi m^* k_B T}}{\pi \hbar} k_B T \right) \left[\mathcal{F}_{1/2}(\eta_{FS}) - \mathcal{F}_{1/2}(\eta_{FD}) \right] . \tag{17.17}$$

Then, using eqn. (17.16) for Q_n in the denominator; we find after some algebra,

$$\boxed{\begin{aligned}
I_{DS} &= W |Q_n(V_{GS}, V_{DS})| \, v_{inj} \left[\frac{1 - \mathcal{F}_{1/2}(\eta_{FD})/\mathcal{F}_{1/2}(\eta_{FS})}{1 + (\frac{\mathcal{T}}{2-\mathcal{T}}) \mathcal{F}_0(\eta_{FD})/\mathcal{F}_0(\eta_{FS})} \right] \\
Q_n &= -q \frac{N_{2D}}{2} \left[\mathcal{F}_0(\eta_{FS}) + (1 - \mathcal{T}) \mathcal{F}_0(\eta_{FS}) + \mathcal{T} \mathcal{F}_0(\eta_{FD}) \right] \\
v_{inj} &= v_{inj}^{\text{ball}} \left(\frac{\mathcal{T}}{2 - \mathcal{T}} \right) \\
v_{inj}^{\text{ball}} &= \sqrt{\frac{2 k_B T}{\pi m*}} \frac{\mathcal{F}_{1/2}(\eta_{FS})}{\mathcal{F}_0(\eta_{FS})} = v_T \frac{\mathcal{F}_{1/2}(\eta_{FS})}{\mathcal{F}_0(\eta_{FS})} \\
\eta_{FD} &= \eta_{FS} - q V_{DS}/k_B T \, .
\end{aligned}} \tag{17.18}$$

These results should be compared with the corresponding expressions for the ballistic IV characteristic as given by eqns. (13.13). Because of scattering, $\mathcal{T} < 1$, so the injection velocity in the presence of backscattering, v_{inj}, is less than the ballistic injection velocity, $< v_{inj}^{\text{ball}}$, which results in a current that is smaller than the ballistic current.

The general expression for I_{DS} can be simplified for small and large drain biases as was done in Exercise 13.1. The linear region currents for

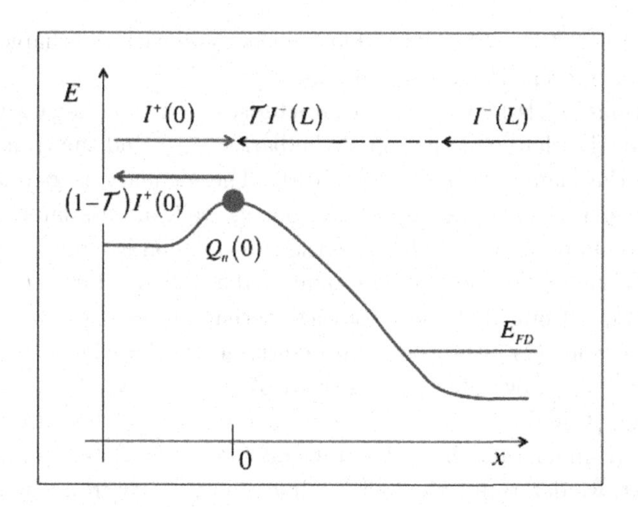

Fig. 17.1 Illustration of the source-injected, backscattered, and drain injected carrier fluxes that contribute to Q_n at the top of the barrier.

Fermi-Dirac and for Maxwell-Boltzmann statistics are

$$I_{DLIN} = W|Q_n(V_{GS})|\mathcal{T}\left(\frac{\upsilon_{inj}^{\text{ball}}}{2k_BT/q}\right)\frac{\mathcal{F}_{-1/2}(\eta_{FS})}{\mathcal{F}_{1/2}(\eta_{FS})}V_{DS} \quad \text{(FD)}$$

$$I_{DLIN} = W|Q_n(V_{GS})|\mathcal{T}\left(\frac{\upsilon_T}{2(k_BT/q)}\right)V_{DS}\,, \quad \text{(MB)} \tag{17.19}$$

where the expression for the ballistic injection velocity from eqn. (17.7) (FD or MB) must be used. Finally, the saturation current is

$$I_{DSAT} = W|Q_n(V_{GS}, V_{DS})|\left(\frac{\mathcal{T}}{2-\mathcal{T}}\right)\upsilon_{inj}^{\text{ball}} \quad \text{(FD)}$$

$$I_{DSAT} = W|Q_n(V_{GS}, V_{DS})|\left(\frac{\mathcal{T}}{2-\mathcal{T}}\right)\upsilon_T\,. \quad \text{(MB)} \tag{17.20}$$

When using Fermi-Dirac statistics, the location of the Fermi level must be known. The Fermi level is found from the known inversion layer charge using the second of eqns. (17.18).

These results should be compared with the corresponding ballistic results in eqns. (17.8) and (17.9). It is interesting to observe that the linear region current is just the linear ballistic current multiplied by the transmission, but the saturation current is the ballistic saturation current multiplied by a factor of $\mathcal{T}/(2-\mathcal{T})$. This difference has to do with the charge balance expression, eqn. (17.16), as we'll discuss in Sec. 17.8.

Equations (17.18) give the IV characteristic of a "Landauer MOSFET" in terms of the charge at the top of the barrier, Q_n, and the transmission. They are the main results of this lecture. These equations give the drain current over the entire range of V_{DS}, but as we'll discuss later, they are difficult to apply in practice because the transmission is a function of V_{DS}.

The IV characteristic would be computed as follows. First, we compute $Q_n(V_{GS}, V_{DS})$ from MOS electrostatics, perhaps using the semi-empirical expression, eqn. (11.14). Next, we determine the location of the source Fermi level, η_{FS}, by solving the second of eqns. (17.18) for η_{FS} given a value of $Q_n(V_{GS}, V_{DS})$. This presents some challenges, because to do so, we need to understand how the transmission, $\mathcal{T}(V_{GS}, V_{DS})$, varies with bias. Next, we determine the ballistic injection velocity from the fourth of eqns. (17.18) and then the injection velocity in the presence of scattering from the third of eqns. (17.18). Finally, we determine the drain current at the bias point, (V_{GS}, V_{DS}) using the first of eqns. (17.18). The main difficulty in using this model is that good models for $\mathcal{T}(V_{GS}, V_{DS})$ do not yet exist. As a result, the semi-empirical virtual source approach is widely used.

In practice, the nondegenerate (Maxwell-Boltzmann) forms of the equations are often used. There is some error — especially above threshold — but the non-degenerate expressions are much simpler, so the trade-off between simplicity and accuracy is often made. For small mass III-V FETs, however, the use of nondegenerate carrier statistics may lead to non-negligible errors.

17.7 The drain voltage-dependent transmission

Although we have developed a simple theory of the nanoscale MOSFET, there are challenges in using the model in practice. The main challenge is that the transmission depends on the drain voltage in a way that is not easy to compute. Figure 17.2 shows why the transmission depends on drain voltage.

As shown on the top of Fig. 17.2, under low bias, the electric field is small across the entire channel. As discussed in Sec. 16.4, the transmission is determined by the length of the low-field region, so for low bias,

$$\mathcal{T}_{LIN} = \frac{\lambda_0}{\lambda_0 + L}. \tag{17.21}$$

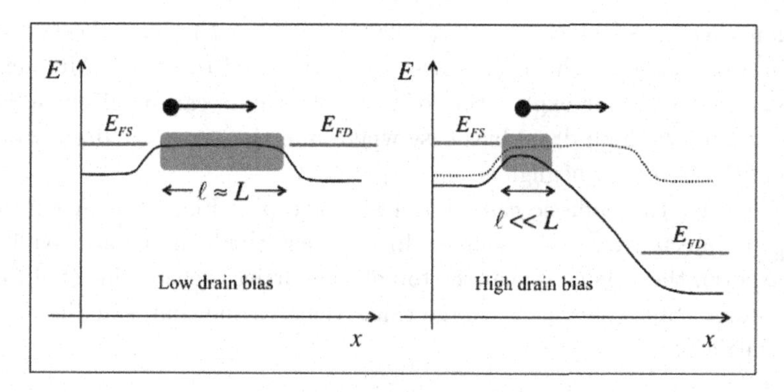

Fig. 17.2 Illustration of why the transmission depends on drain bias and why it is larger for high drain bias than for low drain bias. The mean-free-path in the shaded regions is about λ_0 in both cases.

For high drain bias in a well-designed MOSFET, the low-field region is confined to a short region of length, ℓ, near the beginning of the channel. The high-field part of the channel acts as a near-perfect collector with $\mathcal{T} \approx 1$. As discussed in Sec. 16.4, the transmission of channel in this case is determined by the length of the low-field region, so for high drain bias

$$\mathcal{T}_{SAT} = \frac{\lambda_0}{\lambda_0 + \ell}. \tag{17.22}$$

We conclude that $\mathcal{T}_{SAT} > \mathcal{T}_{LIN}$ because $\ell \ll L$. Under high drain bias, carriers are more energetic in the high-field region, so they scatter more than under low drain bias. Nevertheless, the transmission is higher under high drain bias, so the device delivers a current that is closer to the ballistic limit.

The calculation of the extent of the low-field region as a function of the gate and drain bias requires, in principle, a self-consistent solution to the electrostatic problem in the presence of current flow [1, 2]. When the channel profile, $E_c(x)$, is known, the value of the critical length, ℓ, can be calculated [3-5]. The use of the empirical drain saturation function and injection velocity in the VS model provides an alternative to these calculations.

17.8 Discussion

It might seem confusing that the linear region current as given by eqn. (17.19) is just $\mathcal{T}$ times the ballistic linear current, but the saturation region current as given by eqn. (17.20) is $\mathcal{T}/(2-\mathcal{T})$ times the ballistic saturation current. This occurs because of the need to enforce MOS electrostatics. A clearer explanation of why this occurs can be given by just considering the high drain bias case where injection from the drain to the top of the barrier is negligible.

Consider the ballistic case shown at the top of Fig. 17.3. A current, I^+_{ball}, is injected from the source. In this case (high drain bias, ballistic transport), the only charge at the top of the barrier is charge injected from the source. Since current is charge times velocity, the charge at the top of the barrier is

$$Q_n(x=0) = -\frac{I^+_{\text{ball}}}{W\upsilon_T}\,. \tag{17.23}$$

(Maxwell-Boltzmann statistics are assumed.)

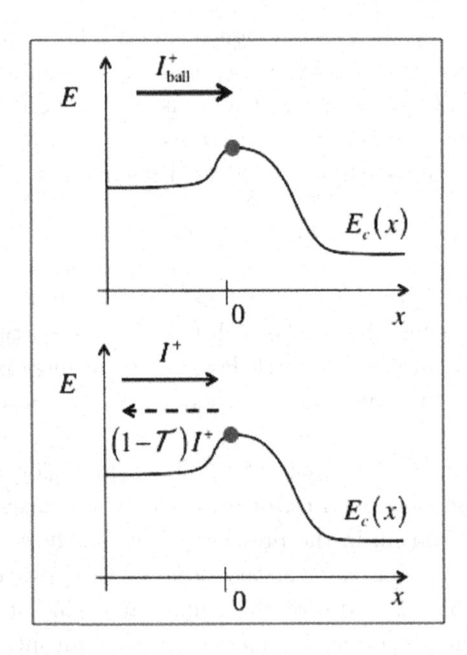

Fig. 17.3 Injected and backscattered currents under high drain bias. Top: The ballistic case. Bottom: In the presence of backscattering.

Next, consider the charge in the presence of scattering. As shown on the bottom of Fig. 17.3, there are two components of the charge; a source-injected component with positive velocity electrons, and a backscattered component with negative velocity electrons. The total charge is

$$Q_n(x=0) = -\frac{I^+ + (1-\mathcal{T}_{SAT})I^+}{W\upsilon_T} = -\frac{(2-\mathcal{T}_{SAT})I^+}{W\upsilon_T}. \qquad (17.24)$$

Now in a well-designed MOSFET, $Q_n(x=0)$ is largely determined by MOS electrostatics and is relatively independent of transport. The charge under ballistic conditions, eqn. (17.23) should be the same as the charge in the presence of scattering, eqn. (17.24). By equating eqn. (17.23) to eqn. (17.24), we find

$$I^+ = \frac{I^+_{\text{ball}}}{(2-\mathcal{T}_{SAT})}. \qquad (17.25)$$

In the presence of scattering ($\mathcal{T} < 1$), so a smaller flux is injected to produce the same $Q_n(x-0)$.

The drain current is $\mathcal{T}$ times the injected current, so for the ballistic case, ($\mathcal{T} = 1$),

$$I^{\text{ball}}_{DS} = I^+ = I^+_{\text{ball}}. \qquad (17.26)$$

and for the general case, ($\mathcal{T} < 1$), we find

$$I_{DS} = \mathcal{T}_{SAT} I^+ = \frac{\mathcal{T}_{SAT}}{(2-\mathcal{T}_{SAT})} I^{\text{ball}}_{DS}. \qquad (17.27)$$

The requirement that MOS electrostatics be enforced results in a saturation current that is $\mathcal{T}/(2-\mathcal{T})$ times the ballistic saturation current.

The question of what mobility means in a nanoscale MOSFET also requires a discussion. According to eqn. (12.44), the mobility is proportional to the mean-free-path. In transport theory, mobility is considered to be well-defined near-equilibrium in a bulk material that is many mean-free-paths long (see Sec. 8.2 in [6]). In modern transistors, the channel length is comparable to a mean-free-path, and under high drain bias, the carriers are very far from equilibrium. Nevertheless, device engineers find that the near-equilibrium mobility is strongly correlated to the performance of nanoscale transistors. How do we explain the relevance of mobility in nanoscale MOSFETs? As shown in Fig. 17.2, the near-equilibrium mean-free-path, λ_0, controls the current under both low and high drain bias. An equilibrium flux of carriers is injected from the source. Under low drain bias, these carriers remain near-equilibrium across the entire channel. Under high drain bias, the carriers gain energy in the drain field, their scattering rate increases, and the mean-free-path decreases. As we have discussed, however, it is the low-field part of the channel that determines the transmission. The carriers are near-equilibrium in the part of the channel that determines the current, so the near-equilibrium mean-free-path controls the current under both low and high drain bias.

We can explain the experimentally observed correlation of nanoscale transistor performance to mobility by arguing that mobility is proportional to the near-equilibrium mean-free-path and the near-equilibrium mean-free-path controls the current of a nanoscale transistor from low to high drain bias. Of course this is only a first order argument. Differences in strain and doping may occur for short channels, and the carriers are not exactly at equilibrium. For very short channels, carriers that enter the channel from the source can excite plasma oscillations near the source, which lower the mean-free-path [7, 8]. This effect has been observed experimentally, but the argument that the high drain bias current is strongly correlated with the near-equilibrium mobility seems to capture the essence of the physics and produces reasonably accurate results in practice.

Exercise 17.1: Analysis of a 25 nm ETSOI N-MOSFET.

To get a feel for some of the numbers involved, it is useful to analyze the measured results of an $L = 25$ nm Extremely Thin Silicon On Insulator (ETSOI) MOSFET [8]. The device is fabricated on a (100) Si wafer and the relevant parameters at 300 K are [9]:

$$\upsilon_{inj} = 0.82 \times 10^7 \text{ cm/s}$$
$$\lambda_0 = 10.5 \text{ nm}.$$

For this problem, we'll need the uni-directional thermal velocity. In Exercise 14.2, we found $\upsilon_T = 1.2 \times 10^7$ cm/s assuming (100) Si at 300K with one subband occupied.

We compute $\mathcal{T}_{LIN}$ from eqn. (17.21):

$$\mathcal{T}_{LIN} = \frac{\lambda_0}{\lambda_0 + L} = \frac{10.5}{10.5 + 25} = 0.33 .$$

To compute $\mathcal{T}_{SAT}$ we solve the third eqn. in (17.18):

$$\mathcal{T}_{SAT} = \frac{2}{1 + \upsilon_T/\upsilon_{inj}} = \frac{2}{1 + 1.2/0.82} = 0.8 .$$

As expected, we find a much higher transmission under high drain bias. To estimate the length of the critical region, ℓ, we solve eqn. (17.22) for

$$\ell = \lambda_0 \left(1/\mathcal{T}_{SAT} - 1\right) = 10.5\,(1/0.82 - 1) = 2.4 \text{ nm}.$$

and find $\ell \ll L$, the bottleneck for current is about 10% of the channel length.

17.9 Summary

In this lecture, we used the Landauer approach introduced in Lecture 12 to compute the IV characteristics of a MOSFET in the presence of scattering. We combined the Landauer expression for current, eqn. (17.1), with the constraint that MOS electrostatics must be satisfied. The result was a fairly simple model for the ballistic MOSFET as summarized in eqns. (17.18). For a MOSFET operating in the subthreshold region, nondegenerate carrier statistics can be employed. Above threshold however, the conduction band at the top of the barrier is close to, or even below the Fermi level, so Fermi-Dirac statistics should be used. Nevertheless, it is common in MOS device theory to assume nondegenerate conditions (i.e. to use Maxwell-Boltzmann statistics for carriers) because it simplifies the calculations and makes the theory more transparent. Also, in practice, there are usually some device parameters that we don't know precisely, so the use of nondegenerate carrier statistics with some empirical parameter fitting often produces reasonable results.

The expressions we have developed provide insight into the physics of the linear and saturation region currents, but they do not provide an accurate model for the drain voltage dependence because we do not have an accurate model for $\mathcal{T}(V_{DS})$. The semi-empirical VS model to be discussed in the next lecture provides additional insight into the linear and saturation region currents as well as a description of the entire IV characteristic.

17.10 References

The detailed transport physics of nanoscale MOSFETs is discussed in:

[1] P. Palestri, D. Esseni S. Eminente, C. Fiegna, E. Sangiorgi, and L. Selmi, "Understanding Quasi-Ballistic Transport in Nano-MOSFETs: Part I — Scattering in the Channel and in the Drain," *IEEE Trans. Electron. Dev.*, **52**, pp. 2727-2735, 2005.

[2] M.V. Fischetti, T.P. O'Regan, N. Sudarshan, C. Sachs, S. Jin, J. Kim, and Y. Zhang, "Theoretical study of some physical aspects of electronic transport in n-MOSFETs at the 10-nm Gate-Length," *IEEE Trans. Electron Dev.*, **54**, pp. 2116-2136, 2007.

The following papers discuss the computation of the critical length for backscattering, ℓ, in the presence of a spatially varying electric field.

[3] Gennady Gildenblat, "One-flux theory of a nonabsorbing barrier," *J. Appl. Phys.*, **91**, pp. 9883-9886, 2002.

[4] R. Clerc, P. Palestri, L. Selmi, and G. Ghibaudo, "Impact of carrier heating on backscattering in inversion layers," *J. Appl. Phys.* **110**, 104502, 2011.

A type of virtual source model that computes computes the bias-dependent transmission (eliminating the need for the empirical drain current saturation function) has recently been reported.

[5] Shaloo Rakheja, Mark Lundstrom, and Dimitri Antoniadis, "A physics-based compact model for FETs from diffusive to ballistic carrier transport regimes," presented at the International Electron Devices Meeting (IEDM), San Francisco, CA, December 15-17, 2014.

The concept of mobility in nanoscale devices is discussed in Sec. 8.2 of

[6] Mark Lundstrom, *Fundamentals of Carrier Transport, 2nd Ed.*, Cambridge Univ. Press, Cambridge, U.K., 2000.

Long-range Coulomb interactions can affect the performance of short channel MOSFETs. For a discussion, see the following papers.

[7] M.V. Fischetti and S.E. Laux, "Long-range Coulomb interactions in small SI devices," *J. Appl. Phys.*, **89**, pp. 1205-1231, 2001.

[8] T. Uechi, T. Fukui, and N. Sano, "3D Monte Carlo simulation including full Coulomb interaction under high electron concentration regimes," *Phys. Status Solidi C*, **5**, pp. 102-106, 2008.

The transistor parameters for Exercise 17.1 were taken the following paper.

[9] A. Majumdar and D.A. Antoniadis, "Analysis of Carrier Transport in Short-Channel MOSFETs," *IEEE Trans. Electron. Dev.*, **61**, pp. 351-358, 2014.

Lecture 18

Connecting the Transmission and VS Models

18.1 Introduction

Equations (17.18) summarize the transmission model for the IV characteristics of MOSFETs. Equations (15.7)-(15.9) summarize the virtual source model for the IV characteristics. The connection between these two models is the topic for this lecture.

We begin, as usual, with the drain current written as the product of charge and velocity,

$$I_{DS} = W \left| Q_n \left(x = 0, V_{GS}, V_{DS} \right) \right| \upsilon(x = 0, V_{GS}, V_{DS}) . \tag{18.1}$$

First, we compute $Q_n(V_{GS}, V_{DS})$ from MOS electrostatics. Next, the average velocity at the top of the barrier must be determined. This is done differently in the transmission and in the VS models.

18.2 Review of the transmission model

We begin this lecture by summarizing the transmission model assuming Maxwell-Boltzmann carrier statistics. The current is given by eqns. (17.18).

The charge at a given bias, (V_{GS}, V_{DS}) is determined by MOS electrostatics. There is no need to know the location of the Fermi level to determine the velocity when Maxwell-Boltzmann carrier statistics are used. The injection velocity is given by

$$\upsilon_{inj} = \upsilon_T \left(\frac{\mathcal{T}}{2 - \mathcal{T}} \right) , \tag{18.2}$$

where the ballistic injection velocity in the Maxwell-Boltzmann limit, υ_T, is given by eqn. (17.7) as

$$\upsilon_T = \sqrt{\frac{2k_BT}{\pi m^*}} . \tag{18.3}$$

The average velocity at a given bias is obtained from

$$\upsilon(x = 0, V_{GS}, V_{DS}) = F_{SAT}\, \upsilon_{inj} , \tag{18.4}$$

where

$$F_{SAT} = \left[\frac{1 - e^{-qV_{DS}/k_BT}}{1 + \left(\frac{\mathcal{T}}{2-\mathcal{T}} \right) e^{-qV_{DS}/k_BT}} \right] . \tag{18.5}$$

Finally, we compute the drain current at the bias point, (V_{GS}, V_{DS}) from eqn. (18.1). Series resistance would be included as discussed in Sec. 5.4.

The difficulty in using the above prescription to calculate the full IV characteristic lies in the difficulty of computing $\mathcal{T}(V_{DS})$. For low V_{DS}, the transmission is known from eqn. (17.21),

$$\mathcal{T}_{LIN} = \frac{\lambda_{LIN}}{\lambda_{LIN} + L} . \tag{18.6}$$

For high V_{DS}, the transmission is given by eqn. (17.22) as

$$\mathcal{T}_{SAT} = \frac{\lambda_{SAT}}{\lambda_{SAT} + \ell} . \tag{18.7}$$

As discussed in Sec. 17.7, $\lambda_{LIN} \approx \lambda_{SAT} = \lambda_0$. The length of the critical region, ℓ, is not easy to compute [1-3], but the Landauer expressions for the linear and saturation region currents are easy to relate to the VS expressions. The linear and saturation region currents for the Landauer MOSFET are given by eqns. (17.19) and (17.20) as

$$\begin{aligned} I_{DLIN} &= W|Q_n|\mathcal{T}_{LIN} \left(\frac{\upsilon_T}{2(k_BT/q)} \right) V_{DS} \\ I_{DSAT} &= W|Q_n|\upsilon_{inj} = W|Q_n| \left(\frac{\mathcal{T}_{SAT}}{2 - \mathcal{T}_{SAT}} \right) \upsilon_T . \end{aligned} \tag{18.8}$$

As we shall see, these equations are easy to relate to the corresponding traditional (diffusive) or VS relations.

18.3 Review of the VS model

The Virtual Source model begins with eqns. (18.1), but then computes the average velocity from

$$\upsilon(x=0,V_{GS},V_{DS}) = F_{SAT}(V_{DS})\upsilon_{sat}\,, \tag{18.9}$$

where the drain voltage dependence of the average velocity is given by the empirical drain saturation function,

$$F_{SAT}(V_{DS}) = \frac{V_{DS}/V_{DSAT}}{\left[1+(V_{DS}/V_{DSAT})^{\beta}\right]^{1/\beta}}\,, \tag{18.10}$$

with

$$V_{DSAT} = \upsilon_{sat}L/\mu_n\,. \tag{18.11}$$

The VS drain current at the bias point, (V_{GS}, V_{DS}), is determined from eqn. (18.1) using the charge from eqn. (15.2) and the average velocity from eqn. (18.9). Series resistance would be included as discussed in Sec. 5.4.

For small drain bias, $F_{SAT} \rightarrow V_{DS}/V_{DSAT}$ and $\upsilon(x=0,V_{GS},V_{DS}) \rightarrow \mu_n V_{DS}/L$. The linear region drain current in the VS model becomes

$$I_{DLIN} = \frac{W}{L}|Q_n(V_{GS})|\,\mu_n V_{DS}\,, \tag{18.12}$$

which is also the result from traditional MOSFET theory. For large V_{DS}, eqn. (18.9) reduces to the traditional velocity saturation expression,

$$I_{DSAT} = W\,|Q_n(V_{GS},V_{DS})|\,\upsilon_{sat}\,. \tag{18.13}$$

The VS model is a semi-empirical model used to fit measured IV characteristics. To fit the measured characteristics of small MOSFETs, the parameters for long channel MOSFETs, μ_n and υ_{sat}, have to be adjusted:

$$\mu_n \rightarrow \mu_{app} \quad \upsilon_{sat} \rightarrow \upsilon_{inj}\,. \tag{18.14}$$

In Lecture 15, we showed that in the ballistic limit, the apparent mobility, μ_{app}, and the injection velocity, υ_{inj}, have clear physical significance. In this lecture, we interpret these two parameters in the presence of carrier scattering.

18.4 Connection

Our goal is to understand the physical significance of the apparent mobility and the injection velocity by relating the VS model to the transmission model.

Linear region: Transmission vs. VS

Using the expression for transmission, $\mathcal{T}_{LIN} = \lambda_0/(\lambda_0 + L)$, we can re-write the transmission expression for the linear current, eqn. (18.8), as

$$\begin{aligned} I_{DLIN} &= \frac{W}{L}|Q_n|\left(\mathcal{T}_{LIN}L\right)\left(\frac{\upsilon_T}{2(k_BT/q)}\right)V_{DS} \\ &= \frac{W}{L}|Q_n|\left(\frac{1}{1/\lambda_0 + 1/L}\right)\left(\frac{\upsilon_T}{2(k_BT/q)}\right)V_{DS}\,. \end{aligned} \tag{18.15}$$

Next, we recall the definition of the mobility, eqn. (12.44),

$$\mu_n = \frac{D_n}{k_BT/q} = \frac{\upsilon_T\lambda_0/2}{k_BT/q}\,, \tag{18.16}$$

and the ballistic mobility, eqn. (12.48),

$$\mu_B = \frac{\upsilon_T L/2}{k_BT/q}\,, \tag{18.17}$$

and use these to re-write eqn. (18.15) as

$$\begin{aligned} I_{DLIN} &= \frac{W}{L}|Q_n|\left(\frac{1}{1/\mu_n + 1/\mu_B}\right)V_{DS} \\ &= \frac{W}{L}|Q_n|\mu_{app}V_{DS}\,, \end{aligned} \tag{18.18}$$

where the *apparent mobility* is defined as

$$\boxed{\frac{1}{\mu_{app}} \equiv \frac{1}{\mu_n} + \frac{1}{\mu_B}}\,. \tag{18.19}$$

To find the apparent mobility, we add the inverse mobility due to scattering to the inverse mobility due to ballistic transport and take the inverse of the sum. This prescription for finding the total mobility due to two independent processes is known as *Mathiessen's Rule* [4].

As discussed in Lecture 12, the ballistic mobility is the mobility obtained when the mean-free-path is replaced by the length of the channel. Carriers scatter frequently in the source and in the drain, so when the channel is ballistic, the distance between scattering events is the length of the channel. By using the ballistic mobility, the linear region current of a ballistic MOSFET can be written in the traditional, diffusive form.

According to eqn. (18.19), the apparent mobility of a MOSFET is less than the lower of the scattering limited and ballistic mobilities. For a long channel MOSFET, $\mu_n \ll \mu_B$, and the apparent mobility is the scattering limited mobility, μ_n. For a very short channel, $\mu_B \ll \mu_n$, and the apparent mobility is the ballistic mobility. Note that the traditional expression for the linear current, eqn. (18.12), could predict a current above the ballistic limit if the channel length is short enough, but if the scattering limited mobility is replaced by the apparent mobility, this cannot happen.

In the linear region, the MOSFET is a gate-voltage controlled resistor (Fig. 18.1). From eqn. (18.18), the channel resistance is

$$R_{ch} = \frac{V_{DS}}{I_{DLIN}} = \frac{L}{W}\frac{1}{|Q_n|\mu_{app}}\,. \tag{18.20}$$

Real MOSFETs have series resistance, so in the linear region

$$I_{DLIN} = \frac{V_{DS}}{R_{ch} + R_S + R_D} = \frac{V_{DS}}{R_{TOT}}\,, \tag{18.21}$$

where R_S and R_D are the source and drain series resistances. By fitting the measured IV characteristic in the linear region to the VS model, both the series resistance and the apparent mobility can be extracted.

To summarize, we have shown that the transmission expression for the linear region current, eqn. (18.8), can be written in the diffusive form, eqn. (18.12), used in the VS model — if we replace the scattering limited mobility, μ_n, in the traditional expression by the apparent mobility, μ_{app}, as in eqn. (18.18).

Saturation region: Transmission vs. VS

According the eqn. (18.8), the factor, $\mathcal{T}_{SAT}/(2 - \mathcal{T}_{SAT})$, is important in saturation. Using eqn. (18.7) for $\mathcal{T}_{SAT}$, we can write

$$\frac{\mathcal{T}_{SAT}}{(2 - \mathcal{T}_{SAT})} = \frac{\lambda_0}{\lambda_0 + 2\ell}\,. \tag{18.22}$$

According to eqn. (18.2), the injection velocity is

$$\begin{aligned} \upsilon_{inj} = \left(\frac{\mathcal{T}_{SAT}}{2 - \mathcal{T}_{SAT}}\right)\upsilon_T &= \frac{\lambda_0 \upsilon_T}{\lambda_0 + 2\ell} \\ &= \frac{1}{1/\upsilon_T + \ell/(\lambda_0 \upsilon_T/2)}\,. \end{aligned} \tag{18.23}$$

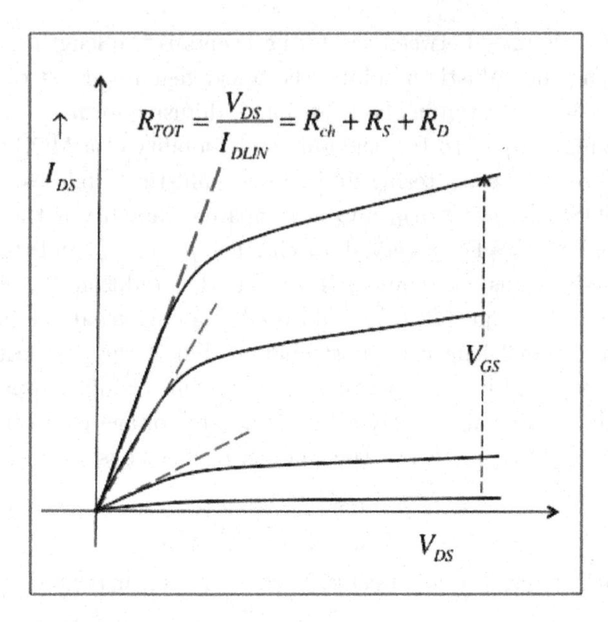

Fig. 18.1 Illustration of how the linear region current is related to the channel and series resistances. For a fixed V_{GS}, the channel resistance is proportional to one over the apparent mobility.

Now, recall the definition of the diffusion coefficient, eqn. (12.18), $D_n = \upsilon_T \lambda_0/2$, which can be used to write the injection velocity as

$$\upsilon_{inj} = \left(\frac{1}{\upsilon_T} + \frac{1}{D_n/\ell} \right)^{-1}, \tag{18.24}$$

or

$$\boxed{\frac{1}{\upsilon_{inj}} = \frac{1}{\upsilon_T} + \frac{1}{D_n/\ell}.} \tag{18.25}$$

According to eqn. (18.25), the injection velocity of a MOSFET is less than the lower of the ballistic injection velocity and D_n/ℓ, which is the velocity at which carriers diffuse across the bottleneck region of length, ℓ. When ℓ is long or D_n small, $D_n/\ell \ll \upsilon_T$, and injection velocity is the diffusion velocity. When ℓ is short or D_n large, $D_n/\ell \gg \upsilon_T$, and the injection velocity is limited by the ballistic injection velocity. The injection velocity cannot be larger than the ballistic injection velocity, but it can be much smaller.

Figure 18.2 is an illustration of what happens in the on-state of a nanoscale MOSFET. Carriers must diffuse across the bottleneck region, but they cannot diffuse faster than the thermal velocity because diffusion is caused by random thermal motion. After diffusing across the bottleneck, they encounter the high field portion of the channel, which sweeps them across and out the drain. The bottleneck region is analogous to the base of a bipolar transistor, and the high-field region is analogous to the collector of a bipolar transistor.

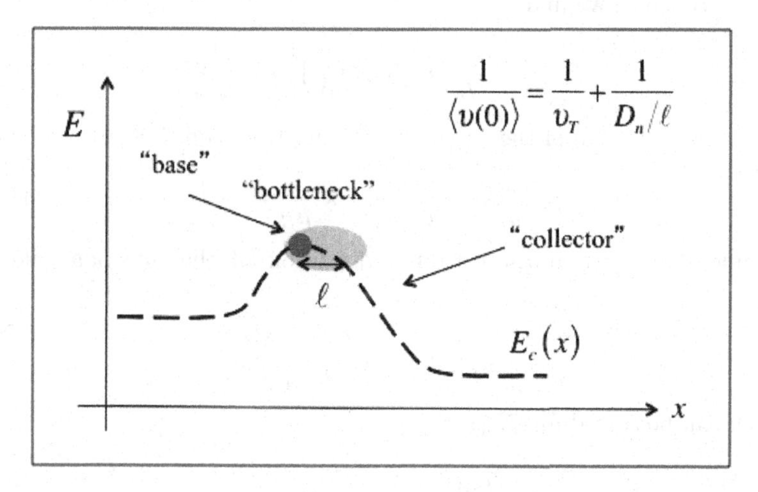

Fig. 18.2 The energy band diagram of a MOSFET in the on-state showing the bottleneck for current flow, where the electric field along the channel is small, and the high-field part the the channel. The bottleneck is analogous to the base of a bipolar transistor, and the high-field region is analogous to the collector.

To summarize, we have shown that the transmission expression for the saturation region current, eqn. (18.8), can be written in the traditional, velocity saturated form, eqn. (18.13), used in the VS model — if we replace the scattering limited velocity, υ_{sat}, in the traditional expression by the injection velocity, υ_{inj}, as defined in eqn. (18.25). The largest that the injection velocity can be is the ballistic injection velocity, υ_T.

Exercise 18.1: Relate the transmission to the parameters of the VS model.

By fitting the VS model to measured data, we determine the apparent mobility and the injection velocity. If we also fit a long channel device, then we can determine μ_n. (The scattering limited mobility might be different in a short channel MOSFET, but as will be discussed in Lecture 19, we can also determine μ_n in the short channel MOSFET.) Assuming that we know μ_{app}, μ_n, and υ_{inj}, show how to determine the transmission in the linear and saturation regions.

Equation (18.8) gives the linear region current in terms of $\mathcal{T}_{LIN}$, and eqn. (18.18) gives the linear region current in terms of μ_{app}. Equating these two expressions, we find:

$$\mathcal{T}_{LIN} = \frac{\mu_{app}}{L}\left(\frac{\upsilon_T}{2k_BT/q}\right)^{-1} = \frac{\mu_{app}}{\mu_B} .$$

Using the definition of the apparent mobility from eqn. (18.19), we find

$$\mathcal{T}_{LIN} = \frac{\mu_{app}}{\mu_B} = \frac{\mu_B\mu_n}{\mu_B+\mu_n} \times \frac{1}{\mu_B} = \frac{\mu_n}{\mu_B+\mu_n} , \tag{18.26}$$

To find $\mathcal{T}_{SAT}$, we begin with the definition of the injection velocity, eqn. (18.2),

$$\upsilon_{inj} = \upsilon_T\left(\frac{\mathcal{T}_{SAT}}{2-\mathcal{T}_{SAT}}\right) ,$$

which can be solved for $\mathcal{T}_{SAT}$

$$\mathcal{T}_{SAT} = \frac{2}{1+\upsilon_T/\upsilon_{inj}} . \tag{18.27}$$

The injection velocity is determined by fitting the VS model to measured data, but the ballistic injection velocity, υ_T, is more difficult to determine. It can be extracted from the measured IV characteristics [5], but it is often computed from the known effective mass and a knowledge of the number of subbands that are occupied.

Exercise 18.2: Mobility and apparent mobility of a 22 nm MOSFET.

Consider an $L = 22$ nm n-channel Si MOSFET at $T = 300$ K biased in the linear region. Assume a (100) oriented wafer with only the bottom subband occupied. Assume that the mobility is $\mu_n = 250\,\text{cm}^2/\text{V}-\text{s}$. What are μ_B, μ_{app}, and $\mathcal{T}_{LIN}$?

For this case, we have seen in eqn. (14.15) that $\upsilon_T = 1.2\times10^7$ cm/s. We find the ballistic mobility from eqn. (18.17) as

$$\mu_B = \frac{\upsilon_T L}{2kT/q} = \frac{(1.2\times10^7)\times(22\times10^{-7})}{2\times0.026} = 508\ \text{cm}^2/\text{V}-\text{s} . \tag{18.28}$$

Since μ_B is comparable to μ_n, this is a quasi-ballistic MOSFET.

The apparent mobility is found from eqn. (18.19) as

$$\mu_{app} = \frac{\mu_n \mu_B}{\mu_n + \mu_B} = \frac{250 \times 508}{250 + 508} = 191\,\text{cm}^2/\text{V} - \text{s}\,.$$

As expected, the apparent mobility is less than the smaller of the ballistic and scattering limited mobilities. Finally, we find the linear region transmission from eqn. (18.26) as

$$\mathcal{T}_{LIN} = \frac{\mu_n}{\mu_B + \mu_n} = \frac{250}{508 + 250} = 0.33\,. \tag{18.29}$$

18.5 Discussion

We have seen in this lecture that one can clearly relate the linear region and saturation region currents of the VS model to the corresponding results from the transmission model. We now understand why the scattering limited mobility that describes long channel transistors needs to be replaced by an apparent mobility that comprehends quasi-ballistic transport. As also shown in this lecture, the saturation velocity in the traditional model corresponds to the injection velocity in the transmission model. The transmission model provides a clear, physical interpretation of the linear and saturation region currents for nanoscale MOSFETs, but the semi-empirical VS model does a better job of describing the shape of the I_D vs. V_{DS} characteristics. This is not a fundamental limitation of the Landauer model; it only happens because of the difficulty of computing $\mathcal{T}(V_{DS})$.

We have discussed three mobilities: 1) The scattering limited mobility, μ_n, 2) the ballistic mobility, μ_B, and 3) the apparent mobility, μ_{app}. Traditional MOSFET theory is expressed in terms of another mobility — the *effective mobility*, μ_{eff}. The term, "effective mobility," is unfortunate but it is the traditional term used for the scattering-limited mobility in MOSFETs [7, 8]. The term, effective, refers to the fact that carriers closer to the surface should have a lower mobility than carriers deeper in the channel because of surface roughness scattering. The effective mobility is the depth-averaged mobility of carriers in the channel. For a Si MOSFET, μ_{eff} is much less than the scattering-limited mobility of carriers in bulk silicon because of surface roughness scattering. For III-V HEMTs, the high bulk mobility is retained because atomically flat interfaces can be produced. In modern MOSFETs, however, quantum confinement is strong, and all carriers in the channel experience surface roughness scattering. Talking of a depth-averaged mobility is not appropriate. For us, $\mu_n = \mu_{\text{eff}}$ simply refers to the scattering limited mobility in a field-effect transistor.

18.6 Summary

In this lecture, we have shown that the transmission model of Lecture 17 can be clearly related to the VS model. By simply replacing the scattering limited mobility, μ_n, in the VS model with the apparent mobility, the correct results for the linear current are obtained from the ballistic to diffusive limits. By simply replacing the high-field, scattering limited bulk saturation velocity, v_{sat}, with the injection velocity, v_{inj}, the correct on-current is obtained. Comparison with measured characteristics showed that nanoscale Si MOSFETs operate well below the ballistic limit but that nanoscale III-V FETs operate quite close to the ballistic limit.

The transmission model suffers from two key limitations. The first is the difficulty of computing I_{DS} vs. V_{DS}, which occurs because of the difficulty of computing $\mathcal{T}(V_{DS})$. The second limitation (which is related to the first) is the difficulty of predicting the on-current, which occurs because of the difficulty of computing the critical length, ℓ, for high drain bias. The result is that it is hard to predict $\mathcal{T}_{SAT}$. Because of these limitations, the transmission and the semi-empirical VS model are often combined with the parameters in the transmission model being determined by fitting the VS model to experimental data, and the physical interpretation of the fitted parameters being provided by the transmission model.

18.7 References

The following papers discuss some of the issues involved in computing transmission in the presence of a spatially varying electric field.

[1] P. Palestri, D. Esseni, S. Eminente, C. Fiegna, E. Sangiorgi, and L. Selmi, "Understanding quasi-ballistic transport in nano-MOSFETs: Part I — Scattering in the channel and in the drain," *IEEE Trans. Electron Dev.*, **52**, pp. 2727-2735, 2005.

[2] P. Palestri, R. Clerc, D. Esseni, L. Lucci, and L. Selmi, "Multi-subband Monte-Carlo investigation of the mean free path and of the kT layer in degenerated quasi ballistic nanoMOSFETs," in Int. Electron Dev. Mtg., (IEDM), Technical Digest, pp. 945-948, 2006.

[3] R. Clerc, P. Palestri, L. Selmi, and G. Ghibaudo, "Impact of carrier heating on backscattering in inversion layers," *J. Appl. Phys.* **110**, 104502, 2011.

Mathiessen's Rule for adding mobilities due to individual processes is discussed in Sec. 4.3.2 of:

[4] Mark Lundstrom, *Fundamentals of Carrier Transport, 2nd Ed.*, Cambridge Univ. Press, Cambridge, U.K., 2000.

As discussed in the following books, the term, effective mobility, is used in conventional MOSFET analysis for the scattering limited mobility of carriers in the inversion layer.

[5] Y. Tsividis and C. McAndrew, *Operation and Modeling of the MOS Transistor*, 3rd Ed., Oxford Univ. Press, New York, 2011. (See Sec. 4.11.)

[6] Y. Taur and T. Ning, *Fundamentals of Modern VLSI Devices*, 2nd Ed., Oxford Univ. Press, New York, 2013. (See Sec. 3.1.5.)

Lecture 19

VS Characterization of Transport in Nanotransistors

19.1 Introduction

Much can be learned about the physics of carrier transport at the nanoscale by carefully examining the *IV* characteristics of well-behaved nanoscale MOSFETs. A number of studies using a variety of methods have been reported (e.g. [1-6]). As described in several publications, the virtual source/transmission model provides a useful tool for studying transport in nanotransistors [7-10]. In this lecture, we'll examine experimental results following the approach of [11, 12]. Both nanoscale Extremely Thin SOI (ETSOI) MOSFETs [13, 14] and III-V HEMTs (High Electron Mobility Transistors) [10, 15] will be examined.

19.2 Review of the MVS/Transmission model

The virtual source and transmission models have been discussed extensively in previous lectures — we summarize the main results here before applying them to experimental data. The specific form of the VS model to be used in this lecture was developed at MIT and will be called the MVS model [16]. The MVS model describes the drain current as the product of charge and velocity [16, 17],

$$I_{DS} = W \left|Q_n \left(x = 0, V_{Gi}, V_{Di}\right)\right| F_{SAT}(V_{Di})\, \upsilon_{inj}\,, \tag{19.1}$$

where $F_{SAT}(V_{Di})\upsilon_{inj}$ is the velocity at the virtual source. The voltages, V_{Gi} and V_{Di} are the intrinsic gate and drain voltages. (The absolute value sign is used because the inversion charge, Q_n, is negative for an n-channel MOSFET.)

In the MVS model, the charge at the virtual source, $Q_n(V_{Gi}, V_{Di})$, is obtained from a semi-empirical expression similar to eqn. (11.14) [16]

$$|Q_n(V_{Gi}, V_{Di})| = m\, C_G(inv) \left(\frac{k_BT}{q}\right) \ln\left(1 + e^{q(V_{Gi} - V_T - \alpha(k_BT/q)F_f)/mk_BT}\right). \tag{19.2}$$

This expression uses an "inversion transition function," F_f [16],

$$F_f = \frac{1}{1 + \exp\left(\frac{V_{Gi} - (V_T - \alpha(k_BT/q)/2)}{\alpha k_BT/q}\right)}, \tag{19.3}$$

which produces an increase in threshold voltage of $\alpha(k_BT/q)$ as the device transitions from subthreshold to strong inversion. Note that $F_f \to 1$ in subthreshold and $F_f \to 0$ in strong inversion. The empirical parameter, α, is typically set to 3.5 [11, 16]. In eqn. (19.2), the threshold voltage depends on drain voltage according to

$$V_T = V_{T0} - \delta V_{Di}\,, \tag{19.4}$$

where V_{T0} is the strong inversion threshold voltage at $V_D = V_{Di} = 0$, and δ is the DIBL parameter in units of V/V. The subthreshold slope parameter in eqn. (19.2), is given by

$$m = m_0 + m' V_{Di}\,, \tag{19.5}$$

where m_0 is the subthreshold parameter at $V_D = V_{Di} = 0$ and $m' = dm/dV_{Di}$ describes the change in m with drain voltage.

The MVS model uses an empirical drain saturation function, which is given by [16]

$$F_{SAT}(V_{Di}) = \frac{V_{Di}/V_{DSATs}}{\left[1 + \left(V_{Di}/V_{DSATs}\right)^{\beta}\right]^{1/\beta}}, \tag{19.6}$$

with

$$V_{DSATs} = \frac{\upsilon_{inj} L_{\text{eff}}}{\mu_{app}}, \tag{19.7}$$

where L_{eff} is the *effective channel length* as discussed by Taur [18]. Note that we have added an s to the subscript SAT in V_{DSATs} to denote the fact that F_{SAT} describes drain current saturation in strong inversion. Under subthreshold conditions, $V_{DSAT} = k_B T/q$ as discussed by Taur and Ning [18]. The MVS model treats this transition between V_{DSAT} in subthreshold and strong inversion heuristically by using the inversion transition function [16],

$$V_{DSAT} = V_{DSATs}(1 - F_f) + (k_B T/q) F_f \,. \tag{19.8}$$

The intrinsic terminal voltages are related to the external terminal voltages according to

$$\begin{aligned} V_{Gi} &= V_G - I_{DS} R_{SD0}/2 \\ V_{Di} &= V_D - I_{DS} R_{SD0}\,, \end{aligned} \tag{19.9}$$

where the series resistance, R_{SD0}, is the sum of the source series resistance, R_{S0}, and the drain series resistance, R_{D0}, which are assumed to be equal and independent of gate or drain voltage.

The MVS model can be fit to the measured transfer characteristics (I_{DS} vs. V_{GS}) and the output characteristics (I_{DS} vs. V_{DS}) to deduce several important device parameters; our analysis will focus on the low V_{DS}, linear region and the high V_{DS}, saturation region.

For small drain bias, $F_{SAT} \to V_{DS}/V_{DSATs}$ and $\upsilon(x = 0, V_{GS}, V_{DS}) \to \mu_{app} V_{DS}/L_{\text{eff}}$. Equation (19.1) in the linear drain current region becomes

$$I_{DLIN} = \frac{W}{L_{\text{eff}}} |Q_n(V_{GS})|\, \mu_{app} V_{DS} = V_{DS}/R_{ch}\,, \tag{19.10}$$

where R_{ch} is the channel resistance. For large V_{DS}, $F_{SAT} \to 1$, and eqn. (19.1) reduces to the traditional velocity saturation expression,

$$I_{DSAT} = W |Q_n(V_{GS}, V_{DS})|\, \upsilon_T\,, \tag{19.11}$$

where

$$\upsilon_T = \sqrt{\frac{2 k_B T}{\pi m^*}} = \upsilon_{inj}^{\text{ball}}\,, \tag{19.12}$$

is the ballistic injection velocity for Maxwell Boltzmann statistics. Note that the ballistic injection velocity can be difficult to compute in practice. Strain and quantum confinement can affect m^*, and eqn. (19.12) assumes only one subband is occupied, which is not always true.

The apparent mobility in the MVS model is given by

$$\frac{1}{\mu_{app}(L_{\text{eff}})} \equiv \frac{1}{\mu_n} + \frac{1}{\mu_B(L_{\text{eff}})}, \tag{19.13}$$

where the scattering limited mobility is

$$\mu_n = \frac{D_n}{k_B T/q} = \frac{\upsilon_T \lambda_0/2}{k_B T/q}, \tag{19.14}$$

and the ballistic mobility is

$$\mu_B(L_{\text{eff}}) = \frac{\upsilon_T L_{\text{eff}}/2}{k_B T/q}. \tag{19.15}$$

The injection velocity under high drain bias is

$$\frac{1}{\upsilon_{inj}} = \frac{1}{\upsilon_T} + \frac{1}{D_n/\ell}, \tag{19.16}$$

where $\ell \ll L_{\text{eff}}$ and

$$D_n = \frac{\upsilon_T \lambda_0}{2}. \tag{19.17}$$

Recall that we have assumed that the mean-free-path in the linear region, λ_{LIN}, is equal to the mean-free-path in the saturation region, λ_{SAT}. While it is not strictly true that $\lambda_{LIN} = \lambda_{SAT} = \lambda_0$, it is physically sensible [19] and is supported by experimental studies [11]. Finally, it is also useful to recall how the parameters in the MVS model are related to the transmission. From eqn. (18.26) for the linear region, we have

$$\mathcal{T}_{LIN} = \frac{\lambda_0}{\lambda_0 + L_{\text{eff}}} = \frac{\mu_{app}}{\mu_B} = \frac{\mu_n}{\mu_B + \mu_n}, \tag{19.18}$$

and from (18.27) for the saturation region, we have

$$\mathcal{T}_{SAT} = \frac{\lambda_0}{\lambda_0 + \ell} = \frac{2}{1 + \upsilon_T/\upsilon_{inj}}. \tag{19.19}$$

The measured injection velocity is related to the transmission according to

$$\upsilon_{inj} = \upsilon_T \left(\frac{\mathcal{T}_{SAT}}{2 - \mathcal{T}_{SAT}} \right). \tag{19.20}$$

This section has summarized the main results that were presented and discussed in earlier lectures. When measured IV characteristics are fit to the MVS model, we will regard the results as measurements of the fixed series resistance, R_{SD0}, the apparent mobility, μ_{app}, and the injection velocity, υ_{inj}. We will also see that the ballistic injection velocity, the scattering limited mobility, the mean-free-path, the critical length, and the linear and saturation region transmissions can all be deduced from measurements.

19.3 ETSOI MOSFETs and III-V HEMTs

The Si MOSFETs to be examined have a simple, well-characterized physical structure that facilitates analysis. As shown in Fig. 19.1, the Si device is a silicon-on-insulator (SOI) structure with an extremely thin SOI layer of thickness, $T_{SOI} = 6.1 \pm 0.4$ nm [11]. The plane of the channel is (100), and the direction of transport is $\langle 110 \rangle$. The gate electrode is polycrystalline silicon, and the oxide is SiON with a Capacitance Equivalent Thickness (CET) of 1.1 nm. The strong inversion gate capacitance, $C_G(inv)$, is obtained from CV measurements on long channel devices [11]. For the devices examined here, $C_G(inv) = 1.98\,\mu\text{F/cm}^2$ for n-FETs [11]. The measured, near-equilibrium mobility for a long channel device is 350 $\text{cm}^2/\text{V}-\text{s}$, which corresponds to a mean-free-path of 15.8 nm.

Neutral stress liners are used in these devices so that the channel is nominally unstrained Si, which simplifies the computation of υ_T. Assuming $m^* = 0.22\,m_0$, we find $\upsilon_T = 1.14 \times 10^7$ cm/s. A process that produces the source/drain extensions during the last high temperature step results in very sharp junctions with low series resistance [13]. The physical length of the gate electrode is determined by CV measurements [12]. Detailed process simulations show that there is 1-2 nm of overlap between the gate electrode and the source/drain extension for n-MOSFETs and p-MOSFETS respectively, so $L_{\text{eff}} = L_G - 2$ nm for n-FETs and $L_{\text{eff}} = L_G - 4$ nm for p-FETs, where L_G is the physical length of the gate electrode. These effective channel lengths were confirmed by a careful analysis of 2D electrostatics [13, 14].

The HEMT is a field-effect transistor in which a wide bandgap III-V semiconductor serves as the "insulator" and a small bandgap III-V semiconductor serves as the channel. The III-V HEMTs to be examined have a high mobility, In-rich channel [10, 15]. As shown in Fig. 19.2, the device is built on an InP substrate. A buffer layer is first grown on the substrate followed by 2 nm of $In_{0.53}Ga_{0.47}As$, 5 nm of InAs, and 3 nm $In_{0.53}Ga_{0.47}As$. The $In_{0.53}Ga_{0.47}As$ layer is lattice-matched to the InP substrate, but there is a mismatch between the lattice spacings of $In_{0.53}Ga_{0.47}As$ and InAs, so the InAs layer is *pseudomorphic* — it is under strain, but the layer is thin enough that the strain can be accommodated without generating crystal defects. On top of this 10 nm thick channel structure is an $In_{0.52}Al_{0.48}As$ barrier layer, which acts as the insulator for this FET. The "T-gate" structure lowers the gate resistance, which is important for RF applications. Heavily doped "cap" layers facilitate low contact resistances.

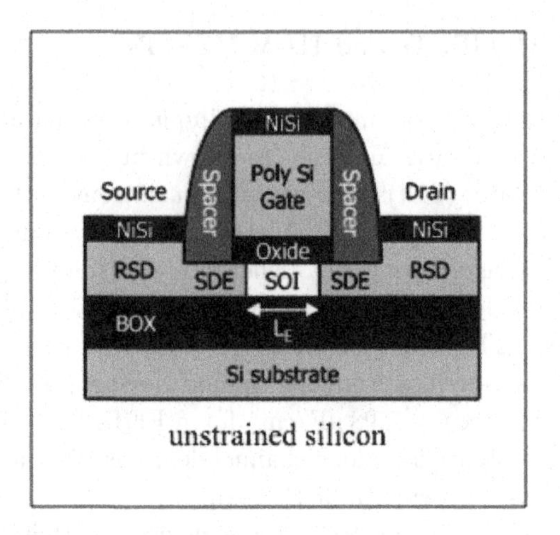

Fig. 19.1 Cross section of the ETSOI MOSFETs analyzed in this lecture. (©IEEE 2014. Reprinted, with permission, from: [11].)

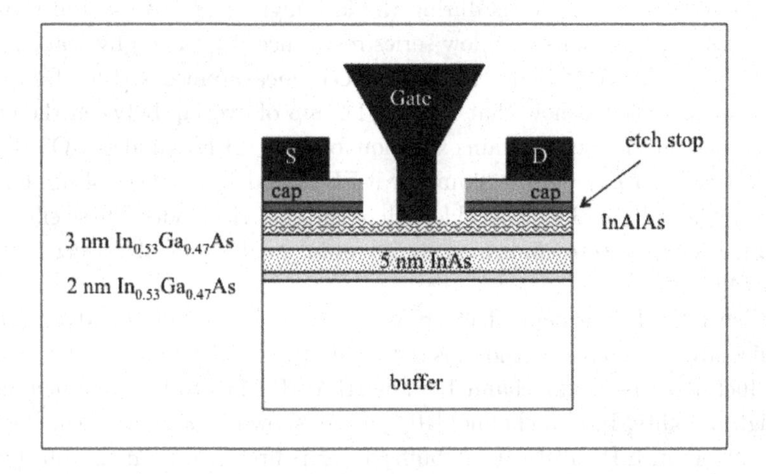

Fig. 19.2 Cross section of the III-V HEMTs analyzed in this lecture. (Adapted from [15].)

The high mobility of the In-rich channel gives this transistor its name — High Electron Mobility Transistor (HEMT). The measured mobility of a long channel device is $12,500\,\text{cm}^2/\text{V}-\text{s}$, which gives a mean-free-path of 153 nm [12]. The channel effective mass is $m_n^* = 0.022m_0$, which gives $\upsilon_T = 3.62 \times 10^7\,\text{cm/s}$ [12]. The 4 nm thick $In_{0.52}Al_{0.48}As$ layer on top of the channel results in an gate capacitance of $C_G(inv) = 1.08\,\mu\text{F/cm}^2$ [12].

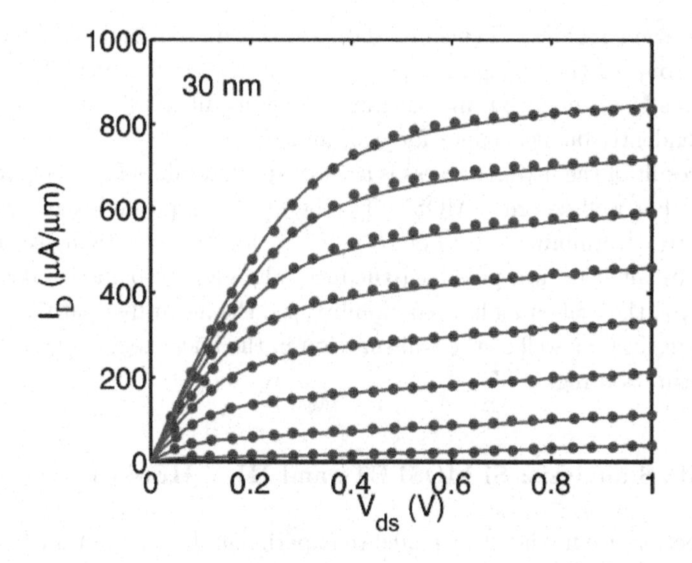

Fig. 19.3 Measured IV characteristics of an $L_{\text{eff}} = 30$ nm ETSOI MOSFET. The points are the measured data (similar to [11]), and the lines are the MVS model fits. The first line is for $V_{GS} = -0.2$ V, and for each line above, V_{GS} increases by 0.1 V. The MVS analysis and plot were provided by Dr. S. Rakheja, MIT, 2014. The data were provided by A. Majumdar, IBM, 2014. Used with permission.

19.4 Fitting the MVS model to measured IV data

Fitting the measured IV characteristics of well-designed MOSFETs typically involves fitting both the transfer and output characteristics. We assume that the physical and effective gate lengths have been independently measured along with the strong inversion gate capacitance. The parameter, α, which controls the transition from weak to strong inversion is set at 3.5 [11, 16]. The parameter, β, in F_{SAT}, is adjusted to match the drain saturation characteristics, but typically falls in a narrow narrow range of $\beta \approx 1.6 - 2.0$ [16]. To fit measured data, such as that shown in Fig. 19.3, four parameters are adjusted. The threshold voltage, V_{T0} is adjusted to fit the measured off-current under low V_{DS}. The DIBL parameter, δ, is adjusted to fit the measured DIBL. (It also affects the output conductance.) The subthreshold slope parameter, m_0, and the punchthrough parameter, m', are adjusted to fit the subthreshold slope under low and high V_{DS}. The apparent mobility, μ_{app}, is adjusted to fit the linear region slope of I_{DS} vs. V_{DS}. The injection velocity, υ_{inj}, is adjusted to fit the measured saturation currents. The series resistance, R_{SD0}, affects both the linear and saturation regions. Typically, data can be fit by hand with only a few iterations, or the fitting process can be automated. Because the series resistance affects the linear and saturation regions differently, it is possible to independently deduce values for μ_{app} and R_{SD0}.

The result of the fitting process is a set of specific values for R_{SD0}, μ_{app}, and υ_{inj}. For well-designed MOSFETs, the fits are typically excellent. In addition to determining values of R_{SD0}, μ_{app}, and υ_{inj}, we will see, that with careful analysis, it is possible to deduce values for the ballistic injection velocity, υ_T, the scattering limited mobility, μ_n, the mean-free-path, λ_0, the critical length, ℓ, as well as the transmission in the linear region, $\mathcal{T}_{LIN}$, and in the saturation region, $\mathcal{T}_{SAT}$.

19.5 MVS analysis: Si MOSFETs and III-V HEMTs

In this section, we fit the MVS model to experimental results for an $L = 30$ nm silicon MOSFET [11] and for an $L = 30$ nm III-V high electron mobility transistor (HEMT) [15]. The fitted MVS parameters will be interpreted according to the transmission model.

Figure 19.3 shows measured *IV* characteristics of an ETSOI MOSFET with $L_{\text{eff}} = 30$ nm [5] along with MVS model fits to the measured data. The MVS fitting parameters are:

Si ETSOI n-MOSFET:

$$
\begin{aligned}
R_{SD0} &= R_{S0} + R_{D0} = 130\ \Omega - \mu\text{m} \\
\mu_{app} &= 220\ \text{cm}^2/\text{V} - \text{s} \\
\upsilon_{inj} &= 0.82 \times 10^7\ \text{cm/s}\,.
\end{aligned}
$$

To interpret these results, we compute the linear and saturation region transmissions. To estimate $\mathcal{T}_{LIN}$ from eqn. (19.18), the ballistic mobility must be known. For the ballistic mobility, we use eqn. (18.17) and find

$$
\mu_B = \frac{\upsilon_T L_{\text{eff}}}{2k_BT/q} = \frac{(1.12 \times 10^7\ \text{cm/s})\,(30 \times 10^{-7}\ \text{cm})}{2 \times 0.026} = 658\ \text{cm}^2/\text{V} - \text{s}\,.
$$

The linear region transmission is estimated from eqn. (19.18) as

$$
\mathcal{T}_{LIN} = \frac{\mu_{app}}{\mu_B} = \frac{220}{646} = 0.34\,.
$$

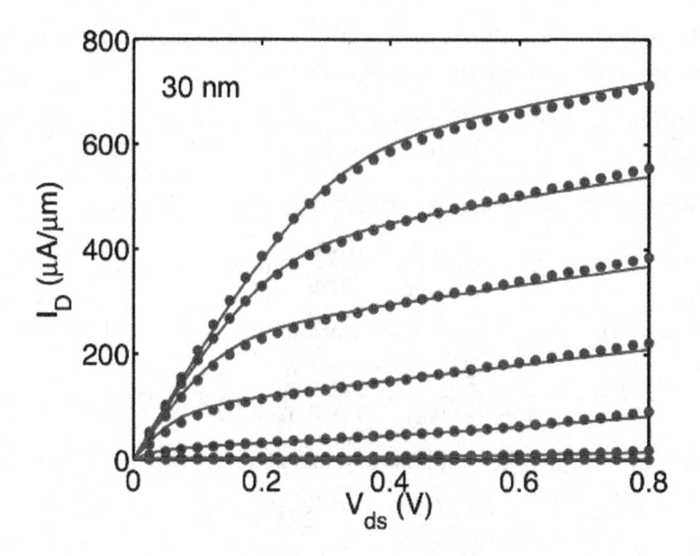

Fig. 19.4 *IV* characteristics of an $L_{\text{eff}} = 30$ nm III-V HEMT. The points are the measured data [15], and the lines are the MVS analysis and plot were provided by Dr. S. Rakheja, MIT, 2014. Data provided by D.-H. Kim. Used with permission.

To estimate the transmission in saturation, we use eqn. (19.20) and find

$$\mathcal{T}_{SAT} = \frac{2}{1 + \upsilon_T/\upsilon_{inj}} = \frac{2}{1 + 1.12/0.82} = 0.85\,.$$

According to the second of eqns. (18.8), we can write the ballistic on-current ratio as

$$B_{SAT} = \frac{I_{DS}(\text{ON})}{I_{DS}^{\text{ball}}(\text{ON})} = \frac{\mathcal{T}_{SAT}}{2 - \mathcal{T}_{SAT}} = 0.72\,.$$

These results, typical for Si MOSFETs, show that the device operates well below the ballistic limit in the linear region and fairly close to the ballistic limit in the saturation region.

Figure 19.4 shows the measured *IV* characteristics of an $L_{\text{eff}} = 30$ nm III-V HEMT [15]. The MVS fitting parameters are:

III-V HEMT:

$$R_{SD0} = R_{S0} + R_{D0} = 400\ \Omega - \mu\text{m}$$
$$\mu_{app} = 1800\ \text{cm}^2/\text{V} - \text{s}$$
$$\upsilon_{inj} = 3.5 \times 10^7\ \text{cm/s}\,.$$

To interpret these results, we compute the linear and saturation region transmissions. For the ballistic mobility, we find

$$\mu_B = \frac{\upsilon_T L_{\text{eff}}}{2k_B T/q} = \frac{(3.62 \times 10^7 \text{ cm/s})\,(30 \times 10^{-7} \text{ cm})}{2 \times 0.026} = 2088 \text{ cm}^2/\text{V} - \text{s}\,.$$

The linear region transmission is estimated from eqn. (19.18) as

$$\mathcal{T}_{LIN} = \frac{\mu_{app}}{\mu_B} = \frac{1800}{2088} = 0.86\,.$$

To estimate the transmission in saturation, we use eqn. (19.20) and find

$$\mathcal{T}_{SAT} = \frac{2}{1 + \upsilon_T/\upsilon_{inj}} = \frac{2}{1 + 3.62/3.50} = 0.98\,.$$

Finally we can estimate the ballistic on-current ratio as

$$B_{SAT} = \frac{I_{DS}(\text{ON})}{I_{DS}^{\text{ball}}(\text{ON})} = \frac{\mathcal{T}_{SAT}}{2 - \mathcal{T}_{SAT}} = 0.96\,.$$

These results, typical for III-V HEMTs, show that the device operates rather close to the ballistic limit in the linear region and essentially at the ballistic limit in the saturation region. This could have been anticipated in two ways. First, the mean-free-path deduced from the scattering-limited mobility was 153 nm — longer than the channel length. Second (and equivalently) the ballistic mobility was lower than the scattering limited mobility.

Although this device operates close to the ballistic limit in terms of on-current, it is important to recall that operation near the ballistic limit only means that the critical part of the channel is short compared to the mean-free-path. Energetic carriers are expected to scatter several times in the high-field region near the drain.

The analysis discussed in this section helps us understand device performance in terms of transmission and the ballistic on-current ratio. As discussed next, a careful analysis of the linear and saturation regions allows us to extract some other useful parameters. Finally, we note that there are some uncertainties in the calculations presented here. The proper effective mass to use depends on the strain in the structure (which may increase or decrease the effective mass) and conduction band nonparabolicity, which increases the effective mass of quantum confined materials. Upper subbands may also be occupied, and the assumption of non-degenerate carrier statistics may not be suitable — especially for III-V FETs. It may be preferable, therefore, to extract υ_T from the measured IV characteristics, as will be discussed in Sec. 19.7.

19.6 Linear region analysis

Analysis of the linear region of a FET can reveal the presence of a ballistic component to the channel resistance, and it can provide a measurement of the scattering limited mobility, μ_n. The MVS fitting procedure allows us to extract a physically meaningful apparent mobility for each channel length. From equations (19.13)-(19.15), we find

$$\frac{1}{\mu_{app}} = \frac{1}{\mu_n} + \left(\frac{\lambda_0}{\mu_n}\right)\frac{1}{L_{\text{eff}}}. \tag{19.21}$$

A plot of $1/\mu_{app}$ vs. $1/L_{\text{eff}}$ should be a straight line with a y-intercept that is one over the scattering limited mobility and a slope that is the ratio of the mean-free-path to the scattering limited mobility. The second term in eqn. (19.21) is just one over the ballistic mobility. The apparent mobility could be channel length dependent if the scattering-limited mobility varies with channel length, but if the plot is linear with a physically sensible slope, then the channel length dependence is most likely due the the ballistic resistance. Figure 19.5 shows results for the III-V HEMT [12]. From this plot, we find $\mu_n = 12,195\,\text{cm}^2/\text{V}-\text{s}$ and $\lambda_0 = 171$ nm. These numbers are very close to those expected from the measured mobility on a long channel FET [12].

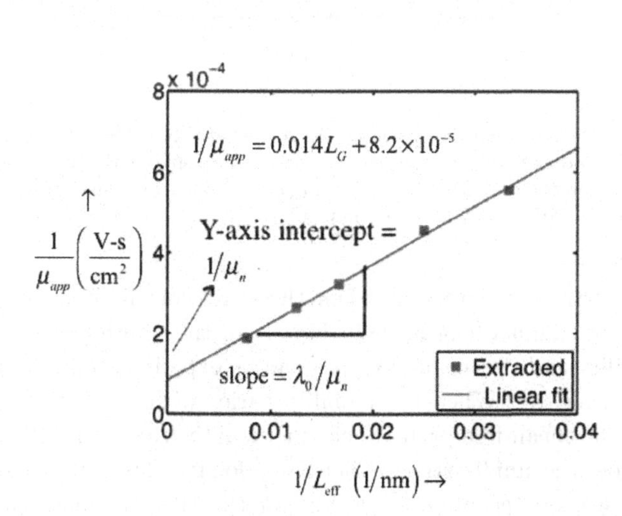

Fig. 19.5 Plot of $1/\mu_{app}$ vs. $1/L$ for the III-V HEMT. From the y-intercept we find the scattering-limited mobility and from the slope of the line, the near-equlibrium mean-free-path. (©IEEE 2014. Reprinted, with permission, from: [12].)

The plot of $1/\mu_{app}$ vs. $1/L_{\text{eff}}$ is not a straight line when the mean-free-path (i.e. the scattering limited mobility) varies with channel length. For such cases, we can deduce the mean-free-path for each channel length by using eqns. (19.13)-(19.15) to find

$$\frac{1}{\lambda_0(L_{\text{eff}})} = \frac{\upsilon_T}{2(k_BT/q)}\frac{1}{\mu_{app}} - \frac{1}{L_{\text{eff}}}\,. \tag{19.22}$$

Figure 19.6 shows the extracted mean-free-path vs. channel length for ET-SOI MOSFETs. Note the decrease in mean-free-path at short channel lengths. This effect may arise from device processing effects, but it has also been predicted to occur because of long-range Coulomb oscillations [20, 21].

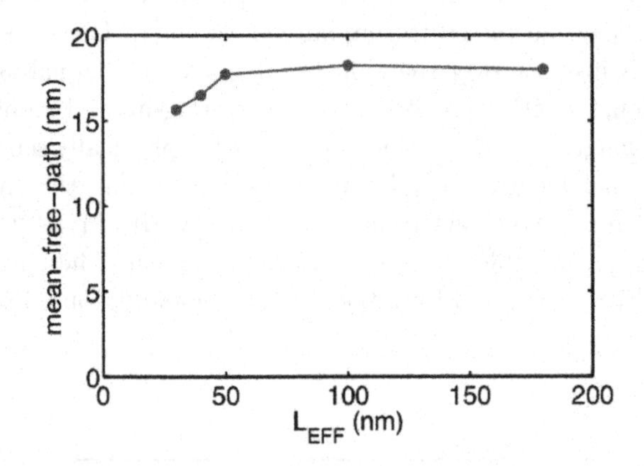

Fig. 19.6 The low bias mean-free-path vs. channel length for ETSOI MOSFETs. For an alternative treatment, which assumes a constant mean-free-path but also includes backscattering from the drain, see [22, 23]. (Figure provided by Xingshu Sun, Purdue University, August, 2014. Used with permission.)

It is interesting to note that when the scattering limited mobility is independent of channel length, as in Fig. 19.5, then both the scattering-limited mobility and the mean-free-path can be experimentally determined without knowing υ_T. When the mobility varies with channel length, we can extract the mean-free-path vs. channel length from eqn. (19.22), but we must know the unidirectional thermal velocity. This can be difficult to compute without accurate knowledge of the effective mass (which is affected by strain and quantum confinement) and the subband populations. As discussed next, however, υ_T can be deduced by analyzing the length dependent injection velocity.

19.7 Saturation region analysis

The magnitude of the injection velocity decreases as the channel length increases. According to eqns. (19.20) and (19.19),

$$v_{inj} = v_T \frac{\lambda_0}{\lambda_0 + 2\ell}, \tag{19.23}$$

which can be written as

$$\frac{1}{v_{inj}} = \frac{1}{v_T} + \frac{2\ell}{\lambda_0 v_T}. \tag{19.24}$$

It is reasonable to assume that ℓ is proportional to L_{eff}. While this is difficult to justify rigorously, a careful analysis of experiments suggests that it is an acceptable approximation in practice [11]. Assuming that $\ell = \xi L_{\text{eff}}$, we can write (19.24) as

$$\frac{1}{v_{inj}} = \frac{1}{v_T} + \frac{2\xi}{\lambda_0 v_T} L_{\text{eff}}. \tag{19.25}$$

A plot of $1/v_{inj}$ vs. L_{eff} should be a straight line. The y-intercept gives the unidirectional thermal velocity and the slope gives ξ, from which we can deduce ℓ. Figure 19.7 shows results for the III-V HEMT [12]. From this plot, we find $v_T = 3.57 \times 10^7$ cm/s and $\xi = 0.09$. This thermal velocity is very close to that expected from the known effective mass, and the critical length is a small fraction of the channel length as expected.

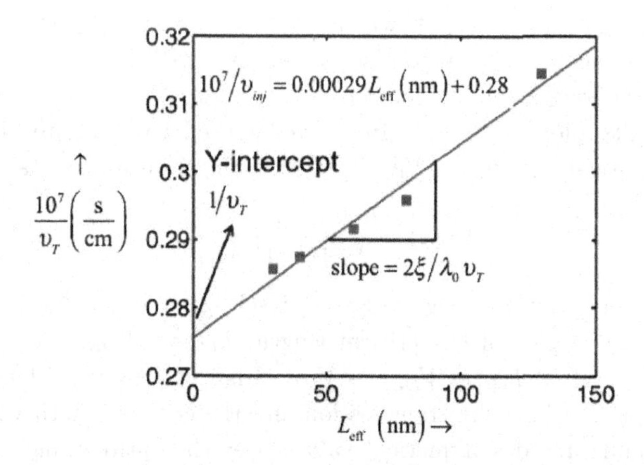

Fig. 19.7 Extraction of the thermal velocity, v_T, for III-V HEMTs by fitting the $1/v_{inj}$ vs. L_{eff} plot with straight line. (©IEEE 2014. Reprinted, with permission, from: [12].)

19.8 Linear to saturation region analysis

One of the challenges in modeling nano-MOSFETs is that we do not have an analytical expression for the drain voltage dependent transmission, $\mathcal{T}(V_{DS})$. Equations (19.18) and (19.19) give $\mathcal{T}$ in the small and large V_{DS} limits. If we had an analytical model for $\mathcal{T}(V_{DS})$, we would not need the empirical drain saturation function, F_{SAT}, as given by eqn. (19.6).

Using the measured IV characteristics of well-behaved nano-MOSFETs, we can extract the experimental $\mathcal{T}(V_{DS})$ characteristic. The process is as follows. First, we fit the measured IV characteristics with the MVS model. Next, we generate intrinsic transistor characteristics by setting $R_{S0} = R_{D0} = 0$ in the MVS model and plotting the resulting IV characteristic. Then we make use of eqn. (17.18) (the Landauer expression for the IV characteristic in terms of the transmission) in the non-degenerate limit to write,

$$I_{DS} = W|Q_n(V_{GS}, V_{DS})|\,\upsilon_T \left(\frac{\mathcal{T}}{2-\mathcal{T}}\right) \left[\frac{1 - e^{-qV_{DS}/k_BT}}{1 + (\frac{\mathcal{T}}{2-\mathcal{T}})e^{-qV_{DS}/k_BT}}\right]. \tag{19.26}$$

The inversion charge, $Q_n(V_{GS}, V_{DS})$, is known from eqn. (19.2) because the parameters in (19.2) have been determined by the MVS fitting to the measured data. Assuming that the ballistic injection velocity, υ_T, is also known, then at any bias point, (V_{GSi}, V_{DSi}), we can fit eqn. (19.26) to the intrinsic IV characteristic and deduce $\mathcal{T}(V_{GSi}, V_{DSi})$. A plot of $\mathcal{T}(V_{GSi}, V_{DSi})$ vs. V_{DSi} at $V_{GSi} = V_{DD}$ is shown in Fig. 19.8 for two different channel lengths. As expected, the transmission decreases as V_{DSi} increases and is smaller for the longer channel length.

The results plotted in Fig. 19.8 can be used to estimate the bias-dependent critical length, $L_C(V_{DSi})$. Writing the transmission as

$$\mathcal{T}(V_{DSi}) = \frac{\lambda_0}{\lambda_0 + L_C(V_{DSi})}, \tag{19.27}$$

we can use the results in Fig. 19.8, set $L_C(V_{DSi} = 0) = L_{\text{eff}}$, and produce Fig. 19.9, a plot of the critical length, L_C, vs. V_{DSi}. We find, as expected, $L_C = \ell \ll L_{\text{eff}}$ as $V_{DSi} \rightarrow V_{DD}$. Figures 19.8 and 19.9 confirm our expectation of how the transmission and the critical length vary with drain bias and provides numerical values over the entire range of drain biases. From the $V_{DSi} = 0$ transmission in Fig. 19.8 and eqn. (19.27) with $L_C(V_{DSi} = 0) = L_{\text{eff}}$, we find the mean-free-path. The result is $\lambda_0(30\,\text{nm}) = 15.4$ nm and $\lambda_0(180\,\text{nm}) = 17.8$ nm.

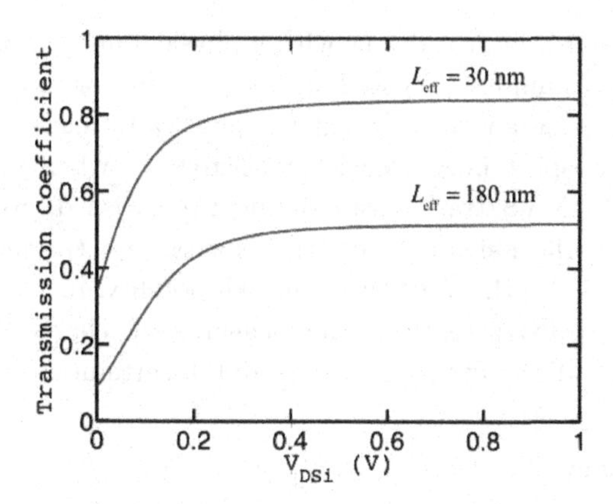

Fig. 19.8 A plot of the extracted transmission vs. drain voltage for ETSOI n-MOSFETs with $L_{\text{eff}} = 30$ nm and $L_{\text{eff}} = 130$ nm. As expected, the transmission is higher under low drain bias than under high drain bias. (Plot produced by Xingshu Sun, Purdue University using ETSOI data supplied by A. Majumdar, IBM.)

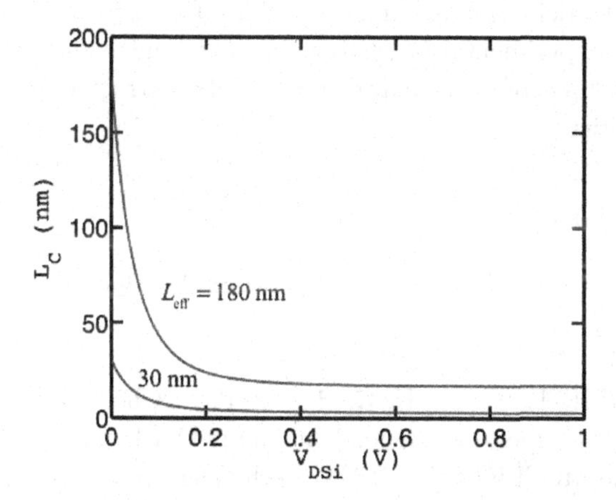

Fig. 19.9 A plot of the deduced critical length for backscattering vs. drain voltage for ETSOI n-MOSFETs with $L_{\text{eff}} = 30$ nm and $L_{\text{eff}} = 130$ nm. The critical length is set to L_{eff} for zero drain bias, and we find $L_C = \ell \ll L_{\text{eff}}$ under high drain bias. (Plot produced by Xingshu Sun, Purdue University using ETSOI data supplied by A. Majumdar, IBM.)

19.9 Discussion

The Virtual Source model provides a semi-empirical description of the *IV* characteristics of well-designed field-effect transistors. By adjusting only a few parameters, excellent fits to the *IV* characteristics can be obtained.

The Landauer approach provides us with a physical interpretation of these parameters. The examples discussed in this lecture show how measured IV characteristics can be analyzed to extract physically relevant information about carrier transport in nanoscale transistors — when the underlying assumption of the MVS/transmission model are satisfied. Key device parameters such as the ballistic injection velocity, υ_T, the mean-free-path for backscattering, λ_0, the scattering limited mobility, μ_n, and the critical length, ℓ can all be extracted from the measured IV characteristic. In the next lecture, we'll discuss the limitations and uncertainties of the model.

19.10 Summary

In this lecture, we showed how to analyze the measured IV characteristics of nano transistors using the MVS/Transmission model. While each type of transistor presents its own challenges, the approach illustrated here provides a starting point for analysis. It should be understood, however, that the MVS model is a model for well-behaved transistors as indicated by excellent fits of experimental IV characteristics to the model. For such well-behaved transistors, physical parameters can be extracted from measured IV characteristics.

19.11 References

The following papers discus various ways to analyze the IV characteristics of nano-MOSFETs.

[1] M. J. Chen, H. T. Huang, K. C. Huang, P. N. Chen, C. S. Chang, and C. H. Diaz, "Temperature dependent channel backscattering coefficients in nanoscale MOSFETs," in IEDM Tech. Dig., pp. 39-42, 2002.

[2] V. Barral, T. Poiroux, M. Vinet, J. Widiez, B. Previtali, P. Grosgeorges, G. Le Carval, S. Barraud, J. L. Autran, D. Munteanu, and S. Deleonibus, "Experimental determination of the channel backscattering coefficient on 10-70 nm-metal-gate double-gate transistors," *Solid-State Electron.*, **51**, no. 4, pp. 537-542, 2007.

[3] M. Zilli, P. Palestri, D. Esseni, and L. Selmi, "On the experimental determination of channel back-scattering in nanoMOSFETs," in IEDM Tech. Dig., pp. 105-108, 2007.

[4] R. Wang, H. Liu, R. Huang, J. Zhuge, L. Zhang, D. W. Kim, X. Zhang, D. Park, and Y. Wang, "Experimental investigations on carrier transport in Si nanowire transistors: ballistic efficiency and apparent mobility," *IEEE Trans. Electron Devices*, **55**, no. 11, pp. 2960-2967, 2008.

[5] V. Barral, T. Poiroux, J. Saint-Martin, D. Munteanu, J. L. Autran, and S. Deleonibus, "Experimental investigation on the quasi-ballistic transport: Part I — Determination of a new backscattering coefficient extraction methodology," *IEEE Trans. Electron Devices*, **56**, no. 3, pp. 408-419, 2009.

[6] V. Barral, T. Poiroux, D. Munteanu, J. L. Autran, and S. Deleonibus, "Experimental investigation on the quasi-ballistic transport: Part II — Backscattering coefficient extraction and link with the mobility," *IEEE Trans. Electron Devices*, **56**, no. 3, pp. 420-430, 2009.

The MVS model has been used to analyze the characteristics of both Si and III-V FETs.

[7] A. Khakifirooz and D. A. Antoniadis, "Transistor performance scaling: the role of virtual source velocity and its mobility dependence," in IEDM Tech. Dig., pp. 667-670, 2006.

[8] A. Khakifirooz and D. A. Antoniadis, "MOSFET performance scaling — Part I: Historical trends," *IEEE Trans. on Electron Devices*, **55**, no. 6, pp. 1391-1400, 2008.

[9] A. Khakifirooz and D. A. Antoniadis, "MOSFET performance scaling — Part II: Future directions," *IEEE Trans. on Electron Devices*, **55**, no. 6, pp. 1401-1408, 2008.

[10] D. H. Kim, J. A. del Alamo, D. A. Antoniadis, and B. Brar, "Extraction of virtual-source injection velocity in sub-100 nm III-V HFETs," in IEDM Tech. Dig., pp. 861-864, 2009.

The analysis approach used in this lecture follows that used in the following two papers. The first paper considers Si MOSFETs (ETSOI MOSFETs) and the second considers both ETSOI MOSFETs and III-V HEMTs. IV.

[11] A. Majumdar and D. A. Antoniadis, "Analysis of Carrier Transport in Short-Channel MOSFETs," *IEEE Trans. Electron. Dev.*, **61**, pp. 351-358, 2014.

[12] S. Rakheja, M. Lundstrom, and D. Antoniadis, "A physics-based compact model for FETs from diffusive to ballistic carrier transport regimes," in IEDM Tech. Dig., 2014.

The ETSOI MOSFETs used for the analysis presented in this lecture are discussed in the following papers.

[13] A. Majumdar, Z. Ren, S. J. Koester, and W. Haensch, "Undoped-body, extremely-thin SOI MOSFETs with back gates," *IEEE Trans. on Electron Devices*, **56**, no. 10, pp. 2270-2276, 2009.

[14] A. Majumdar, X. Wang, A. Kumar, J. R. Holt, D. Dobuzinsky, R. Venigalla C. Ouyang, S. J. Koester, and W. Haensch, "Gate length and performance scaling of undoped-body, extremely thin SOI MOSFETs," *IEEE Electron Device Lett.*, **30**, no. 4, pp. 413-415, 2009

The III-V HEMTs used for the analysis presented in this lecture are discussed in the following paper.

[15] D. H. Kim and Jesus del Alamo, "30-nm InAs Pseudomorphic HEMTs on a InP Substrate With a Current-Gain Cutoff Frequency of 628 GHz," *IEEE Electron Device Letters*, **29**, no. 8, pp. 830-833, 2008.

The MIT Virtual Source Model is a physics-based compact model based on the Landauer model and suitable for use in circuit simulation. The first paper below described the model, and the second citation is a location from which the model can be downloaded.

[16] A. Khakifirooz, O. M. Nayfeh, and D. A. Antoniadis, "A Simple Semiempirical Short-Channel MOSFET Current Voltage Model Continuous Across All Regions of Operation and Employing Only Physical Parameters," *IEEE Trans. Electron. Dev.*, **56**, pp. 1674-1680, 2009.

[17] Shaloo Rakheja; Dimitri Antoniadis (2013), "MVS 1.0.1 Nanotransistor Model (Silicon)," `https://nanohub.org/resources/19684`.

For a discussion of the meaning of effective channel length, see Sec. 4.3 in Taur and Ning. Section 3.1.3.2 discusses the subthreshold current and shows that $V_{DSAT} = k_BT/q$ in subthreshold.

[18] Y. Taur and T. Ning, *Fundamentals of Modern VLSI Devices*, 2nd Ed., Oxford Univ. Press, New York, 2013.

The following paper argues that the mean-free-paths for backscattering under low and high drain bias are approximately the same. Reference [11] provides experimental evidence that this is true.

[19] M. S. Lundstrom, "Elementary scattering theory of the Si MOSFET," *IEEE Electron Dev. Letters*, **18**, pp. 361-363, 1997.

Long-range Coulomb interactions can cause the mobility to decrease in short channel MOSFETs. For a discussion, see the following papers.

[20] M. V. Fischetti and S. E. Laux, "Long-range Coulomb interactions in small SI devices," *J. Appl. Phys.*, **89**, pp. 1205-1231, 2001.

[21] T. Uechi, T. Fukui, and N. Sano, "3D Monte Carlo simulation including full Coulomb interaction under high electron concentration regimes," *Phys. Status Solidi C*, **5**, pp. 102-106, 2008.

Two recent papers discuss the anomalous apparent mobility vs. channel length behavior observed in some MSOFETs. By anomalous, we mean, not described by eqns. (18.19) and (18.17)

[22] D. A. Antoniadis, "On apparent electron mobility in Si nMOSFETs from diffusive to ballistic regime," *IEEE Trans. Electron Dev.*, **63**, no. 7, pp. 2650-2656, 2016.

[23] K. Natori, H. Iwai, and K. Kakushima, "Anomalous degradation of low-field mobility in short-channel metal-oxide-semiconductor field-effect transistors," *J. Appl. Phys.*, **118**, no. 23, p. 234502, 2015.

Lecture 20

Limits and Limitations

20.1 Introduction

As the dimensions of high-performance transistors for digital logic continue to shrink, some questions arise. "What are the fundamental limits of MOSFETs?" "How close to these limits can semiclassical models be used?" These questions will be briefly addressed in this lecture. Even when the semiclassical model can be used, questions about the validity of the simplifying assumptions that make the transmission model for MOSFETs tractable must be asked. Some questions are straightforward (e.g. how good is the assumption of gate voltage independent series resistances) and others involve more subtle discussions of complex transport physics. Some of these questions will also be addressed in this lecture.

20.2 Ultimate limits of the MOSFET

In Lecture 3, we presented a simple model of the MOSFET as a barrier-controlled device and summarized it in Fig. 3.7. This simple barrier model can be used to establish some fundamental limits for transistors operating as digital switches. Although the approach is a little different here, we arrive at the same expressions for the fundamental limits as in [1]. The arguments are simple and heuristic but sufficient to estimate the fundamental limits.

Figure 20.1 summarizes our simple model for the MOSFET as a barrier controlled logic switch (the same model device would apply to a bipolar transistor as well). The off-state is shown on the left. The large energy barrier prevents electrons in the source from flowing out the drain (a large drain voltage is assumed). The on-state is shown on the right. A large gate voltage pushes the barrier down, electrons flow from the source, across the channel, and into the drain. Ballistic transport is assumed in the channel, so the carriers deposit their energy in the drain, where they relax to the bottom of the conduction band through strong inelastic scattering.

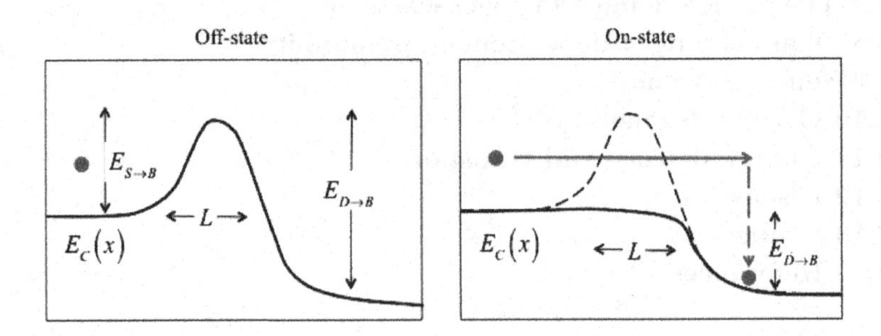

Fig. 20.1 Simple, barrier model for a MOSFET as a digital switch. Left: the Off-state. Right: the On-state. The energy barrier for electrons from the source to the top of the energy barrier is $E_{S\to B}$ and energy barrier for electron from the drain to the top of the energy barrier is $E_{D\to B}$.

As shown in Fig. 20.2, this simple model can be used to establish the minimum energy for a switching event. The large gate voltage in the on-state eliminates the energy barrier between the source and the channel, but a barrier, $E_{D\to B}$, from the drain to the top of the barrier in the channel exists because of the positive drain voltage. After electrons have thermalized in the drain, there is some probability, $\mathcal{P}$, that they will be thermionically emitted over the barrier and return to the source. If this happens, a switching event did not occur. By requiring that the probability, $\mathcal{P}$, is less than one-half,

$$\mathcal{P} = e^{-E_{D\to B}/k_B T} < \frac{1}{2}, \tag{20.1}$$

we find the minimum energy barrier as

$$E_{\min} = k_B T \ln 2, \tag{20.2}$$

which is 0.017 eV at room temperature. Electrons that enter the drain dissipate their kinetic energy by inelastic scattering, so the minimum switching energy is $E_S|_{\min} = k_B T \ln 2$. The argument used here should be viewed as a simple, heuristic argument. More fundamental considerations [2, 3] and careful analysis [4] lead to the same conclusion for the minimum switching energy.

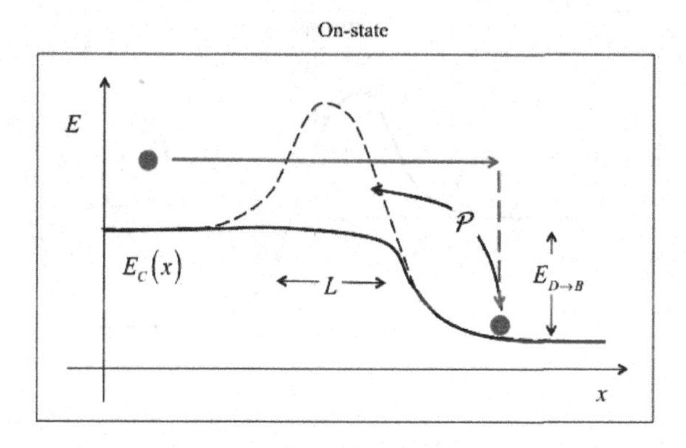

Fig. 20.2 Illustration of a switching event in the on-state. The probability of a switching event is $1 - \mathcal{P}$, where $\mathcal{P}$ is the probability that an electron from the drain can be thermionically re-emitted back to the source.

Next, let's estimate the minimum gate length. As shown in Fig. 20.3, when the device is in the off-state, the barrier must be high enough and thick enough to prevent electrons from the source to flow to the drain. To estimate the barrier height needed, consider thermionic emission. The height of the energy barrier in the Off-state must be at least $E_{\min} = k_B T \ln 2$ to ensure that the probability that an electron surmounts the barrier is less than one-half. The minimum thickness of the barrier (the length of the channel) is determined by quantum mechanical tunneling through the barrier. The probability that electrons from the source tunnel through the barrier can be estimated with a standard quantum mechanical approximation (the WKB approximation) [1]. Requiring that this probability be less than one-half for the device to be considered to be off, we find

$$\mathcal{P} = e^{-2\sqrt{2m^* E_{S\to B}}L/\hbar} < \frac{1}{2}, \tag{20.3}$$

which can be solved for the channel length, L, to find

$$L > \frac{\hbar}{\sqrt{2m^* E_{S\to B}}}. \tag{20.4}$$

By using $E_{S\to B} = E_S|_{\min}$, we find the minimum channel length to be

$$L_{\min} = \frac{\hbar}{\sqrt{2m^* E_S|_{\min}}}. \tag{20.5}$$

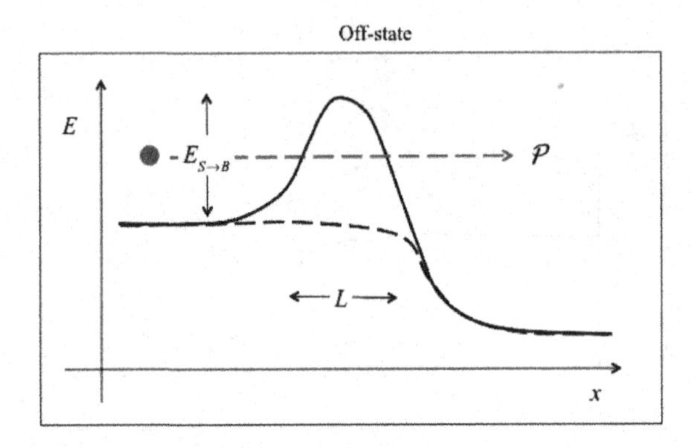

Fig. 20.3 Illustration of the off-state and the probability, $\mathcal{P}$, that an electron from the source can quantum mechanically tunnel through the barrier to the drain.

Finally, we can estimate the switching time of the device. In the on-state, electrons simply flow across the channel at the ballistic velocity. The minimum transit time across the channel is

$$\tau_{\min} = \frac{L_{\min}}{\upsilon_T}, \tag{20.6}$$

Using eqn. (20.5) for L_{min}, $\sqrt{2k_BT/\pi m^*}$ for υ_T, and discarding some factors on the order of unity, we find

$$\tau_{\min} = \frac{\hbar}{E_S|_{\min}}. \tag{20.7}$$

Having estimated the minimum switching energy, channel length, and switching time, we can evaluate them at room temperature (assuming $m^* = m_0$) to find

$$\boxed{\begin{aligned} E_S|_{\min} &= k_B T \ln 2 && = 0.017\,\text{eV} \\ L_{\min} &= \frac{\hbar}{\sqrt{2m^* E_S|_{\min}}} && = 1.5\,\text{nm} \\ \tau_{\min} &= \frac{\hbar}{E_S|_{\min}} && = 40\,\text{fs}\,. \end{aligned}} \tag{20.8}$$

The minimum switching energy of a single transistor as estimated by eqn. (20.8) is far below the switching energy in a typical CMOS circuit, which can be estimated from $E_S = C_S V_{DD}^2$, where C_S is the average capacitance being switched. In a typical circuit, $C_S \approx 1$ fF and $V_{DD} \approx 1$ V, which gives a switching energy that is a few hundred thousand times the fundamental limit. This occurs because the typical capacitance at a node being switched is far greater than the intrinsic gate capacitance of a single transistor. The additional capacitance is due to parasitics (e.g. parasitic gate to drain capacitance), to the capacitance of the wiring, and because a single logic gate typically drives a number of output gates (so-called *fanout*). On the other hand, a typical circuit node does not switch on every cycle, so this number should be multiplied by an *activity factor* that is much less than one.

Another consideration is *noise margin*; the probability of an error, $\mathcal{P}$, must be orders of magnitude smaller than one-half. The minimum channel length and power dissipation was determined by requiring the on-off ratio to be two. Realistic circuits require an *on-off ratio* of roughly 10^4, so channel lengths and switching energies will always be well-above the fundamental limit. Nevertheless, current day channel lengths of about 20 nm are within about an order of magnitude of the fundamental lower limit. The device transit time is also within an order of magnitude of the estimated lower limit. The nature of CMOS circuits, however, is such that the average switching energy is likely to always be orders of magnitude higher than the fundamental limit for a single device.

These rough estimates of the fundamental limits are instructive and indicate that some of them are being approached with CMOS technology. A key question for device researchers is: "Is there a fundamentally better switching device than a MOSFET?". Note that the fundamental limit for the switching energy of a MOSFET can be obtained from some very general arguments that do not assume a specific device [2]. It is also interesting to note that the lower limit size can be obtained from the uncertainty relation, $\Delta p \Delta x \geq \hbar/2$, and the lower limit for device speed can be obtained from $\Delta t \Delta E \geq \hbar/2$ [1]. These considerations suggest that there may not be a digital switching device that is fundamentally better than a MOSFET.

20.3 Quantum transport in sub-10 nm MOSFETs

The practical limits of transistor downscaling can be explored by numerical device simulation. Figure 20.4 shows results obtain by simulating quantum transport in a nanowire Si MOSFET. Electrostatic control for this model device is excellent, so the scaling limits are determined by quantum mechanical tunneling of electrons from the source through the source to channel barrier in the off-state. The plots show the energy-resolved current. At $L = 12$ nm (upper left), the off-state leakage current flows almost entirely over the top of the barrier. This transistor is operating as a classical, barrier-controlled device. When the channel length decreases to 10 nm (upper right), a small fraction of the current begins to tunnel through the barrier. At 10 nm, however, good transistor performance is still obtained — the device continues to operate as a classical barrier-controlled device. For $L = 7$ nm (bottom left), a substantial fraction of the off-state current flows by tunneling under the barrier. The performance of the device (e.g. its subthreshold slope) degrades. Finally, at $L = 5$ nm, most of the off-state current is due to tunneling through the barrier. At this channel length, it is difficult to modulate the current by controlling the barrier height because the barrier is transparent to electrons.

The results shown in Fig. 20.4 suggest that our semicassical, transmission model for the MOSFET should continue to describe Si devices down to channel lengths of 10 nm or so. To scale devices further, heavier effective masses may be needed to suppress tunneling [5]. Scaling to channel lengths of 5 nm presents many challenges — both practical (such as the increasing importance of parasitic resistance and capacitance at very short cannel lengths) and fundamental, such as direct tunneling through the barrier [7]. Numerical studies of effective mass engineering by strain and channel orientation suggest, however, that it may be possible to realize good performance — even down to 5 nm channel lengths [7]. It is clear however, that the practical and fundamental limits of MOSFET down-scaling are being approached.

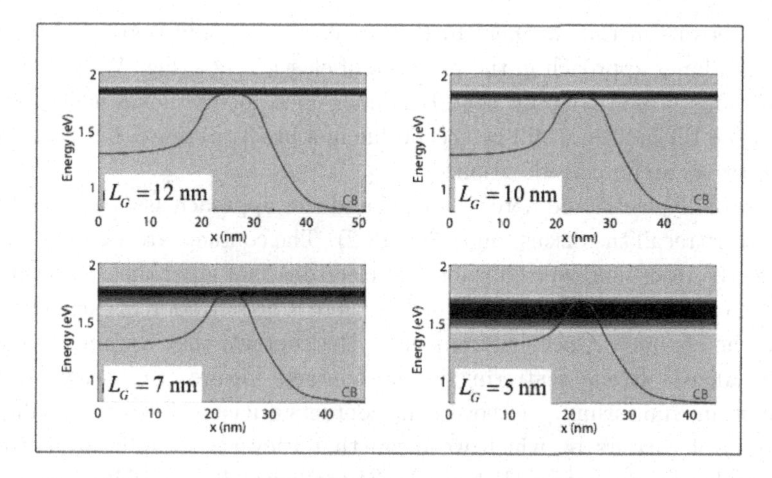

Fig. 20.4 The energy-revolved current as computed by a quantum transport simulation for a silicon nanowire MOSFET in the off-state under high drain bias. The nanowire is $< 110 >$ oriented and is 3 nm in diameter. (Simulations performed by Dr. Mathieu Luisier, ETH Zurich and used with permission, 2014).

20.4 Simplifying assumptions of the transmission model

Based on the Landauer approach to carrier transport, we have developed in these lectures a transmission model for the IV characteristics of nanoscale MOSFETs. Under an appropriate set of simplifying assumptions, the Landauer approach can be derived from a fully quantum mechanical treatment of dissipative quantum transport [8]. Alternatively, it can be derived from the semiclassical Boltzmann Transport Equation (BTE) under an appropriate set of simplifying assumptions [9, 10]. As discussed in the previous section, transistors with channel lengths above about 10 nm can generally be described semiclassically. Accordingly, we focus on the assumptions that underlie the semiclassical version of the Landauer approach. For a more careful exposition of the Landauer approach, see volumes 1 and 2 in this series [11, 12].

First, note that the transmission model uses the Landauer approach to express the terminal current as a linear combination of contact Fermi functions. This follows mathematically from BTE, if and only if the non-linear scattering terms from the exclusion principle can be excluded. The non-linear exclusion principle terms drop out in two cases: 1) Elastic scattering, and 2) Non-degenerate carrier statistics. In this work, we have emphasized 2), but in the on-state, Fermi-Dirac statistics may be required to describe the electrons in the channel. In that case, we can only rigorously justify the Landauer approach in the presence of elastic scattering. When neither conditions 1) nor 2) apply, then Landauer does not mathematically follow from the BTE; it may still be acceptable in a practical sense, but each case requires a careful consideration.

Secondly, we should note that the Landauer approach assumes idealized contacts (recall the discussion in Sec. 12.2). The contacts are assumed to be perfectly absorbing, which means that electrons that enter the contact from the channel are completely absorbed — there is no reflection of carriers back into the channel. Once electrons enter the contact, they are immediately thermalized; strong scattering in the contacts ensures that they always remain in equilibrium. Moreover, the contacts are considered to be infinite sources of carriers by which we mean that they can supply any current demanded by the gate and channel without being depleted. Real contacts can deviate from this ideal.

In the approach discussed in these notes, the third set of assumptions has to do with the transmission, which is described by a bias-independent mean-free-path (the near-equilibrium mean-free-path) and a bias-dependent critical length (Sec. 16.4). Scattering in a nanotransistors is complicated, and it is not obvious that such a simple description is adequate.

A fourth consideration has to do with electrostatic self-consistency. In our simple treatment, we do not spatially resolve the electrostatic potential within the device, we focus on the top-of-the-barrier and include a DIBL parameter to account for two-dimensional electrostatics. The validity of this approach needs to be considered.

A fifth consideration involves the use of Fermi-Dirac statistics. We have assumed Boltzmann statistics for most of our discussion. Fermi-Dirac statistics can be included; it complicates the model but can be important for III-V FETs [13, 14].

Finally, the sixth consideration has to do with the assumption that the inversion layer charge is controlled by electrostatics and not by transport. This is not true in general, and can become important for III-V devices [13, 14].

Several more issues could be raised. For example, we have assumed a simple, isotropic energy band, but the nonparbolicity of the conduction band can be important, as can the multiple valleys in the conduction band of Si (recall Secs. 9.2 and 9.3) and the warping of the valences bands. These issues can be important and need to be considered on a case by case basis, but we will focus in the next few sections on the issues identified above — beginning with the derivation of the Landauer approach from the BTE.

20.5 Derivation of the Landauer Approach from the BTE

In this section, we derive the Landauer Approach, eqn. (12.2), from a simple form of the BTE. The goal is to provide some understanding of the assumptions underlying eqn. (12.2). Consider the field-free semiconductor slab shown in Fig. 20.5 in which two large contacts in equilibrium (not shown) inject fluxes of charge carriers, $F_1(E)$ and $F_2(E)$, into the slab. At the right, a $+x$-directed flux, $\mathcal{T}(E)F_1(E)$, emerges due to the injected flux at the left. At the right, there is also a $+x$-directed flux, $(1-\mathcal{T}(E))F_2(E)$, due to the part of the injected flux, $F_2(E)$, that backscatters. We assume elastic scattering within the slab so that the transmission from the left to right is the same as from the right to left.

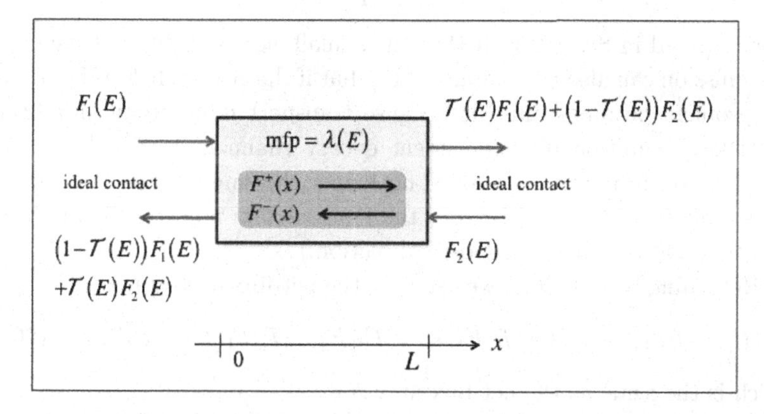

Fig. 20.5 A semiconductor slab with carrier fluxes, $F_1(E)$, and $F_2(E)$ injected from equilibrium contacts (not shown). Inside the slab, there is a $+x$-directed flux, $F^+(x)$, and a $-x$-directed flux, $F^-(x)$.

Within the slab, there is a positively-directed flux, $F^+(x)$, and a negatively-directed flux, $F^-(x)$. The positively-directed flux decreases when it back-scatters to a negatively-directed flux, and it increases when the negatively-directed flux backscatters to a positively-directed flux. Similar considerations apply to the negatively-directed flux. Accordingly, we can write:

$$\begin{aligned} \frac{dF^+(E)}{dx} &= -\frac{F^+}{\lambda} + \frac{F^-}{\lambda} \\ \frac{dF^-(E)}{dx} &= -\frac{F^+}{\lambda} + \frac{F^-}{\lambda}, \end{aligned} \tag{20.9}$$

where we assume that the two fluxes flow in a single energy channel (i.e. elastic scattering). The two equations have the same signs because F^- is taken to be positive when directed in the $-x$-direction. Equations (20.9) are a simple, steady-state BTE in which velocity space is discretized in one positively-directed velocity and one negatively-directed velocity. As discussed in Secs. 12.4 and 16.5, λ is the mean-free-path for backscattering. The quantity, dx/λ, is the probability per unit length that a positive (negative) flux backscatters to a negative (positive) flux. It is straightforward to solve (20.9) subject to the given boundary conditions and show that (see Sec. 6.3 in [12]):

$$\mathcal{T}(E) = \frac{\lambda(E)}{\lambda(E) + L}, \tag{20.10}$$

as was stated in Sec. 12.4. If there is a small electric field in the slab, the transmission can also be computed [15], but if the electric field is large, then the problem becomes difficult because transport is far from equilibrium, and the assumption of independent energy channels breaks down. The transmission from left to right is no longer the same as the transmission from right to left. (In this case, the transmission approaches one in one direction and zero in the opposite direction.)

Returning to Fig. 20.5, we see that the net flux at $x = L$ is

$$F(E) = \mathcal{T}F_1(E) + (1 - \mathcal{T})F_2(E) - F_2(E) = \mathcal{T}\left[F_1(E) - F_2(E)\right], \tag{20.11}$$

which is the same as the net flux at $x = 0$.

At the left, the injected current between E and $E + dE$ is

$$I_1(E)dE = qF_1(E)dE = qv_x^+ \frac{D(E)}{2} f_1(E)dE, \tag{20.12}$$

where v_x^+ is the velocity in the $+x$ direction, $D(E)$ is the density of states, and the factor of 2 comes from the fact that only half of the states have a velocity in the $+x$ direction. The equilibrium Fermi function of contact 1 is $f_1(E)$. Similarly, the current injected at the right is

$$I_2(E)dE = qv_x^+ \frac{D(E)}{2} f_2(E)dE. \tag{20.13}$$

(We assume that $|\upsilon_x^+(E)| = |\upsilon_x^-(E)|$.) Next, we define the number of modes (or channels for conduction) at energy, E, as

$$M(E) \equiv \frac{h}{4}\upsilon_x^+ D(E)\,. \tag{20.14}$$

(It is easy to check the dimensions and show that M is dimensionless). Using this result in the expressions for $I_1(E)$ and $I_2(E)$, we find the net current as

$$I(E) = I_1(E) - I_2(E) = \frac{2q}{h}\mathcal{T}(E)M(E)\left[f_1(E) - f_2(E)\right]\,. \tag{20.15}$$

The total current is found by integrating over all of the energy channels to find

$$I = \int I(E)dE\,. \tag{20.16}$$

The final result is

$$I = \frac{2q}{h}\int \mathcal{T}(E)M(E)\left[f_1(E) - f_2(E)\right]dE\,, \tag{20.17}$$

which is eqn. (12.2), the Landauer expression for the current. This simple derivation is sufficient to show where eqn. (12.2) comes from, but for a deeper discussion of the Landauer approach, the reader should consult Datta [11].

20.6 Non-ideal contacts

Contacts limit the performance of devices. Series resistance is always a concern, but other effects can occur as well. These other effects are not typically a problem for Si MOSFETs, but they can be a problem in III-V and GaN FETs [13, 14, 16].

At the top of the barrier, the current can be written as $I_D = qn_s(0)\left\langle \upsilon_x(0)\right\rangle$. The value of the charge density at the top of the barrier is controlled by gate electrostatics, but if the source is not doped heavily enough, then it cannot supply the charge needed at the top of the barrier. The source depletes, large electric fields result, and device performance degrades. This effect has been called source exhaustion [17]. Another effect can also occur. The channel is typically thinner than the source, and it can be difficult for electrons from the source to get into the channel. This effect has been called *source starvation* and can be important in III-V FETs [18]. Paradoxically, this is a case for which scattering can actually improve the performance of a device. Simulations of realistic contact structures show that the performance in the presence of scattering is better than for the ballistic case because scattering helps funnel electrons into the channel [19].

Non-ideal source effects have been modeled by including a gate-voltage-dependent series resistance. This can be done empirically [13] or more physically by including ungated FETs in the source/drain regions adjacent to the channel [16].

As will be discussed in Sec. 20.8, carriers can also backscatter from the source and drain. This effect, observed in some transistors, can also be considered to be a contact effect [31].

20.7 The critical length for backscattering

Computing the transmission in a field-free-slab is fairly easy and, as shown by (20.10), the result is simple. In the channel of a MOSFET, however, there can be a strong electric field that varies rapidly in space, and computing the transmission involves careful consideration of so-called non-local semiclassical transport effects such as velocity overshoot (see Sec. 8.6 in [15]). We have argued that the result can be written in the form

$$\mathcal{T}(E) = \frac{\lambda_0(E)}{\lambda_0(E) + L_C}, \tag{20.18}$$

where λ_0 is the near-equilibrium mean-free-path and $L_C \ll L$ is a critical length for backscattering.

As discussed in Sec. 16.4, this equation is physically sensible. As discussed in Sec. 19.8, the experimentally extracted transmissions behave according to (20.18) and show that $L_C \to L$ for low drain bias and $L_C \to \ell$ where $\ell \ll L$ for high drain bias. Equation (20.18) can be derived, if we assume near-equilibrium transport, but under high drain bias, transport is far from equilibrium for most of the channel. Because transport is far from equilibrium, the use of the near-equilibrium mean-free-path in (20.18) can be questioned. The argument is that the scattering that causes electrons to return to the source occurs very near the top of the barrier before the electrons have been significantly heated. It seems surprising that such a simple equation could describe such a complicated problem, but Monte Carlo simulations that treat far from equilibrium transport show that (20.18) can, in fact, work rather well [20].

The on-current of a MOSFET is proportional to the injection velocity, which is given by (18.23) as

$$\upsilon_{inj} = \left(\frac{\mathcal{T}_{SAT}}{2 - \mathcal{T}_{SAT}} \right) \upsilon_T = \frac{\lambda_0 \upsilon_T}{\lambda_0 + 2\ell} \,.$$

The injection velocity is determined by the transmission in saturation or, equivalently, by the high-bias critical length, ℓ. With the MVS model, we fit the injection velocity to measured data. From the measured data, the critical length, ℓ can be deduced [21], but to predict the on-current, we need to predict ℓ.

The length, ℓ, is approximately the distance over which the potential increases by k_BT/q from its value at the top of the barrier [20], but this is only a rough estimate. Assuming non-degenerate, near-equilibrium conditions, one can derive an expression for ℓ in terms of the channel potential, $V(x)$ [22]. An analytical expression that does not assume near-equilibrium conditions can also be derived [23].

Careful studies of carrier backscattering in nanoscale MOSFETs using Monte Carlo simulations to treat non-local transport self-consistently with the Poisson equation have been reported [24-27]. The study reported in [22] confirmed that the scattering that returns electrons to the source occurs very near the top of the barrier, but the critical length is somewhat longer than the distance over which the potential increases by k_BT/q. The critical length also depends on the shape of the potential profile, which is influenced by self-consistent electrostatics. As a result, ballistic simulations of the potential profile cannot be used to predict the critical length. The authors of [24] conclude that the assumption that $\ell \ll L$ for high drain bias is a good one, but the precise calculation of ℓ requires self-consistent simulations that treat the various scattering processes realistically.

20.8 Channel length dependent mfp/mobility

The transmisson model described in Lecture 17 is written in terms of the ballistic injection velocity, which depends on the bandstructure, and the mean-free-path (mfp) for back-scattering, λ, which depends on bandstructure, scattering physics, and on how electrons are distributed in momentum space. For high drain bias, the carrier scattering rate and mfp vary greatly along the channel as the carriers gain energy from the channel electric field. A key assumption of our model is that the appropriate mfp to use when computing the transmission is the near-equilibrium mean-free-path (i.e. $\lambda \approx \lambda_0$) because the scattering that controls the transmission occurs very near the source before the carriers have had a chance to gain significant energy.

In Lecture 18, we related the transmission model to the VS model by defining a quantity that has the units of mobility (eqn. (18.16))

$$\mu_n = \frac{\upsilon_T \lambda_0/2}{k_B T/q},$$

Strictly speaking, mobility is a quantity that is well-defined only near-equilibrium and only in the bulk (see Sec. 8.2 in [15]), but it is convenient to write the transmission model in traditional form and to express the mfp in terms of a mobility. If the velocity of the injected flux is υ_T and if the near-equilibrium mfp at the top of the barrier, λ_0, is the same as in a very long channel device, then the mobility in (18.16) is the same mobility that would be measured in a long channel MOSFET. In these lectures, we have often used the long channel mobility to estimate the near-equilibrium mfp, λ_0, in a nanoscale FET.

When we expressed the transmission model in the VS form, we found that the drain current in the linear region was proportional to the apparent mobility (eqn. (18.19))

$$\frac{1}{\mu_{app}} = \frac{1}{\mu_n} + \frac{1}{\mu_B}.$$

The apparent mobility depends on both the real, scattering limited mobility, μ_n, and on the ballistic mobility, μ_B, where (as given by eqn. (18.17))

$$\mu_B \equiv \frac{\upsilon_T L/2}{k_B T/q}.$$

We see that the apparent mobility which is easily extracted from the *IV* characteristic, decreases for short channel lengths because the ballistic mobility decreases with channel length.

For some transistors, the length dependence of the apparent mobility seems to be entirely determined by the ballistic mobility (Fig. 19.5), but for others it appears that λ_0 decreases at short channel lengths (Fig. 19.6). The cause(s) for the decrease in mfp for short channel lengths is not yet fully understood. Some studies indicate that charged defects, perhaps unintentionally introduced during device processing are the cause [29]. Other studies point out that *long range Coulomb scattering* is a possible cause. In this case, electrons in the channel interact with the sea of electrons in the source and drain and excite plasma oscillations [30]. This additional scattering process should increase in strength as the channel length decreases.

The use of metal gates instead of polysilicon gates should help to screen out these long range Coulomb interactions, but this is a fundamental process that should be present in all FETs. Clearly understanding the cause of the mfp reduction at short channel lengths — how much is fundamental and how much is process-related and changeable will be important as channel lengths shrink below 20 nm.

It is important to realize that in the apparent mobility

$$\frac{1}{\mu_{app}(L)} = \frac{1}{\mu_n(L)} + \frac{1}{\mu_B(L)},$$

both the scattering limited mobility, μ_n and the ballistic mobility, μ_B depend, in principle on the channel length. The scattering limited mobility, $\mu_n(L)$, may be less than the corresponding mobility in a long channel device.

Equations (18.19) and (18.17) can be combined to write the channel length dependent apparent mobility as

$$\mu_{app}(L) = \frac{\mu_n}{1 + \frac{\mu_n}{aK_B L}},$$

where

$$K_B = \frac{2k_B T/q}{\upsilon_T},$$

and a is a numerical factor. In the theory we have presented in these notes, $a = 1$, and the transistor data that we analyzed in Lecture 19 could be described with $a = 1$. It now seems clear, however, that $a = 1$ does not describe all transistors (see [31] and references therein). For some transistors, the apparent mobility vs. channel length characteristic can only be described with $a < 1$. For example, a recent FinFET analysis needed $a \approx 0.4$ [31]. Natori has argued that this behavior can be explained by assuming that carriers backscatter not only in the channel, but also at the channel to drain transition [32]. According to Natori's model,

$$a = 1 - R_D,$$

where R_D is the drain backscattering coefficient. To explain the $a \approx 0.4$ observed for the FinFETs in [31], a very significant R_D of about 0.4 is needed. It is possible that details of the channel to drain access regions explain why some transistors display $a \approx 1$ and others show $a < 1$. More studies will be needed to clarify our understanding of this issue.

20.9 Self-consistency

We have argued that scattering deep in the channel — far from the top of the barrier — does not matter much because these carriers cannot surmount the top of the barrier and return to the source. Scattering does, however, slow down the carriers and because there is a steady injection of carriers from the source, the population of electrons builds up within the channel. The increased electron density in the channel couples to the Poisson equation and changes the electrostatic potential everywhere — including near the top of the barrier. The result, as shown in Fig. 20.6, is that the shape of the potential profile changes, so the critical length for backscattering changes.

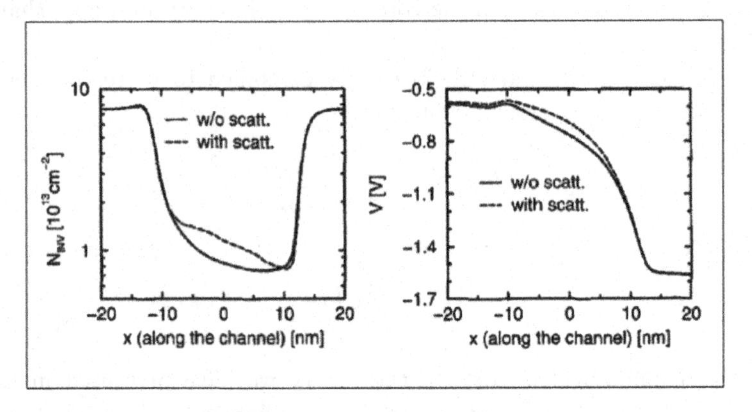

Fig. 20.6 Illustration of the effect of scattering on the self-consistent electrostatics of a nanoscale MOSFET. Left: The electron density vs. position with and without scattering. Right: The conduction band edge vs. position with and without scattering. (©IEEE 2004. Reprinted, with permission, from: P. Palestri, D. Esseni, S. Eminentet, C. Fiegna, E. Sangiorgi, and L. Selmi, "A Monte-Carlo Study of the Role of Scattering in Decananometer MOSFETs," Tech. Digest, Intern. Electron Dev. Mtg, pp. 605-608, 2004.)

Figure 20.6 shows some results of self-consistent numerical simulations. At the left, we see the electron density versus position for the case of a ballistic channel and when scattering is included. As expected, scattering increases the electron density in the channel. Figure 20.6 shows on the right that the added negative charge in the channel causes the conduction band to "float up" and broaden. The result is that the critical length for backscattering increases, which means that the transmission decreases, which lowers the current. We conclude that scattering deep in the channel can affect the current [20, 24]. For a well-designed transistor, however, this effect is small because in a well-designed transistor, the potential at and near the top of the barrier is largely controlled by the gate voltage and not by the drain voltage or by the potential deep in the channel. This can be seen from the fact that for well-behaved transistors, 2D electrostatics in subthreshold (when there is little charge in the channel) and above threshold (where there is a lot of charge in the channel) can be described by the same DIBL parameter (i.e. the parallel shift of the subthreshold charactersitics and the output conductance in the saturation region are both a consequence of 2D electrostatics and both can be described by the same DIBL parameter).

20.10 Carrier degeneracy

Our use of the transmission model in Lectures 18 and 19 assumed Boltzmann statistics for carriers. This seems to work well for Si MOSFETs (e.g. [21]) and also reasonably well for III-V FETs as shown in Lecture 19. For III-V FETs, however, the small effective masses increase the importance of carrier degeneracy and a more physical model is obtained when Fermi-Dirac statistics are included [13, 14]. As shown by eqns. (17.18), the ballistic injection velocity increases with increasing $|Q_n|$ when Fermi-Dirac statistics are included. The relation between mobility and mean-free-path also changes when Fermi-Dirac statictics are used (i.e. eqn. (12.44) is replaced by (6.33) in [12]),

$$\ll \lambda \gg = \frac{2(k_B T/q)\mu_n}{\upsilon_T} \times \frac{\mathcal{F}_0(\eta_F)}{\mathcal{F}_{-1/2}(\eta_F)} \, .$$

Recall that υ_T is the non-degenerate, uni-directional thermal velocity.

The gate capacitance in strong inversion is also lowered when Fermi-Dirac statistics are employed because the quantum capacitance discussed in Sec. 9.2 is reduced when Fermi-Dirac statistics are employed. For a careful treatment of carrier degeneracy in an extended VS model, see [15, 16]. Some useful extensions to Fermi-Dirac statistics of the nondegenerate expressions presented in these notes can be found in [33].

20.11 Charge density and transport

The drain current is proportional to the product of charge and velocity. In the MVS model, the charge at the top of the barrier is determined by

MOS electrostatics using the semi-emipircal expression, eqn. (19.2), which depends only on gate and drain voltages. The injection velocity depends on the transmission, as given by eqn. (19.20). The separation of charge and transport is, however, an approximation.

As illustrated in Fig. 17.1, the charge at the top of the barrier consists of a negative velocity component and a positive-velocity component and is related to the transmission by eqn. (17.18) as

$$Q_n = -q\frac{N_{2D}}{2}\left[\mathcal{F}_0(\eta_{FS}) + (1-\mathcal{T})\mathcal{F}_0(\eta_{FS}) + \mathcal{T}\mathcal{F}_0(\eta_{FD})\right] . \tag{20.19}$$

In the diffusive limit ($\mathcal{T} \ll 1$), both positive and negative velocity states at the top of the barrier are occupied at all drain biases, but in the ballistic limit ($\mathcal{T} \rightarrow 1$) under high drain bias, only positive velocity states are occupied. The value of the transmission determines the location of the Fermi level (η_{FS}), which determines the ballistic injection velocity, $\upsilon_{inj}^{\text{ball}}$.

In the MVS model, we use eqn. (19.2) to determine Q_n from the gate and drain biases and then eqn. (20.19) to determine η_{FS} from which the ballistic injection velocity can be determined. In principle, however, the value of Q_n itself depends on $\mathcal{T}$. As discussed in Sec. 9.5, the gate capacitance in inversion is the series capacitance of an insulator capacitance and a semiconductor capacitance. For an ETSOI device, the semiconductor capacitance is just the quantum capacitance, C_Q. In the diffusive limit ($\mathcal{T} \ll 1$), both positive and negative velocity states are occupied, and the quantum capacitance in the degenerate limit is proportional to the density-of-states as given by (9.49). In the ballistic limit ($\mathcal{T} \rightarrow 1$), only positive velocity states are occupied, so the quantum capacitance in the degenerate limit is proportional to one-half of the density-of-states. For III-V FETs, this effect can be important because light effective mass results in a small C_Q, which significantly lowers the gate capacitance. Because III-V FETs operate close to the ballistic limit, the already small quantum capacitance, which is reduced by a factor of two under high drain bias, can be an important factor in III-V FETs and should be accounted for in order to do justice to the physics [13, 14].

20.12 Discussion

My goal in these lectures has been to convey the essential physics of nanoscale FETs as illuminated by detailed numerical simulations and experiments. This "essentials only" approach is useful for understanding and interpreting the results of simulations and experiments and as a basis for the development of semi-empirical compact models for FETs, such as the MVS model. This is true as long as the approach correctly captures the essential physics. Detailed numerical studies that support the transmission model have been reported (e.g. [24-27]). These simulations confirm the understanding that the scattering that limits the on-current occurs in a short region near the virtual source and that a device may deliver a current close to the ballistic on-current even in the presence of a good deal of scattering — as long as it does not occur in the critical, bottleneck region, but they also show that quantities like the specific length of the critical layer and the specific velocities of the forward and reverse-directed flux, can only be quantitatively predicted with detailed simulations [24]. The simulations of [24] also confirmed that the current injected from the source under ballistic conditions, I^+_{ball}, is larger than the current injected from the source in the presence of channel backscattering, I^+ — for the reasons discussed in Sec. 17.8.

Simulation results are shown in Figs. 20.7 and 20.8 [23, 25]. These simulations treat electron transport self-consistently with the Poisson equation. They include quantum confinement effects and a detailed treatment of the relevant scattering processes. Figure 20.7 shows how the 2D k-states in the channel of an $L = 25$ nm MOSFET under high gate and drain bias are occupied. In the source, there is a symmetric, near-equilibrium distribution of occupied k-states — the source acts as a good Landauer contact. At the top of the barrier, the distribution of occupied states is strongly asymmetric with positive k-states mostly occupied; only a few negative velocity k-states are occupied as a result of backscattering. Deeper in the channel, the radius of occupied states expands as the electrons are accelerated by the electric field, but the distribution becomes more and more asymmetric with most of the occupied states being those along the direction of the electric field. Finally, at the drain, we see again a symmetric, thermal distribution of occupied states.

Figure 20.8 shows simulations of the electron distribution vs. velocity along the transport direction — this time for a 14 nm MOSFET under high gate and drain bias [25]. Two cases are considered with (dashed lines) and without (solid lines) scattering in the channel. Again, an equilibrium distribution is observed in the source, and a highly asymmetric (approximately hemi-Maxwellian) distribution at the top of the source to channel barrier is observed. For the ballistic case (which is much like the example of Fig. 14.3), there are no negative velocity electrons at the top of the

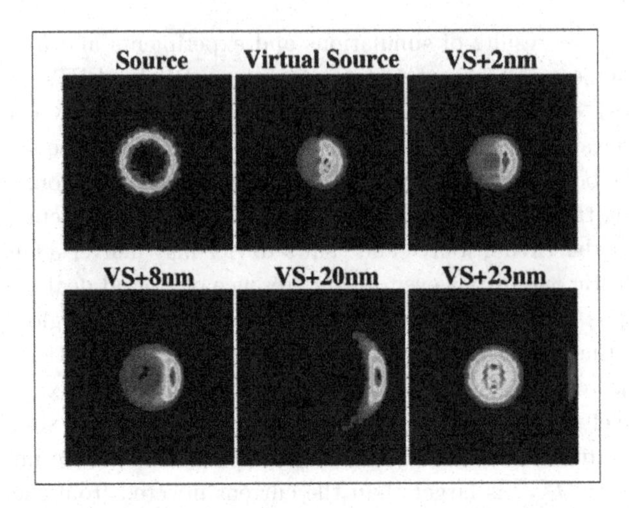

Fig. 20.7 Occupation of 2D k-states in the channel of an $L = 25$ nm MOSFET as obtained from the numerical simulations of [23]. The device is in the on-state, and the distributions are shown at six different locations between the source and drain. (©IEEE 2007. Reprinted, with permission, from: [23].)

barrier, but when there is scattering in the channel, a small population of negative-velocity electrons is observed. Deeper in the channel a ballistic peak of electrons develops as electrons are accelerated in the electric field. Very similar features are observed in fully ballistic simulations of nanotransistors [28]. Simulations like those in [24-27] support the conceptual picture of the essential physics that we have developed in these lectures.

Other studies also using very detailed numerical simulation have, however, raised some concerns [34, 35]. These studies emphasize the fundamental nature of long-range Coulomb interactions, but they also note that the increasing use of metal gates is likely to screen these interactions and reduce, but not eliminate the effect. They also discuss the importance of source starvation, which we discussed briefly in Sec. 20.6. It is especially important to treat these effects for III-V FETs, and techniques to do so in a VS framework have been developed [13, 14, 16]. These authors also point out that the potential barrier at the virtual source is not fixed; it is affected by transport and it may not be possible to maintain the equilibrium charge at the top of the barrier when current flows. Some aspects of this effect were discussed in Secs. 14.3, 17.6 and 17.8, and the most recent version of the MVS model attempts to treat these effects by including a better description of the charge in the presence of transport [13, 14]. The

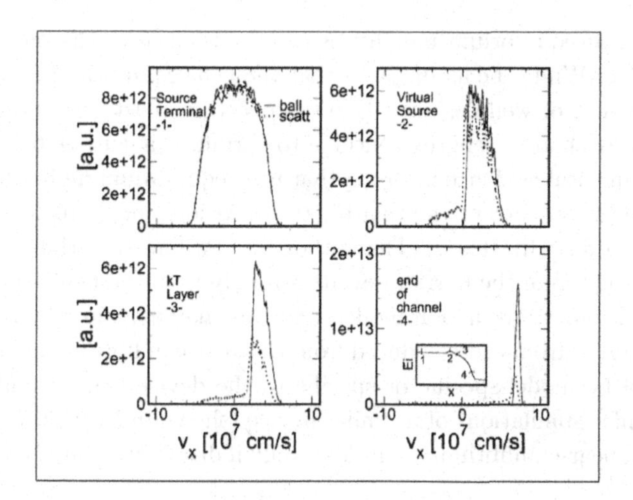

Fig. 20.8 Occupation of k-states in the channel of an $L = 14$ nm MOSFET as obtained from the numerical simulations of [25]. The device is in the on-state and the distributions are shown at four different locations between the source and drain. The distributions are plotted as a function of velocity along the direction of the channel. Solid lines assume no scattering in the channel; the dashed lines treat scattering everywhere in the device. The noise in the results comes from the stochastic process used to solve the BTE. (©IEEE 2007. Reprinted, with permission, from: [23].)

authors of [34, 35] also point out that the Poisson equation couples the electron density through the channel to the potential at the beginning of the channel, so scattering everywhere affects the length of the critical layer. (Recall the discussion in Sec. 20.9) The authors of [24] also made this point, and it was mentioned in the first paper on the transmission approach to MOSFETs [20], but while it is true in general, for well designed MOSFETs, good electrostatic design minimizes the influence of this effect.

The study in [34] provides an interesting discussion of scattering in the critical layer and its relation to the low-field mobility. The authors point out that our expression for transmission can have predictive power only if a prescription is given for calculating the mean-free-path and critical length. This is a valid criticism and one that makes the Landauer or VS model a semi-empirical one that must be fit to data. Obviously a predictive model would be preferred, that is what numerical simulations can sometimes do. The basic transmission model of the MOSFET is not predictive; its value value lies in providing a conceptual framework for understanding.

The simulations of [34] find that the scattering time in the critical layer is much different than in the bulk and that as a result, the authors conclude that the near-equilibrium mobility is of no relevance to the on-current of a MOSFET. While the assumption that the near-equilibrium mfp controls the on-current of well-designed FETs is one that must be continually re-examined as channel lengths continue to shrink, this author believes that the experimental evidence is strong that near-equilibrium mobility is correlated with the on-current and that the transmission model provides a simple explanation for why this is. The authors of [34] find that the occupied k-states at the top of the barrier deviate strongly from the hemi-Maxweillian assumed in our transmission model, and this may explain why their mfps are so different from those deduced from the near-equilibrium mobility. But this could be model-specific or specific to the device being simulated because similar simulations of a similar device (shown in Figs. 20.7 and 20.8) do show a near-equilibrium, hemi-Maxwellian distribution at the top of the barrier.

Simulations like those presented in [24-28] and [34, 35] have been enormously useful in elucidating the physics of transport in nanoscale MOSFET. They also help us understand what the transmission model gets right and what its limitations are. They have played an important role in the evolution of the transmission model, which is an on-going process because as channel lengths continue to scale down, new physical effects become important. The authors of [34] conclude that the transmission model is a useful, qualitative guide to understand the essential physics of nanoscale MOSFETs, but that it cannot yield quantitative predictions for $L < 50$ nm. The authors of [24] agree. This author would not disagree, but he has been impressed with the ability of the MVS model to produce excellent fits to a wide variety of Si, III-V, and other FETs with channel lengths down to at least 30 nm using only a few, physically sensible fitting parameters. This seems to suggest that the MVS model (and the transmission model that it is based upon) is getting some important things right and the the extracted parameters have physical significance. Finally, we note that recent extensions to the VS model have made it possible to predict the entire IV characteristic from the near-equilibrium mobility and a few key device parameters [13, 14, 36].

20.13 Summary

In this lecture, some of the physical effects that occur in nanoscale FETs were discussed. When one looks closely at what happens inside a small field-effect transistor using detailed simulations, things are very complicated, but the transmission model outlined in these lecture notes provides a simple, physically sound understanding of the essential physics of nanoscale FETs in terms of a only a few, physically meaningful parameters. It is, in fact, not uncommon in science that the macroscopic behavior of a system that appears to be enormously complicated at the microscale can often be described in terms of a few simple parameters [37]. The nanoscale MOSFET is an example of this phenomenon.

The transmission model suffers from an important limitation — it is difficult to compute $I_D(V_{GS}, V_{DS})$ for arbitrary voltages because of the difficulty of computing $\mathcal{T}(V_{GS}, V_{DS})$. (We have only discussed $\mathcal{T}$ in the small and large V_{DS} limits.) As a result, it is difficult to predict the on-current because of the difficulty of computing the critical length, ℓ, for high drain bias. Because of this limitation, key parameters in the Landauer/VS model are determined by fitting the model to experimental data, and the physical interpretation of the fitted parameters is provided by the transmission model. It should be noted, however, that recent advances are leading to more predictive VS models [13, 14, 36].

Technology developers rely on sophisticated computer simulations to design and optimize devices. These numerical simulations treat the flow of electrons and holes (either semi-classically or quantum mechanically) under the influence of their self-consistent electrostatic potential. The transmission model and the related VS model describe the essential physics of good transistors. They can be used to analyze and interpret the results of experiments and detailed simulations, and they can form the kernel of a physics-based compact model for use in circuit design. Device researchers need both types of models - detailed simulations that include as much physics as possible and that solve the governing equations with the fewest possible approximations, and they need simple, conceptual models like the transmission model that get to the heart of the problem as simply as possible.

20.14 References

The approach used in this lecture to establish the fundamental limits for MOSFETs as digital switches is similar to the approach of Victor Zhrinov and colleagues.

[1] V. V. Zhirnov, R. K. Cavin III, J. A. Hutchby, and G. I. Bourianoff,

"Limits to Binary Logic Switch Scaling — A Gedanken Model," *Proc. IEEE*, **91**, pp. 1934-1939, 2003.

The classic paper on the need to dissipate energy in digital computation was written by Rolf Landauer.

[2] R. Landauer, "Irreversibility and Heat Generation in the Computing Process," in *IBM J. Research and Development*, pp. 183-191, 1961.

In subsequent work, Charles Bennett and Rolf Landauer showed that there is, in fact, no lower limit to the energy needed to switch a bit if special techniques known as reversible computing are employed. Attempts to implement this idea have proven to be challenging.

[3] Charles Bennett and Rolf Landauer, "The fundamental limits of digital computation," *Scientific American*, **61**, pp. 48-57, 1985.

As shown by Meindl, the Landauer limit of $k_B T \ln 2$ energy dissipation per bit can also be obtained by analyzing a CMOS inverter circuit.

[4] J. D. Meindl and J. A. Davis, "The Fundamental Limit on Binary Switching Energy for Terascale Integration (TSI)," *IEEE J. Solid State Circuits*, **35**, no. 10, pp. 1515-1516, 2000.

Current device research makes use of quantum mechanical transport simulations to explore the limits of MOSFETs. Some examples are listed below.

[5] Jing Wang and Mark Lundstrom, "Does Source-to-Drain Tunneling Limit the Ultimate Scaling of a MOSFET?" International Electron Devices Meeting Tech. Digest, pp. 707-710, San Francisco, CA, Dec. 2002.

[6] Mathieu Luisier, Mark Lundstrom, Dimitri A. Antoniadis, and Jeffrey Bokor, "Ultimate device scaling: intrinsic performance comparisons of carbon-based, InGaAs, and Si field-effect transistors for 5 nm gate length," presented at the International Electron Device Meeting, Dec., 2011.

[7] S. R. Mehrotra, Sung Geun Kim, T. Kubis, M. Povolotskyi, M. S. Lund-

strom, G. Klimeck, "Engineering Nanowire n-MOSFETs at $L_g < 8$ nm," *IEEE Trans. Electron Dev.*, **60**, no. 7, pp. 2171-2177, 2013.

The connection between the so-called NEGF approach to quantum transport and the Landauer approach is discussed by Datta.

[8] S. Datta, "Steady-state Quantum Kinetic Equation," *Phys. Rev. B*, **40**, Rapid Communications, pp. 5830-5833, 1989.

The connection between the Boltzmann Transport Equation and the Landauer approach is discussed in the following two papers.

[9] M. A. Alam, Mark A. Stettler, and M. S. Lundstrom, "Formulation of the Boltzmann Equation in Terms of Scattering Matrices," *Solid-State Electron.*, **36**, pp. 263-271, 1993.

[10] Changwook Jeong, Raseong Kim, Mathieu Luisier, Supriyo Datta, and Mark Lundstrom, "On Landauer vs. Boltzmann and Full Band vs. Effective Mass Evaluation of Thermoelectric Transport Coefficients," *J. Appl. Phys.*, **197**, 023707, 2010.

The Landauer approach to carrier transport at the nanoscale is discussed in more depth in Vols. 1 and 2 of this series.

[11] Supriyo Datta, *Lessons from Nanoelectronics*, 2nd Ed., PART A: Basic Concepts, World Scientific Publishing Company, Singapore, 2017.

[12] Mark Lundstrom, *Near-Equilibrium Transport: Fundamentals and Applications*, World Scientific Publishing Company, Singapore, 2012.

Extensions of the MVS model to III-V FETs are described in the following two papers.

[13] Shaloo Rakheja, Mark Lundstrom, and Dimitri Antoniadis, "An Improved Virtual-Source-Based Transport Model for Quasi-Ballistic Transistors Part I: Capturing Effects of Carrier Degeneracy, Drain-Bias Dependence of Gate Capacitance, and Non-linear Channel-Access Resistance," *IEEE Trans. Electron. Dev.*, **62**, no. 9, pp. 2786-2793, 2015.

[14] Shaloo Rakheja, Mark Lundstrom, and Dimitri Antoniadis, "An Improved Virtual-Source-Based Transport Model for Quasi-Ballistic Transistors Part II: Experimental Verification," *IEEE Trans. Electron. Dev.*, **62**, no. 9, pp. 2794-2801, 2015

Chapter 9 Sec. 9.4.2 in the following text discussed the transmission in the presence of an electric field.

[15] Mark Lundstrom, *Fundamentals of Carrier Transport, 2nd Ed.*, Cambridge Univ. Press, Cambridge, U.K., 2000.

The non-ideal source effects, source exhaustion and source starvation, are discussed in the following papers.

[16] Ujwal Radhakrishna, Tadahiro Imada,, Toms Palacios, and Dimitri Antoniadis, "MIT virtual source GaNFET-high voltage (MVSG-HV) model: A physics based compact model for HV-GaN HEMTs," *Phys. Status Solidi C*, **11**, No. 3-4, pp. 848-852, 2014.

[17] Jing Guo, Supriyo Datta, and Mark Lundstrom, Markus Brink, Paul McEuen, Cornell, Ali Javey, Hongjie Dai, Hyoungsub Kim, and Paul McIntyre, "Assessment of MOS and Carbon Nanotube FET Performance Limits using a General Theory of Ballistic Transistors," Intern. Electron Devices Meeting Tech. Digest, pp. 711-714, San Francisco, CA, Dec. 2002.

[18] M. V. Fischetti, L. Wang, B. Yu, C. Sachs, P. M. Asbeck, Y. Taur, and M. Rodwell, "Simulation of Electron Transport in High-Mobility MOSFETs: Density of States Bottleneck and Source Starvation," Intern. Electron Devices Meeting Tech. Digest, pp. 109-112, Washington, DC, Dec. 2007.

[19] R. Venugopal, S. Goasguen, S. Datta, and M. S. Lundstrom, "A Quantum Mechanical Analysis of Channel Access, Geometry and Series Resistance in Nanoscale Transistors," *J. Appl. Phys.*, **95**, pp. 292-305, Jan. 15, 2004.

Very simple arguments for the simple treatment of backscattering as given by (20.18) are discussed in the following paper.

[20] M. S. Lundstrom, "Elementary Scattering Theory of the Si MOSFET," *IEEE Electron Dev. Lett.*, **18**, pp. 361-363, 1997.

The extraction of the critical length for backscattering from measured data is discussed by Majumdar and Antoniadis.

[21] A. Majumdar and D. A. Antoniadis, "Analysis of Carrier Transport in Short-Channel MOSFETs," *IEEE Trans. Electron. Dev.*, **61**, pp. 351-358, 2014.

The following two papers discuss the computation of the critical length for backscattering, ℓ, in the presence of a spatially varying electric field. The first paper assumes near-equilibrium conditions, and the second does not.

[22] Gennady Gildenblat, "One-flux theory of a nonabsorbing barrier," *J. Appl. Phys.*, **91**, pp. 9883-9886, 2002.

[23] R. Clerc, P. Palestri, L. Selmi, and G. Ghibaudo, "Impact of carrier heating on backscattering in inversion layers," *J. Appl. Phys.* **110**, 104502, 2011.

Detailed numerical simulations of MOSFETs that treat off-equilibrium transport in the presence of a self-consistent potential are described in the following papers.

[24] P. Palestri, D. Esseni S. Eminente, C. Fiegna, E. Sangiorgi, and L. Selmi,, "Understanding Quasi-Ballistic Transport in Nano-MOSFETs: Part I — Scattering in the Channel and in the Drain," *IEEE Trans. Electron. Dev.*, **52**, pp. 2727-2735, 2005.

[25] L. Lucci, P. Palestri, D. Esseni L. Bergagnini, and L. Selmi, "Multisubband Monte Carlo study of transport, quantization, and electron-gas degeneration in ultrathin SOI n-MOSFETs," *IEEE Trans. Electron. Dev.*, **54**, pp. 1156-1164, 2007.

[26] J. Lusakowski, M. J. Martin Martinez, R. Rengel, T. Gonzalez, R. Tauk, Y. M. Meziani, W. Knap, F. Boef, and T. Skotnicki, "Quasiballistic transport in nanometer Si metal-oxide-semiconductor field-effect-

transistors: Experimental and Monte Carlo analysis," *J. Appl. Phys.*, **101**, 114511, 2007.

[27] H. Tsuchiya, K. Fujii, T. Mori, and T. Miyoshi, "A Quantum-corrected Monte Carlo study on quasi-ballistic transport in nanoscale MOSFETs," *IEEE Trans. Electron Dev.*, **53**, pp. 2965-2971, 2006.

[28] J.-H. Rhew, Zhibin Ren, and Mark Lundstrom, "A Numerical Study of Ballistic Transport in a Nanoscale MOSFET," *Solid-State Electronics*, **46**, pp. 1899-1906, 2002.

The reduction of mean-free-path at short channel lengths is currently a topic of research. The first paper below presents evidence that this reduction is due to processing-induced charged defects. The second paper describes a long-range Coulomb scattering process that might play a role.

[29] V. Barrel, T. Poiroux, S. Barrund, F. Andrieu, O. Faynot, D. Munteanu, J.-L. Autran, and S. Deleonibus, "Evidences on the physical origin of the unexpected transport degradation in ultimate n-FDSOI devices," *IEEE Trans. Nanotechnology*, **8**, pp. 167-173, 2009.

[30] M. V. Fischetti and S. E. Laux, "Long-range Coulomb interactions in small Si devices. Part I: Performance and Relaibility," *J. Appl. Phys.*, **89**, pp. 1205-1231, 2001.

Two recent papers discuss the anomalous apparent mobility vs. channel length behavior observed in some MOSFETs. By anomalous, we mean, not described by eqns. (18.19) and (18.17)

[31] D. A. Antoniadis, "On apparent electron mobility in Si nMOSFETs from diffusive to ballistic regime," *IEEE Trans. Electron Dev.*, **63**, no. 7, pp. 2650-2656, 2016.

[32] K. Natori, H. Iwai, and K. Kakushima, "Anomalous degradation of low-field mobility in short-channel metal-oxide-semiconductor field-effect transistors," *J. Appl. Phys.*, **118**, no. 23, p. 234-502, 2015.

Extensions of some of the expressions used in these notes to include Fermi-Dirac statistics can be found in the the following paper.

[33] M. Lundstrom and X. Sun, "Some useful relations for analyzing nanoscale MOSFETs operating in the linear region." [Online]. Available: http://arxiv.org/abs/1603.03132, 2016.

In addition to the studies of [22-26], other detailed numerical studies of nano MOSFETs have examined the validity of the transmission model and reached more skeptical conclusions as to its usefulness.

[34] M. V. Fischetti, S. Jin, T.-W. Tang, P. Asbeck, Y. Taur, S. E. Laux, M. Rodwell, and N. Sano, "Scaling MOSFETs to 10 nm: Coulomb effects, source starvation, and virtual source model," *J. Comp. Electronics*, **8**, no. 2, pp. 60-77, 2009.

[35] M. V. Fischetti, S. T. P. O'Regan, S. Narayanan, C. Sachs, S. Jin, J. Kim, and Y. Zhang, "Theoretical study of some physical aspects of electronic transport in nMOSFETs at the 10-nm gate length," *IEEE Trans. Electron Dev.*, **54**, no. 9, pp. 2216-2136, 2007.

A new version of the VS model that can predict, not fit, IV characteristics has recently been reported.

[36] Shaloo Rakheja, Mark Lundstrom, and Dimitri Antoniadis, "A physics-based compact model for FETs from diffusive to ballistic carrier transport regimes," presented at the International Electron Devices Meeting (IEDM), San Francisco, CA, December 15-17, 2014.

It may seem surprising that the very complicated physics of nanoscale FETs can be simply described in terms of only a few parameters. The following paper explains how phenomena that are complex at the microscale can be described at the macroscale in terms of only a few parameters.

[37] B. B. Machta, R. Chachra, M. K. Transtrum, and J. P. Sethna, "Parameter Space Compression Underlies Emergent Theories and Predictive Models," *Science*, **342**, pp. 604-606, 2013.

Appendix A

Listing of Exercises

Applications of the theory presented in these notes were illustrated by several exercises. This appendix lists those exercises and the corresponding sections in which they can be found.

Sec. 4.5 Saturated region: Classical pinch-off

Exercise 4.1: Linear to saturation square law IV characteristic.
Exercise 4.2: Electric field vs. position in the channel.

Sec. 5.4 Series resistance

Exercise 5.1: Analysis of experimental data.

Sec. 6.6 The body effect

Exercise 6.1: Some typical numbers.

Sec. 7.5 Approximate gate voltage — surface potential relation

Exercise 7.1: Some typical numbers.

Sec. 8.4 The mobile charge above threshold

Exercise 8.1: Inversion layer capacitance and thickness.

Sec. 9.3 The mobile charge (ETSOI)

Exercise 9.1: Intrinsic electron sheet density.
Exercise 9.2: Semiconductor potential at the beginning of inversion.
Exercise 9.3: Inversion layer capacitance and capacitance equivalent thickness.

Sec. 12.3 Large and small bias limits (Landauer Approach)

Exercise 12.1: Prove that the area under the Fermi window is one.
Exercise 12.2: Derive the unidirectional thermal velocity.

Sec. 13.6 Charge-based current expressions

Exercise 13.1: Show that eqn. (13.13) gives the correct linear and saturation region currents.
Exercise 13.2: Derive the IV characteristics, analogous to eqns. (13.13), for a ballistic nanowire MOSFET.

Sec. 14.4 Ballistic injection velocity

Exercise 14.1: Ballistic injection velocity in the fully degenerate limit.
Exercise 14.2: Ballistic injection velocity for a realistic MOSFET.

Sec. 15.4 Connection (ballistic and VS models)

Exercise 15.1: Show that the VS fitting parameters used in Fig. 15.1 are the expected values.

Sec. 16.6 Discussion

Exercise 16.1: Mean-free-path and transmission in a 22 nm MOSFET.

Sec. 17.8 Discussion

Exercise 17.1: Analysis of a 25 nm ETSOI N-MOSFET.

Sec. 18.4 Connection (transmission and VS models)

Exercise 18.1: Relate the transmission to the parameters of the VS model.
Exercise 18.2: Mobility and apparent mobility of a 22 nm MOSFET.

Index

Printed in the United States
By Bookmasters